P9-DUN-511

plastics as design form

thelma r.newman

To my men
Jack
Jay
Lee

foreword

"It's only *made of* plastic."

For quite some time this statement has been an implied indictment of an entire group of materials that represent over one hundred years of development and exploration. Perhaps no group of materials has been so severely maligned as those of the plastics field.

Without question, the lack of education in connection with plastics and poor presentation of information have brought about consumer antipathy toward the products of the industry. However, it is also true that many artists and designers have been slow to explore plastics as materials of great integrity.

The person not participating in the use of plastics materials has a total lack of understanding of the field. Indeed, the term "plastic" is constantly misapplied. Actually, the word is a description of properties that many materials display. In addition to the synthetic materials, clay, wax, rubber, and others manifest plastic characteristics. Terms such as thermoplastic resins, thermosetting resins, polymerization, polyfluorocarbons, cellulosics, polyvinylidene are not a part of the everyday vocabulary of most people. Many of us have even forgotten "celluloid," the first synthetic plastic, which appeared over one hundred years ago.

Fortunately, the products of the industry have progressed from the celluloid kewpie doll that was a major aspect of the identity of the plastics industry in the mind of the general public. But much of this progress has not been readily apparent to the vast majority of consumers, and the integrity of plastics as art and design form still needs careful exploration and presentation.

Dr. Newman, in this publication, does not repeat her treatment of the field previously explored in Plastics as an Art Form. *She proceeds in* Plastics as Design Form *to complement and fill in where her first book left off.*

There is yet a great contribution to be made by the innovative designer who involves himself with plastics. The possible changes which can be brought about in our environment excite the imagination.

Works by many of the artists whose creative use of plastics has helped to nullify the often adverse view of the field are included in the over seven hundred photographs in this volume. This book will both motivate and instruct, as the author nudges the designer and the plastics industry to accept a profound responsibility for "the plastic presence," as the first chapter is titled.

Donald L. Wyckoff
Executive Vice President
American Crafts Council

preface

Plastics as Design Form *is meant to complement and not to repeat content of* Plastics as an Art Form. *If a subject is amply covered in* Plastics as an Art Form, *it is not duplicated here. When I wrote the first book, plastics was still very much an unknown entity to the artist-craftsman-designer. A decade later, after having compiled and digested mountains of research and experienced a broad range of successes and mistakes, I have translated from the technical realm into lay terms another volume of information.*

This book was designed to speak to a broad sector of interest—to the designer, architect, manufacturer, artist, educator, and craftsman, in fact, to all those people who have something to do with designing and making useful forms. It speaks on a professional level, omitting "artsy-craftsy" projects and emphasizing pace-setting designs, for the most part. The content deliberately assaults weak design standards that nurture plastics as fakes. Instead, design approaches that could lead to innovative conceptualizations for the use of plastics as plastics constitute the underlying philosophy. The Bauhaus school tried to raise the level of applied art, nee design form, to the level of studio art. It nearly did. Bauhaus influences are still very much with us today.

A broader range of possibility is presented here, broader than Bauhaus standards and going beyond my own purist prejudices for essential design. For men are many-object creatures, having divergent tastes; there are so many possibilities that our vision should encompass a global approach.

My search for innovative design form in plastics as plastics covered two continents. Designers, architects, artists, and manufacturers everywhere were exceedingly generous in sharing their discoveries and giving me photographs. There are still a few people and companies who look upon their products as proprietary articles and refuse to divulge any information. When that happens, as it did for Plastics as an Art Form, *I either pretermit information or I try to rediscover the process that would lead to the product. Sometimes I find another route other than the procedures taken to create the original design form. This brings outcries: "You are wrong. I did not do it that way!" But is it wrong if another solution is discovered? If these artists, designers, and manufacturers would adopt a more broad-minded approach toward responsible sharing, we would, I submit, be farther along toward understanding about plastics.*

Despite search for a range of possibility, examples lean heavily toward furniture and lighting. That is because, in these two areas, designers have had more opportunities and freedom to innovate. But processes are translatable. The route we take may travel over other territory, but if it leads to understanding the meaning of plastics, then we have reached our goal. And if, along the way, we have developed a facility for working with plastics, then that is an added bonus. Instructions as to how you may work with plastics are encompassed in text and photos.

For the most part, technical information is confined to the text and supporting diagrams and charts. Contrary to the usual approach, content is also embodied in the photographs and in captioning of photographs. In fact, much of the design and step-by-step process information is described in that manner. Photographs, then, do not function here as illustrations mainly to support a text, but perform as visual content reinforced with captions. And along the way, built into the text, is specific health and safety information.

Plastics is a giant industry encompassing hundreds of formulations and dozens of processing techniques. Because of the size of the industry, there are natural and practical limitations to what could and should be covered in a book. The outlook in Plastics as Design Form *reaches out more into industrial procedures than* Plastics as an Art Form, *but stays within the design studio for developing basic process concepts. As with all texts and reference books, information is here for you to use directly, or to start you off toward further research. Material contained in the Appendices and Bibliography will suggest sources of additional investigation.*

acknowledgment

A source book with so broad a scope as this owes its existence to the experiments, successes, and failures of countless people and companies. All the resources listed under courtesy credits co-operated generously by providing pictures and information. It is their gift to all who read these pages, and a contribution that will help make these new technologies for design more readily useable.

Very deep gratitude, with love, goes to my husband Jack, who encouraged and facilitated this effort, and to my sons Jay and Lee, who helped in the studio, in research, and at the typewriter from the beginning to deadline day. To Norm Smith, for his impeccable photo processing and advice goes my appreciation. Thanks also to secretary Carmela Scorese, who pounded away at the typewriter with dedication. My research in Italy was aided considerably by Miss Vanna Becciani's invaluable help.

Special thanks also is due the following people and companies for their extra assistance in unstintingly providing information and/or photos, and/or tours of their facilities:

American Cyanamid Company
Artemide and Mr. Canal
Artmonger's Manufactory, Inc.
Atelier International
Scott Bader Co., Ltd.
Bayer-Leverkusen
Miss Vanna Becciani of the Association of Industrial Designers in Italy
Francesco Binfare of Cassina Center in Italy
Virginia Burdick Associates
Julia Busch
CASABELLA
Dr. Guilio Castelli of Kartell-Binasco
Stanley Cook and Company
Dr. Albert G.H. Dietz

DOMUS
E.I. du Pont de Nemours & Co., Inc.
Duro Plastic of New York
Eastman Chemicals Products, Inc.
General Electric Silicones Division
Mrs. A. Baserga and F. Weiler of Gommagomma spa
Robert T. Henson of Flexible Products Co.
Just Plastics, Inc.
Louis E. Keyes of Cadillac Plastics
Ugo La Pietra
Paolo Lomazzi
Edward MacEwen of PPG Industries
Marbon Division of Borg Warner
Mobay
MODERN PLASTICS
Monsanto
Museum of Contemporary Crafts
Mrs. Giannina Cortopassi, Dr. Alberto Li Giotti, and Mr. Mario Magnoni of Oppenheimer of Italy
Plasteco
PLASTICS TECHNOLOGY
The Polymer Corporation
PROGRESSIVE ARCHITECTURE
Dom Andrisani, Ed Jansen, and Harold Kirkpatrick of Reichhold Chemical Co.
Ben Allen, John R. Gill, James Knight, Robert Neu, and Richard Strebel of Rohm and Haas Co.
Celia Sebiri
Neal Small Designs
Stendig, Inc.
Mrs. Muriel Willet of Willet Studios
The Spiro Zakas Association
John Zerning

T.R.N.

Photographs are by the author unless credits are indicated.

contributing
designers
architects
artists
engineers

Eero Aarnio
Archizoom Group
Sergio Asti
Gae Aulenti
Milo Baughman
Mario Bellini
Cini Boeri
Olaf Von Bohr
Phillip Borden
Paolo Caliari
Casoni & Casoni
Achille Castiglioni
Giorgina Castiglioni
Pier Giacomo Castiglioni
Wendell Castle
Gino Colombini
Giànni Colombo
Joe C. Colombo
Sam Davis
DeCursu, DePas, D'Urbino,
Beretta, Lomazzi
Design Group
Albert G. H. Dietz
Felix Drury
Anna Castelli Ferrieri
Frattini
David Garber
Ignazio Gardella
Giorgio Gaviraghi
Karl Gerstner
Daniel Grataloup
Ted Hallman
Casper Henselman
Albert Herbert
Isao Hosoe
Jude Johnson
Vladimir Kagan
Alexander Kower
Carolyn Kriegman
Alan Landis
Aldo Lanza
Ugo La Pietra
Michael Lax
Sam Lebowitz
Les Levine
James L. Littlejohn
Paolo Lomazzi
Jeffrey B. Low
Bix Lye
Vico Magistretti
Angelo Mangiarotti
Jack Marshall
Jeanne Martin
Eugene Massin
Sergio Mazza
Clement Meadmore
Henry Miller
Andrew Ivar Morrison
Peter Muller-Munk Association
Jay Newman
Lee Newman
Thelma R. Newman
Peter Nicholson
Filippo Panseca
Verner Panton
Henry Pearson
Charles O. Perry
Project One
Alberto Rosselli
Antonio Rossin
Eero Saarinen
Claudio Salocchi
Carlos Sansegundo
Richard Sapper
Ben Shahn
Neal Small
Giorgio Soavi
Walter Stein
Merle Steir
Elisa Stone
Giotto Stoppino
J. R. Strignano
Superstudio
Nancy Thompson
Wilhelm Vest
Al Vrana
Ian Walker
Kenneth Winebrenner
Spiro Zakas
Marco Zanuso
John Zerning
Nell Znamierowski

contents

Foreword (by Dr. Donald L. Wyckoff) vii
Preface ix
Acknowledgment xi
Contributing Designers, Architects, Artists, and Engineers xiii

CHAPTER 1.

THE PLASTIC PRESENCE 1

CHAPTER 2.

PLASTIC FACTS AND GENERIC EXTENTS 30

History of Plastics in a Capsule 30
That Plastic Quality 31
Specific Families 31
A Primer of Processes 39
Injection Molding 39
Compression Molding 40
Extrusion 40
Calendering 41
Blow Molding 43
Casting 43
Coating 43
Rotational Molding 43
Filament Winding 43
Forming (Thermoforming) 43
Hand Layup 43
Laminating and High Pressure Laminating 48
Pulp Molding 51
Solvent Molding 51
Et Cetera 51
Bonding and Joining Techniques 52
General Application Considerations 58
Selection Factors 58
Fillers 59
Fibrous Materials 59
Spheres 59
Inorganic Fillers 59
In Conclusion 62
Processing Aids 63
Colorants 63

Surface Coloring 63
Flammability 63
Toxicity 64
Permanence 66

CHAPTER 3.

DESIGN PARAMETERS AND THE SIXTH SENSE 67

Design Standards 67
Measurements 68
General Design Considerations 70
The Development of a Principle 75
Definition of Surface 75
Application of HP Structures 83
Cocoon Construction System 84
Analysis of This Design Approach 84
Design Parameters 89
Stress and Deformation as Design Considerations 89
Plastic Memory 92
Elongation 92
Tensile Strength 92
Flexural Strength 94
Coefficient of Linear Thermal Expansion 94
Unpredictable Variables 94
Some Predictable Variables 95
Structural Design Possibilities 96
Designs Omitted 96

CHAPTER 4.

LIGHT AND TRANSPARENCY 122

Light 123
Light in Transparent Form (A Focus on Polyester) 143
Light and Mylar 149
Literal Light and Metallized Acrylic 149

CHAPTER 5.

DESIGNING WITH RIGID SHEET, BLOCK, ROD, AND TUBE 153

General Fabrication Considerations 153
Basic Equipment and Supplies 154
Cutting 154
Circular Sawing 154
Band Saws 155
Jig Saws and Sabre Saws 156
Hand Saw and Scriber 156
Routing and Shaping 156
Drilling 156
Turning 158
Repairing Cracks 158
Bonding and Fastening 158
Screws and Tapping 158

Cementing 161
Dip or Soak Cementing 163
Capillary Method 166
Mortar Joint Cementing with Viscous Cements 166
Other Types 167
Other Methods 167
Ultrasonic Welding 167
Cementing for ABS 170
Cementing for Polycarbonates 170
Forming 170
Cold Forming 170
Heat Forming 172
Strip Heating 176
Drape Forming 177
Plug and Ring Forming 180
Free Blowing 180
Vacuum Forming 180
Vacuum Snap-Back Forming 181
Finishing 182
Scraping 184
Sanding and Ashing 184
Polishing and Buffing 185
Flame Polishing 186
Cleaning of Acrylic 187
Decorating 187
Engraving 187
Silk Screening 187
Spray Painting 187
Brush Painting 189
Dyeing 189
Hot-Stamping 189
Vacuum Metallizing 189
Frosting 189
Specific Design Considerations 189
New Varieties of Acrylic 190

CHAPTER 6.

A GALLERY OF DESIGNS WITH ACRYLIC 193

Environments 193
Windows 201
Furniture 204
Lamps 207
Containers 210
Ornaments 212
Play Forms and Other Things 214

CHAPTER 7.

DESIGNING WITH FOAM, FILM, AND FIBER 217

Polyurethane 218
Rigid Foams 220
Pour-in-Place Rigid Foams 220

Frothing Rigid Foams 220
Spraying Rigid Foams 221
Hand-Casting Rigid Urethane Parts 221
Flexible Urethane Foams 221
Working with Flexible Foam Slabs 221
Cutting 234
Coating and Covering Foams 236
Bonding Foams 238
Adhesive Types—Advantages and Disadvantages 239
Safety Factors in Working with Urethanes 239
Film 264
Sealing Films 265
Fiber 265

CHAPTER 8.

DESIGNING WITH CASTING AND LAMINATING LIQUIDS 266

Mold Making 266
Preparation of the Master or Pattern 266
Pattern Mounting 266
Mold Release 268
Making of RTV Molds 269
Making of a Polyurethane Elastomer Mold 273
Preparation of the Pattern 273
Mixing the Polyurethane Elastomer 273
Making of Fiberglass Reinforced Polyester Molds (FRP) 273
Working with "Unfilled" Polyester Resin 274
Catalyst-Resin Proportions 279
The Catalyst 280
Casting Procedure 280
Colorants and Fillers 280
Molds 280
Surface Tackiness 280
Solvents 283
Fastening 283
Epoxies 283
Working with Putty-Type Epoxies 283
Working with Liquid Epoxies 283
Curing Agents 284
Silicone Crystal-Clear Casting 284
Water-Extended Polyester Resin (WEP) 284
Working with WEP 284
Reinforced Polyester (RP) 284
Basic Processes with RP 302
Hand Layup 302
Spray-up 303
Vacuum-Bag Molding 303
Pressure-Bag Molding 303
Autoclave Molding 303
Filament Winding 303
Centrifugal Casting 303

Continuous Pultrusion 304
Matched Die Molding 304
Durability of FRP 304
Premix 304
Finishing Materials 304
Casting with Acrylic Resins 305
Casting Using a Sirup 305
Health Hazards 307

CHAPTER 9.

PLASTIC PROSPECTIVE 308

Plastics and Ecology 308
A Visual Language for Plastics 309
Some Design Concepts for Futurity 309
Technology and Potential 311

APPENDICES

A. NOMOGRAPH TO DETERMINE MATERIAL COSTS 315

B. SOURCES OF SUPPLY 317

C. SHOPS THAT WILL WORK WITH SMALL USERS AND MAKE PROTOTYPES 329

D. TRADE ASSOCIATIONS AND TRADE PUBLICATIONS 331

GLOSSARY 333

BIBLIOGRAPHY 341

INDEX 343

tables

Properties of Thermosetting Resins 34
Properties of Thermoplastics 36
Process Selection Chart 40
Bonding and Joining 53
Typical Solvents for Solvent Cementing 58
Inorganic Fillers 59
Machining Difficulties and Causes - Check List 159
Acrylic Clinic 190
Guidelines to Mold Material Selection 274
Variation in Gel Time with Temperature and Catalyst Concentration for a Typical Rapid Room-Temperature Curing Polyester Resin 283
Visual Defects in Reinforced Plastics 286
Comparison Chart 304

plastics as design form

1

the plastic presence

The plastics industry is now in its adolescence—just as marvelous, incorrigible, and promising as any adolescent. The field has grown, in the past hundred years, into a pervasive, ungainly giant. Wherever one looks, whatever one touches, indeed tastes and smells, becomes part of our plastic presence. Something formed of these divergent polymers pervades and even permeates every sector of life from skin and bone to food and furniture. Surgeons are using plastic parts as substitutes for human members. We eat a so-called ice cream and a so-called pudding concocted of assorted polymers. Tooth filling, button, stocking, pillow, chair, eyeglasses, typewriter, bathroom, raincoat—the list is endless.

Experience with plastics has accrued to a considerable fund with notable refinements in design and construction techniques, just as there continue to be misapplication and astoundingly dreadful designs. At the same time that plastics are on a threshold of innovation, as anyone can see from the photos here, they have also branched out into a multimillion-dollar industry that assaults every good design value. New forms have been built that never could have been built before; one example is integral skin urethane furniture. Urethane provides within the same solid piece and through a *single process* varying densities of hardness and a finished covering or skin, with no added adhesives, fasteners, or supports. A concept such as this is enough to send the designer's mind off in countless creative directions and purge from memory historical binds of what traditional materials were able to do. Yet this same material, with its variable density, long life, easily formable potential, has now become the great imitator of hand-carved wooden furniture. Our polymer chemists and engineers ingeniously designed new processes to make fakes feasible. It is too bad that this same technological prowess could not be put to work, instead, to use plastics as plastics. [1]

Besides just plain bad taste and the excuse to meet demands of a market that buys objects in bad taste, there are a few other negatives mitigating against the liberation of plastics from the shackles of traditional materials and ways of thinking. Professor Albert G.H. Dietz of Massachusetts Institute of Technology writes that codes, prediction of performance, fire resistance, cost, unfamiliarity on the part of designers, and unfamiliarity with the building industry (for one) on the part of plastics suppliers are some reasons why plastics are not used more or are used poorly. With experience would come prediction of performance and users would eventually become more familiar with plastics' applications (reason enough to justify writing this book). But building codes and stratification of labor in the building industry into demarcated jobs, such as the carpenter, paper hanger, welder, mason, et cetera, automatically delimit the uses of plastics. Plastics cannot be placed into the procrustean bed of these disciplines. A man who hangs a sheet of plastic as a wall covering would need to use the saw of the carpenter; if he melted plastic around a beam, he would need to use a heat gun, but not a welder's torch, as well as a saw; if he was going to cast concrete into a styrofoam mold, again he would be cutting across trade-union lines.[2] Codes echo fixed notions that classify myriad, diverse plastics into known categories of traditional materials.

On the design level, too, the architect using plastics may have to think like a sculptor or a designer; a designer would have to think in terms of the environment as an architect would or even as a lighting engineer. Indeed, plastics are *today's* materials cutting across old boundaries in interdisciplinary process.

It was a paradox that accidentally forced interdisciplinary thinking in Italy and converted the Italian architect into an industrial designer. A new kind of leadership was born and a renaissance of design in Italy was to emerge.

After World War II, Italy had to start again. There was a hungry young generation, anxious to throw off shackles of the old, forming a ready market. There were too many architects, with not enough buildings to build. There was also an inner group of these architects who had friends who were manufacturers owning small factories. An ambience developed, and there

Eero Saarinen's chairs for Knoll Associates appeared over thirteen years ago. His classic has opened new directions in designing with plastics, and has stimulated a host of imitators. (*Courtesy, Knoll Associates, Inc.*)

[1] One engineering article claims that plastics are used as plastics when the construction of a chair suits plastics. The design may be the same imitation wooden chair, though.

[2] This concept is developed more fully by Armand Winfield in "A Case Study: The Plastic House," *Progressive Architecture,* October 1970, pp. 80–87.

was no class of technocrats to interfere. The architect-designers got there before the engineers and were able to charge with new life and vitality the design ideas that were to emerge in the next quarter of a century. A fresh wind was blowing in Italy.

These architect-designers were a new breed. They "discovered" plastics and thought of plastics in a structural—space sense, evolving new ways of looking at the material. Both architects and their manufacturer friends learned about plastics together. The architect-designer was given free reign. Not only was this a new relationship with industry; because architect-designers led the way without the mutilation of design for the sake of price, or market, or machine, this association raised the level of their fabrications, as we observe if we survey the creative scope of products selling in our department stores and if we study the representative photos in this book. Nowhere else today, in a collective sense, is the architect-designer able and even willing to initiate untried materials. Not only was there a ready market in Italy, but also a hungry one in the United States.

Has the public attitude about those "cheap" plastics (usually toys) changed? Is the public ready to accept expensive plastics as "investment" items such as furniture, homes, and cars?

A recent survey commissioned by *Modern Plastics* magazine and conducted by Opinion

Another pace-setter design is by Mario Bellini of Milan, Italy, 1968. It is an automatic phonograph that works like a toaster. A record can be inserted in the slot and the phonograph plays in all positions. The manufacturer is S. P. A. Luigi Cozzi dell'Aquila. (*Courtesy, Mario Bellini*)

Marco Zanuso and Richard Sapper created this injection molded acrylic television set in 1970 for Brionvega. When not in use, the set looks like a dark box, but when operating the screen projects through the transparent smoke gray front. (*Courtesy, Zanuso & Sapper*)

Research Corp., Princeton, N.J., stated that "plastics' image isn't at all what we had been led to believe. For every market," they continued, "the image is more positive than negative, and in some cases it's almost a love affair." [3] SPI reported that there is still a substantial confusion about what is plastic and what is not. People have difficulty identifying plastic parts from natural, traditional materials. But, according to the O.R.C. report, younger people like plastics better than older people. "This in turn suggests that efforts by manufacturers to hide the plastic nature of their product may indeed be a short-sighted merchandising technique. Far from affecting sales negatively, the word 'plastics' is more likely to be a positive sales tool. Some consumer advertising has apparently recognized this fact already and is stressing the plastic nature of the product," the survey concluded.

The trend is upward. Young people, in particular, want to express their own personalities and mark their ideas on their own age. Their life style is different. Plastics fills this need for uniqueness. What other material is so unfettered by tradition than plastics? These materials can answer demands never before made and, therefore, reflect in a new way a neoteric mode of living.

Designers are asking new questions about man—his behaviors, his patterns—and are basing designs on the results of their studies. Robert Zeidman, an industrial designer (Robert Zeidman Associates), said: "The designer no longer plunges into designing an object. He is vitally concerned about the human element—man's behaviors, structure, and needs. Before any design is committed to paper, studies are made to underpin design solutions." Joe Colombo, an Italian designer, supported this: "What man needs is a microcosm from which a logical progression accrues to what his environment will look like." An environment, according to Designer Colombo, is an accruement of man-forms and *not* a macrocosm which would adapt man to a sys-

[3] "What *is* Plastics' Image Anyway?" *Modern Plastics,* July 1970, pp. 66–70.

tem. Architect Magistreti of Milan, Italy, added another dimension: "Man's meaning is most important. His form, for instance, is more important than technical aspects. A chair will have to follow the curves of man's body. What materials will do is important but technology is at the designer's service, it is a means."

Agreed, we need to know what a material will do. What are its limits; what is its potential? We cannot start with formal preconcepts, only with human conditions to determine human-material relationships and contiguities.

Design with plastics needs to be emancipated from formal preconcepts—rigidity of forms dictated by grains in wood, linear arrangements of structures in steel, planes descriptive of boards. Curved forms, generally, describe best what plastics will do both from a formal, aesthetic vantage point and from a technical sense. Wood made this transition via molds in laminated forms of the 1930's and 1940's. These were forced shapes for wood, but for plastics they predicted a direction. New materials bring new designs. The four-legged wooden chair looks the way it does because the raw material dictated, in part, what it would do and tradition filled in as a guide. A chair made of plastic, however, could have many more legs, or none. It could be a ball, cube, cone, hard, soft. Plastics' potential is to be innovative, flexible, and mobile.

Ours is a mobile society. People move about cross-country, cross-oceans faster and faster. Furniture and even homes need to have "traveling potential." Collapsible, redisposable, stacking, packing, take-apart, blow-up furniture and environments fill this criteria for mobility. Lomazzi designed the blow-up chair of see-through vinyl and knock-down chairs and bookcases of Formica. Joe Colombo created a storage system of infinite combinations using six parts and four support members. Olaf Von Bohr designed a component library wall for Kartell that can be assembled in minutes; and, more miraculously, each piece can be formed of ABS via injection molding in four minutes. Yale Professor Ralph Drury demonstrated in the Pittsburgh "Three Rivers" project that the real potential of urethane foams is that large living spaces can be created *in situ* with relatively few materials. Curvilinear insulated spaces can be sprayed up over any support including blow-up forms. The result is a rigid shelter. One problem is that people, so used to living with post and lintel, horizontals and verticals, in boxes, cannot get used to living in nonplanar areas.

A whole range of plastics such as acrylics, polycarbonates, polyesters, and other types can be more transparent than glass. And they are easily formable. Transparency is a new reality that translates opaque design principles in a new way. Shapes of things that are transparent have a morphological quality—form and structure are one, but both can also be seen at the same time. Transparent plastics can change the quality of spaces. The sense of density of forms disappears and in its place is color, texture, light reaching beyond and through the perimeters of a form to fill an entire room with dancing patterns, echoes of the mother form, variations, and transformations. Carlos Sansegundo, Neal

Tomorrow's needs for stackable furniture taking up very little space have been answered by Paolo Lomazzi's design for a child's chair. It is injection molded of ABS and comes in bright colors. (*Courtesy, Paolo Lomazzi*)

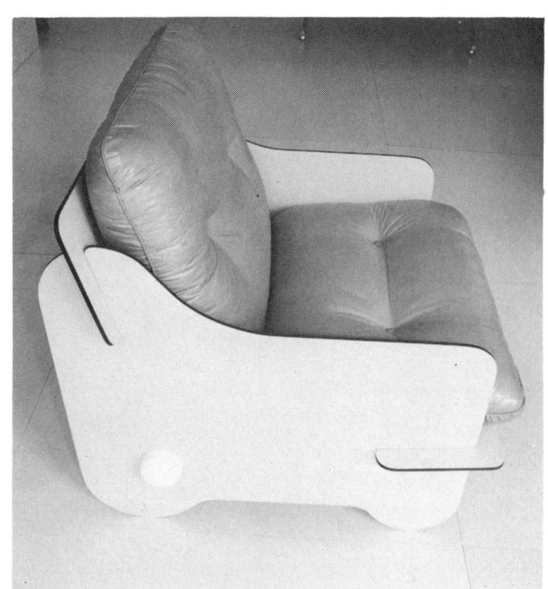

These are knockdown units of Formica that come apart for storage or moving. They are made by Zanotta of Italy. The upholstery of vinyl, the various parts slide into place in grooves and the only mechanical fastener used for the chair is an enlarged plastic "screw." The bookcase sides fit together via a notched injection molded piece. Pegs and fasteners are dramatized to accent the design. (*Courtesy, Paolo Lomazzi*)

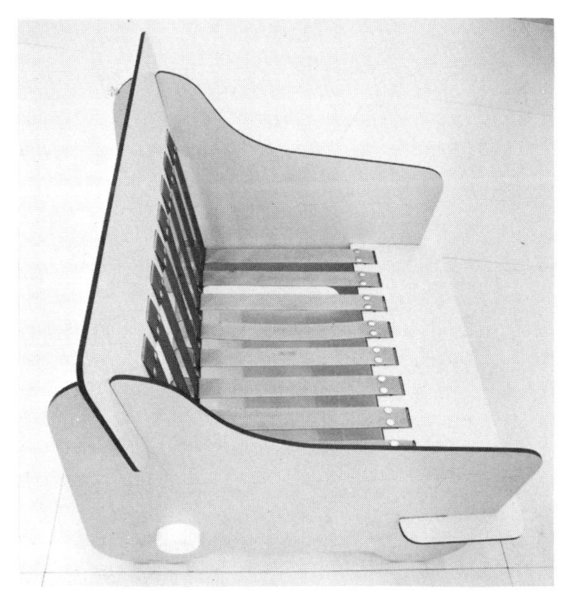

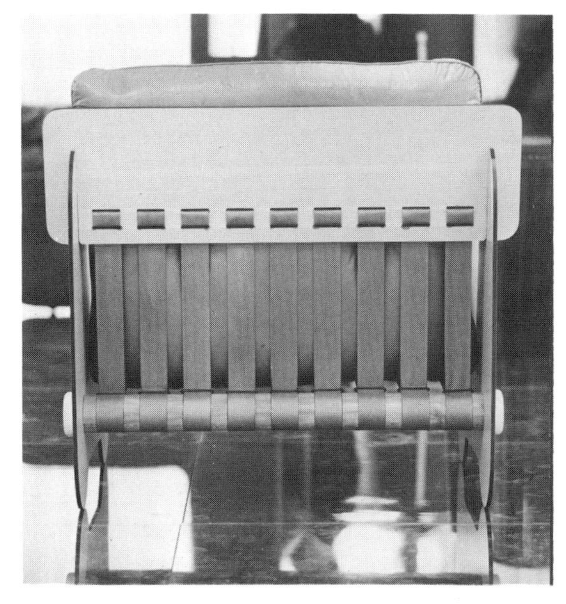

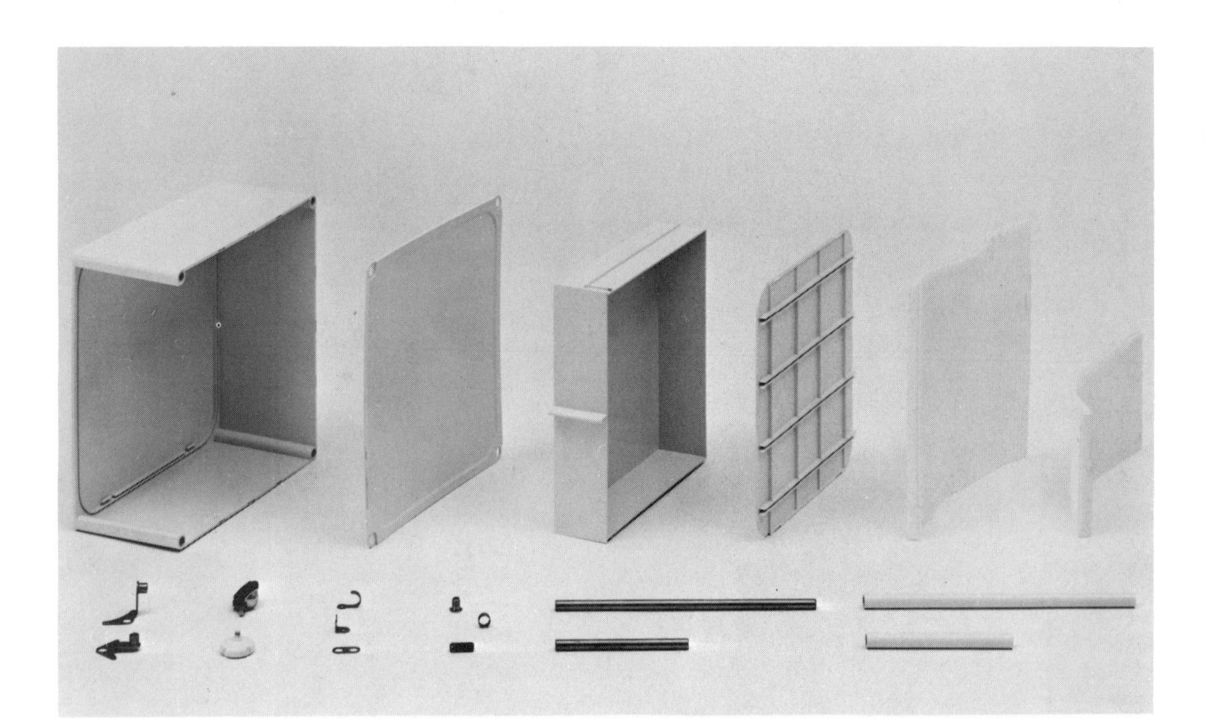

Infinite combinations are possible with Joe Colombo's ABS injection molded design for a storage system. Six basic module pieces, four supporting rods, and attaching hardware make for various combinations. (*Courtesy, Joe C. Colombo*)

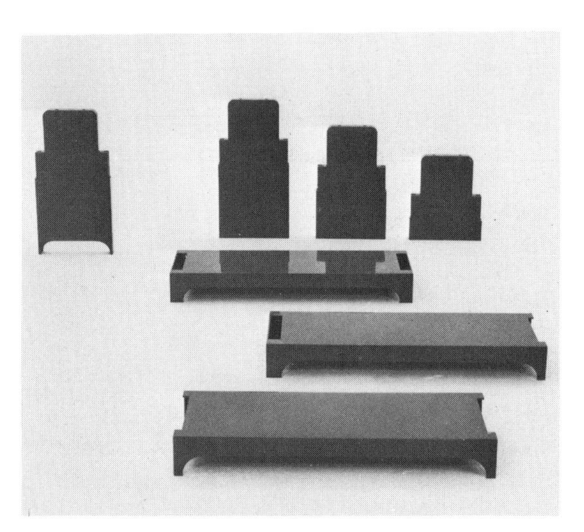

Olaf Von Bohr designed this library component system for Kartell-Binasco in 1969. Parts are injection molded of Marbon's ABS Cycolac. It takes four minutes to make each part and just a few more minutes to assemble. The surface has a permanent, washable, colorful finish. (*Courtesy, Kartell-Binasco*)

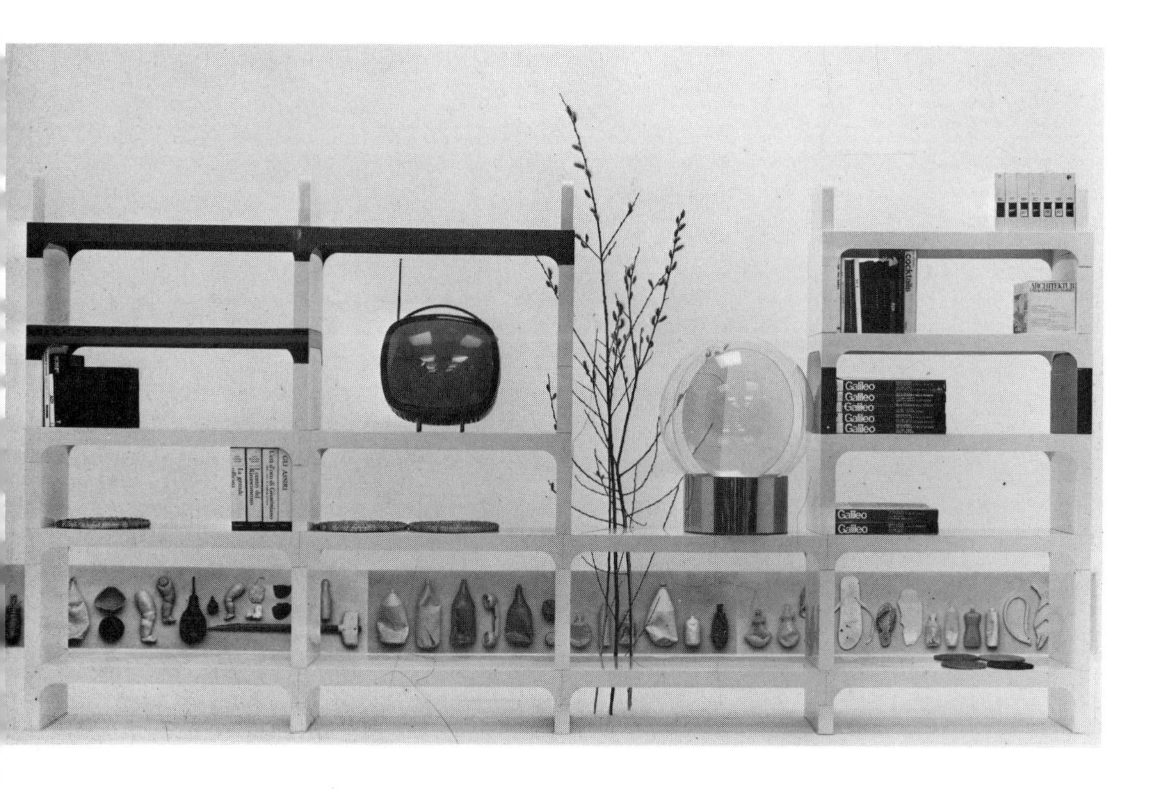
cocktails
Galileo
Galileo
Galileo
Galileo
Galileo
Galileo
Galileo

High intensity polyethylene reinforced with fiberglass makes for a durable, colorful, inexpensive seat. A frame of chip-resistant enamel or chrome permits stacking. Because they nest ¾″ apart, a stack of 40 chairs stores in a space 18″ wide, 20″ deep, less than 7′ high. (*Courtesy, Canadian Seating Co., Ltd.*)

Neal Small is a master at utilizing acrylic's design potential. His light table holds a 50-watt reflector bulb that lights the white acrylic indentation. These units vary in size from a 16″ cube to cubes 16″ high, 20″ x 20″ or 24″ x 24″. (*Courtesy, Neal Small Designs, Inc.*)

Small, and Spiro Zakas folded, bent, and sliced acrylic into a range of objects, lighting, and furniture, and have inspired a flood of imitators. Ugo LaPietra designed pattern into light forms that project their shapes around the room.

Synthetic yarns and fibers yield yet another potential. Fibers woven into gauzelike films can be stretched like body stockings over armatures, expanding at every rib and contracting into curves between supports. Nell Znamierowski utilizes heat-reacting qualities of synthetic fibers such as acrylic to design texture and color effects in rugs and wall hangings via unique commercial weaving processes.

Neal Small diagonally slices a cast acrylic tube for the design stripe pattern in this 12″ vase. There is no waste because of counter changing colors. (*Courtesy, Neal Small Designs, Inc.*)

Spiro Zakas slices and bends a 4′ x 8′ acrylic sheet for this table of ¾″ thickness. (*Courtesy, The Spiro Zakas Association*)

The clear Plexiglas chair, by Carlos Sansegundo, has a seat and swivel back that is vacuum formed. Fasteners are of acrylic so that they expand and contract with the piece. He orders the pieces cut to size from Rohm & Haas Company and then refines and puts together the parts. 33″ x 13″ x 10″. (*Courtesy, Carlos Sansegundo*)

These are just a few excursions into possibility. Excluded from our concern here is a vast area of design using plastics: plastic foam laminated curtain walls, plastic pipe for plumbing, floor coverings, wall coverings, insulation, household implements, articles of clothing, et cetera. Plastics can be very precious and expensive, but it also can be inexpensive when a chair can be made completely in five minutes.

The public is catching up to the vast range of qualities and uses for plastics. But plastics still has to prove itself and does so in baby steps rather than in giant possibilities. Wood, for example, does not have to justify its worth. If a design fails, it is usually because of joint failure and it is the designer's fault even if the wood cracks and splinters in some other part than a joint. If a plastic piece fails, whether it is a joint failure or a structural weakness, plastics (as a category) is blamed, even though the ultimate failure is due to poor designing and processing. Responsible designing and manufacturing is the only answer. Lack of standards, penny pinching because of competition can destroy a market. Industries such as Monsanto, Reichhold, Mo-

This table by Carlos Sansegundo fits together in four sections of smoke-gray and transparent Plexiglas. Mr. Sansegundo reports that acrylic scratches no more than wood or metal. A scratched piece is still beautiful and is very easy to maintain. He plays with edge lighting in his designs. This piece is 48″ x 48″ x 16″. (*Courtesy, Carlos Sansegundo*)

Acrylic is strip heated and bent in a jig for this magazine end table design by Andrew Ivar Morrison for Stendig, Inc. Opaque acrylic does not show scratches easily because light does not pipe through the scratches. (*Courtesy, Stendig, Inc.*)

Plexima, a Scandinavian design much copied, utilizes the light piping qualities of acrylic and nylon extruded "cord." (*Courtesy, Koch & Lowy*)

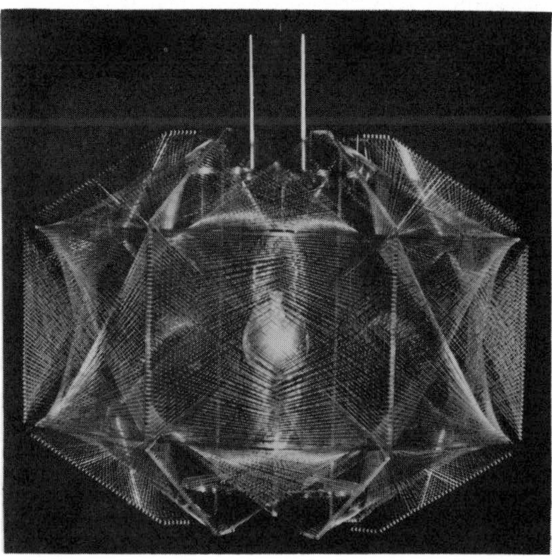

Lomazzi's model for an inexpensive child's chair is of hollow construction, made of polypropylene for blow molding. (*Courtesy, Paolo Lomazzi*)

bay, and Rohm and Haas have committed themselves to experimentation and the development of models. This is exemplary. We need more of this sense of responsibility and more leadership in good design.

Just a handful of designs included in this collection are projections. All the other examples are produced and marketed here and abroad. There is a third illusive category, though. Somewhere in the text lies chemical and engineering potential that goes beyond our conceptualization. When life style and technology meet, we will see a surge of design innovation barely conceived of today because plastics are capable of doing much more than our present limited vision can imagine.

Polyethylene spheres are rotationally molded for Neal Small's two designs—light table and stacking floor lamp. The forms are massive and dominate a setting. The light table is 16″ high and the lamp is 47″ high, 32″ wide, and uses 16″ spheres. (*Courtesy, Neal Small Designs, Inc.*)

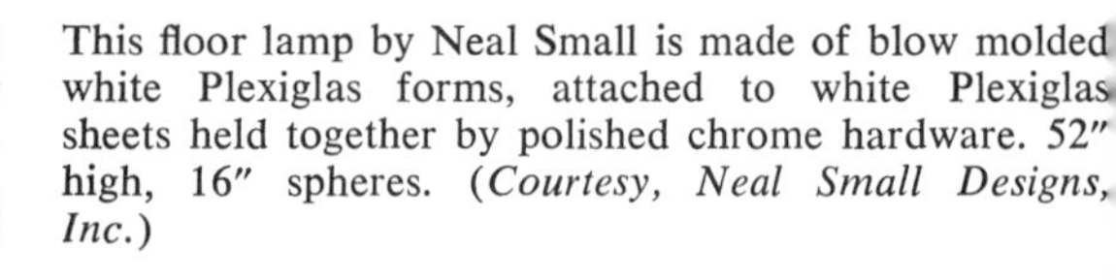

This floor lamp by Neal Small is made of blow molded white Plexiglas forms, attached to white Plexiglas sheets held together by polished chrome hardware. 52″ high, 16″ spheres. (*Courtesy, Neal Small Designs, Inc.*)

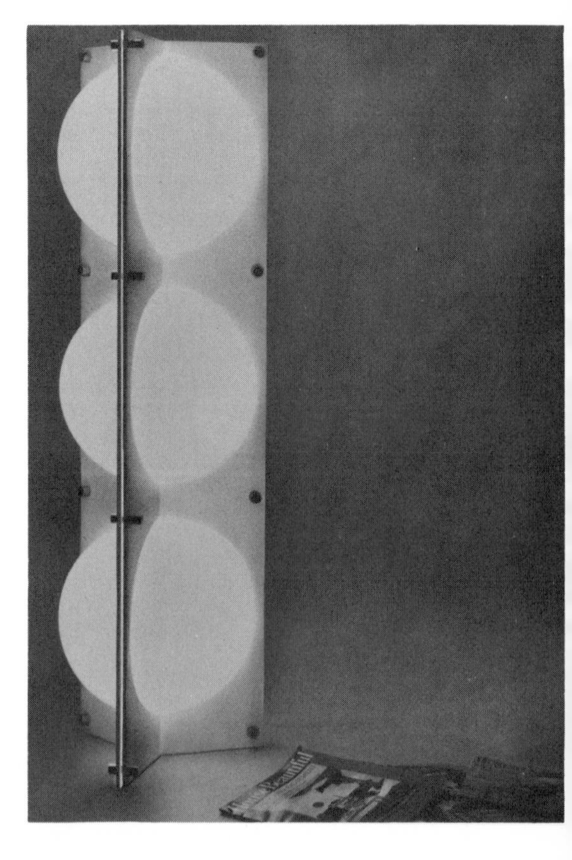

Vico Magistretti's lamp (1969) for Artemide is injection molded of melamine. He uses counteropposed curves in a sensitively proportioned design that is highly functional as well as attractive. (*Courtesy, Vico Magistretti*)

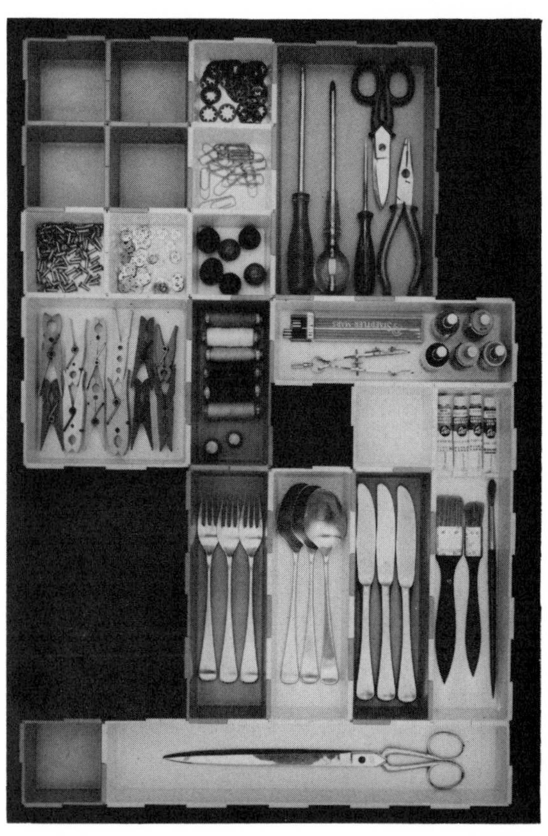

Colorful polystyrene, injection molded by Kartell—Olaf Von Bohr's design for a modular gadget organizer—can build to any size and for many purposes. (*Courtesy, Kartell-Binasco*)

Sweeping Shovel by Gino Colombini for Kartell. Polystyrene, injection molded. (*Courtesy, Kartell*)

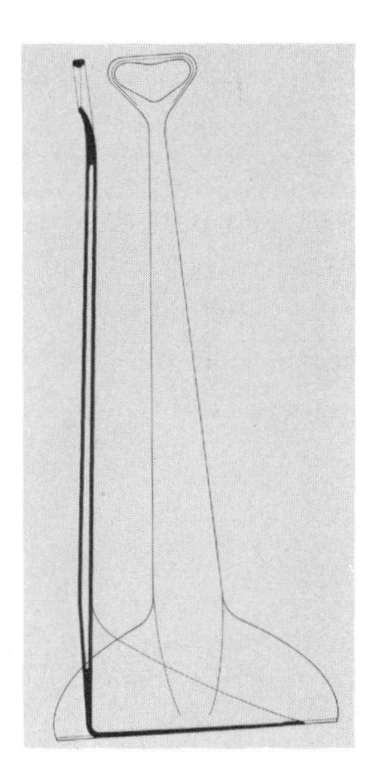

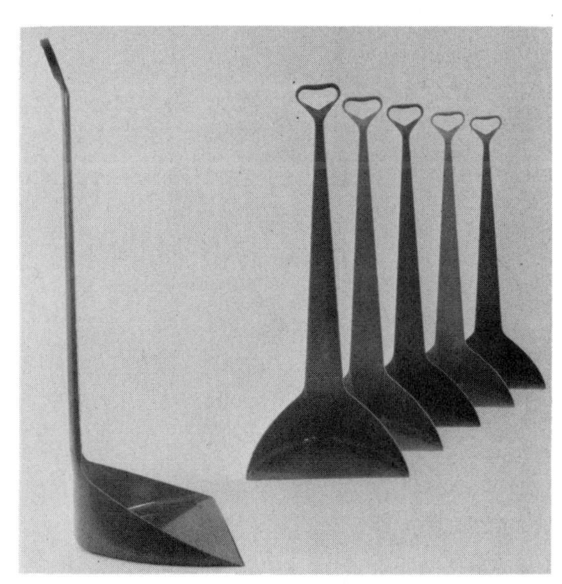

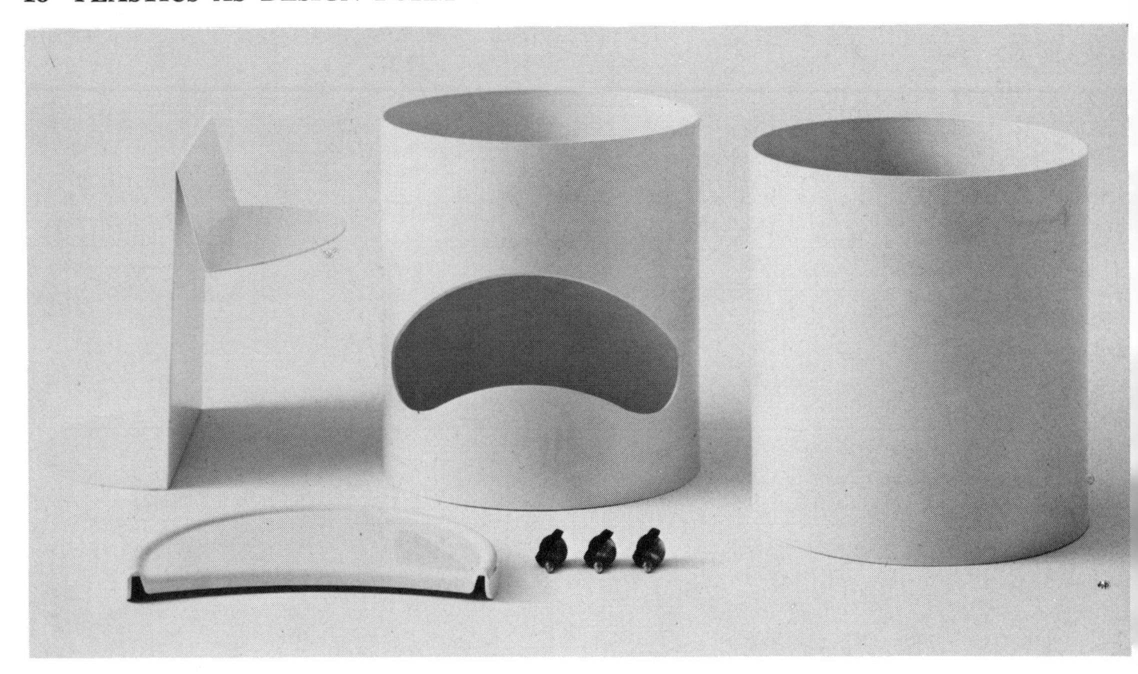

Interchangeable, cylindrical container system, injection molded of ABS. Designed by Joe Colombo, 1969. (*Courtesy, Joe C. Colombo*)

Architect Sergio Asti designed this melamine ashtray for Mebel. (*Courtesy, Sergio Asti*)

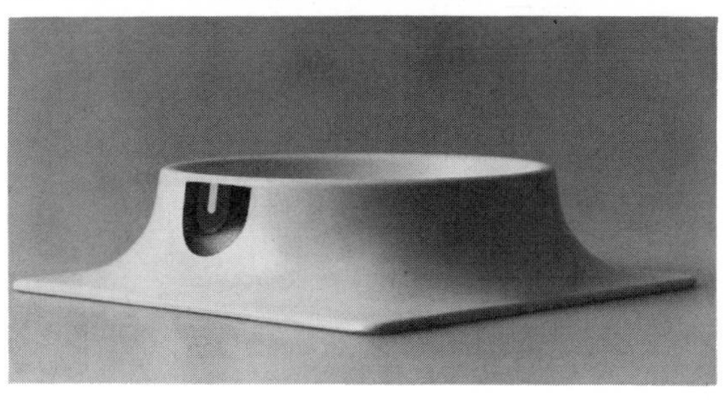

Cylindrical, mobile storage system, injection molded of ABS. Designed by Anna Castelli Ferrieri for Kartell, 1970. (*Courtesy, Kartell-Binasco*)

Chair, injection molded of ABS. Designed by Architect Gae Aulenti for Kartell, 1970. (*Courtesy, Kartell-Binasco*)

Nesting stools, injection molded of ABS. Time for fabricating: three minutes. Designed by Giotto Stoppino for Kartell. (*Courtesy, Kartell-Binasco*)

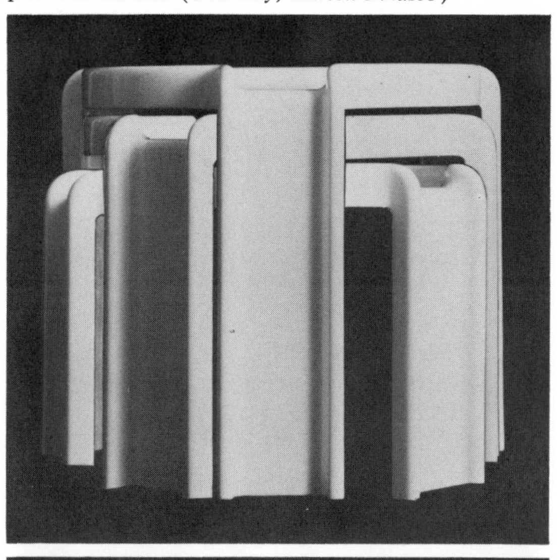

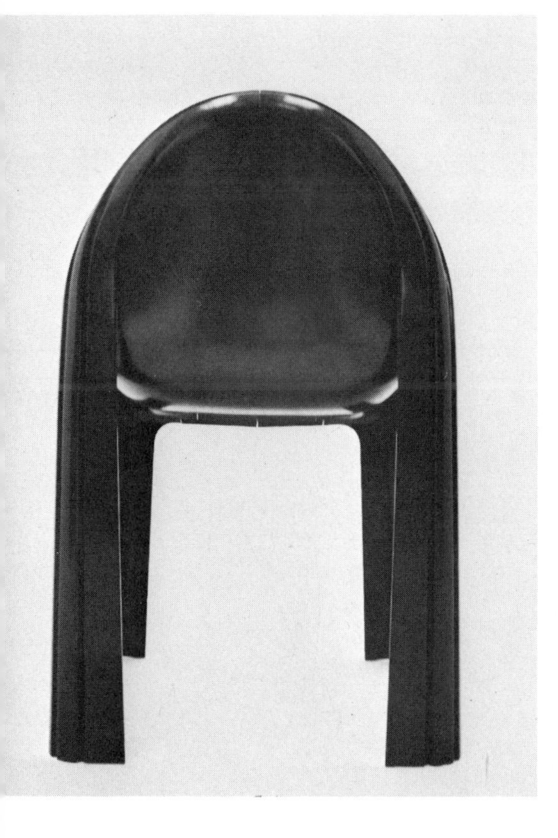

Stacking tables, injection molded of brightly colored ABS. Legs are fluted to provide extra strength, as the convolutions in a scallop shell. Edges are curved to relieve stresses and avoid cracking. Designed by Vico Magistretti for Moreddi. (*Courtesy, Moreddi*)

The blow-up environment has the great potential of quickly—within a few days—spanning large areas without interfering supports. The same system that maintains air pressure also provides heat or cooling. Polyvinyl chloride impregnated fiberglass is the covering used in Architect Paolo Lomazzi's design. (*Courtesy, Paolo Lomazzi*)

This is a design projection by Vladimir Kagan for Monsanto. It is a table that could help to make moving easier. The table folds up and uses the same principle as a collapsible drinking cup. (*Courtesy, Chemstrand Co.*)

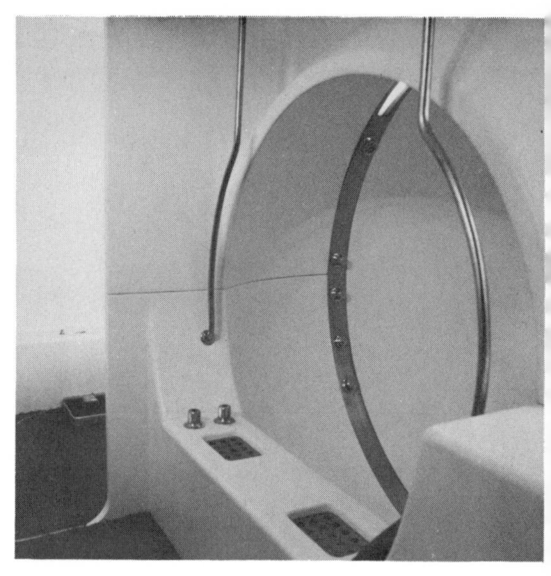

A prefabricated bathroom for the *Visiona* environment by Joe C. Colombo. (*Courtesy, Joe C. Colombo*)

An experimental environment made of various plastics and containing all the accoutrements for living, for *Visiona,* an annual project of the Bayer Co. Designer Joe C. Colombo. (*Courtesy, Bayer-Leverkusen*)

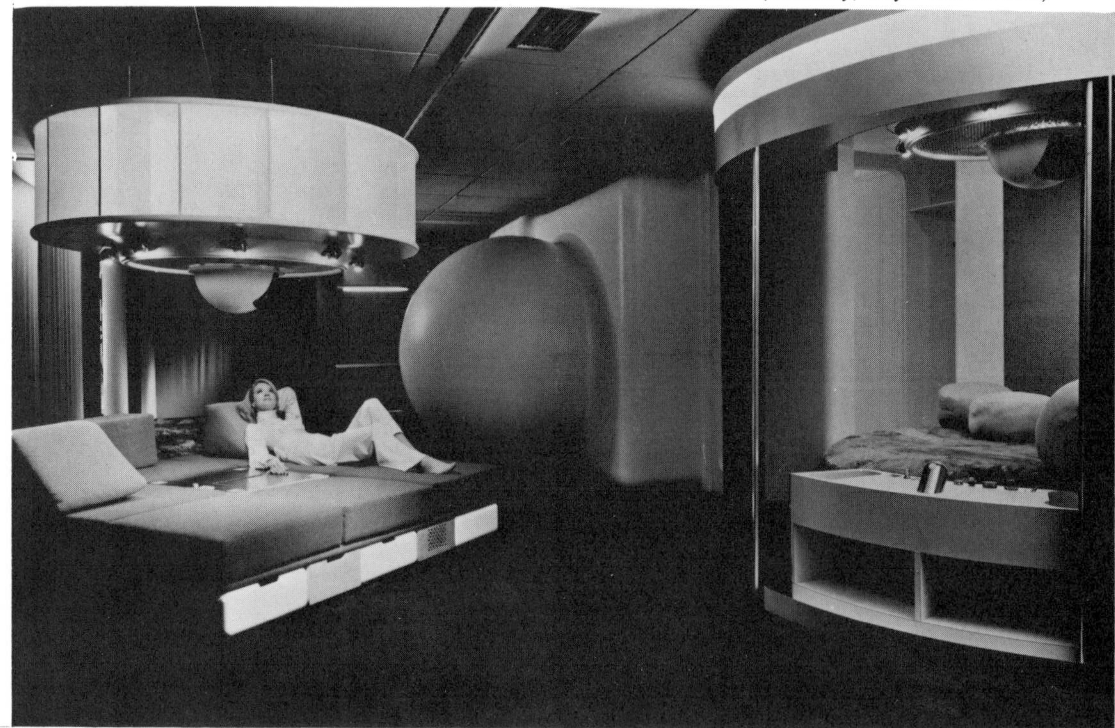

Mass-produced basic box components made of lightweight maintenance-free plastics could be fully assembled in the factory and delivered by truck to the site. There they could be stacked on top of one another or placed side-by-side to make either multiroom dwellings or multistory units. The design by Albert Herbert was a projection for Monsanto. (*Courtesy, Chemstrand Co.*)

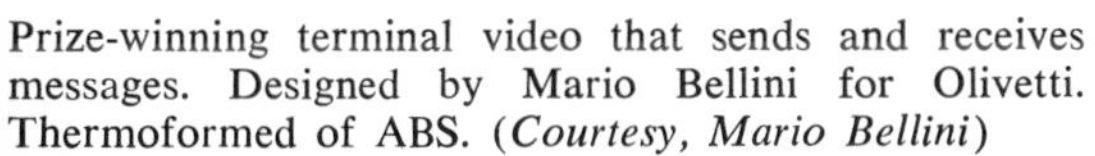

Prize-winning terminal video that sends and receives messages. Designed by Mario Bellini for Olivetti. Thermoformed of ABS. (*Courtesy, Mario Bellini*)

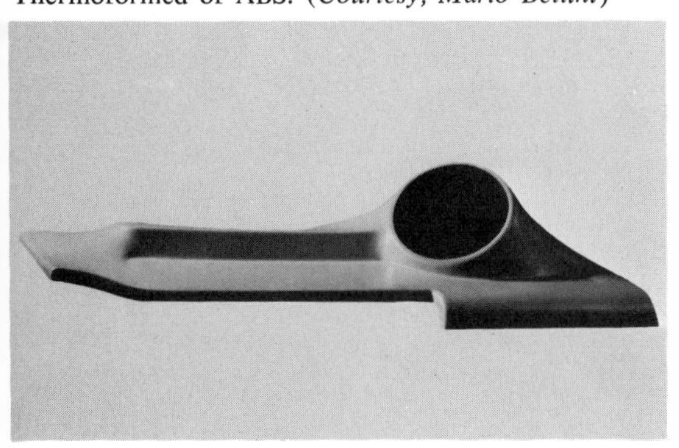

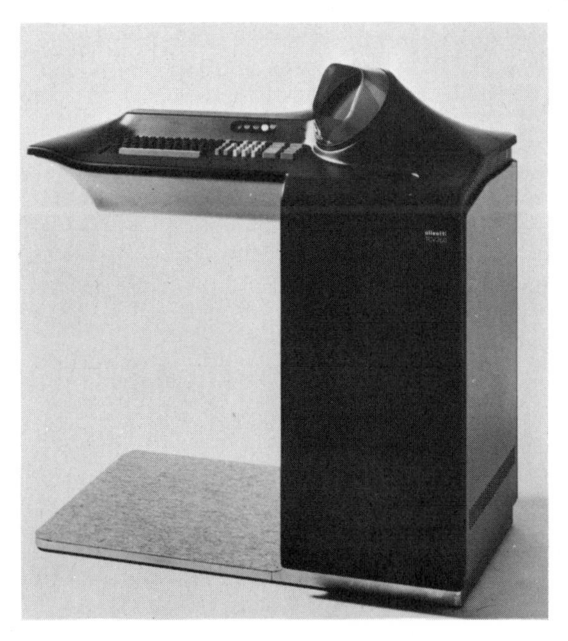

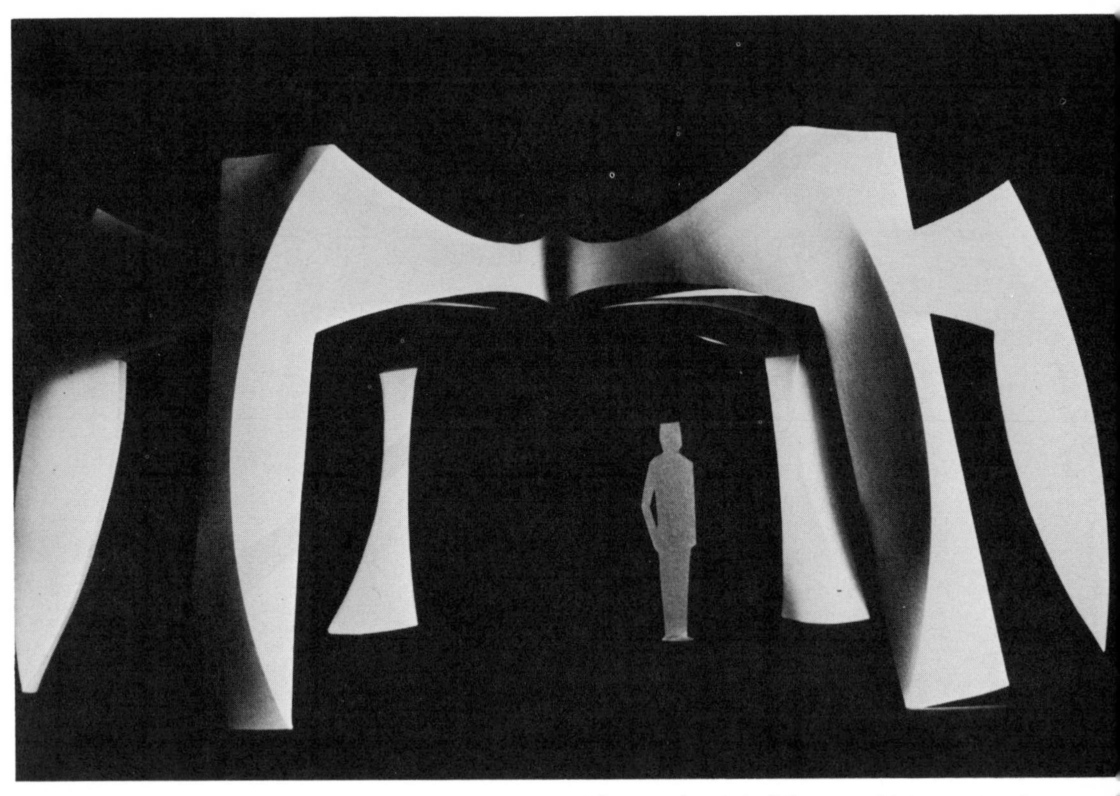

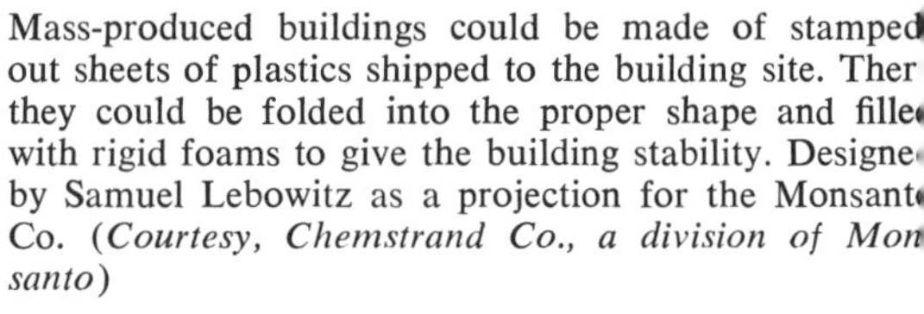

Mass-produced buildings could be made of stamped out sheets of plastics shipped to the building site. Ther they could be folded into the proper shape and fille with rigid foams to give the building stability. Designe by Samuel Lebowitz as a projection for the Monsant Co. (*Courtesy, Chemstrand Co., a division of Mon santo*)

These are two design projections based on a hyperbolic parabolic by Engineer John Zerning of England. In both cases large areas can quickly be spanned with FRP molded in vinyl-cocoon molds. (*Courtesy, John Zerning*)

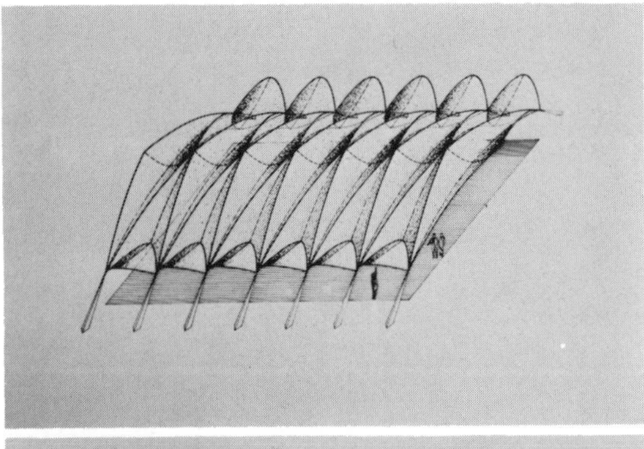

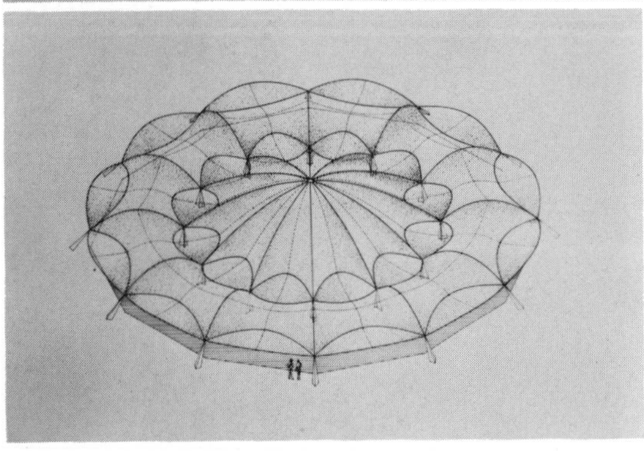

Carolyn Kriegman's body ornament is made of acrylic. (*Courtesy, N.J. State Arts Council. Photographer, Joseph Crilley*)

Petals of transparent primary colors unfold and fold again for self-discovery experiences in color mixing. (*Courtesy, Creative Playthings*)

This is an acrylic model of a design to be executed in blow molded transparent polyvinyl chloride for a temporary building, possibly for a World's Fair. Architect Ugo LaPietra collaborated with Architects Lomazzi, D'Urbino, DePas, and Beretta, 1969. (*Courtesy, Lomazzi, D'Urbino, DePas, Beretta*)

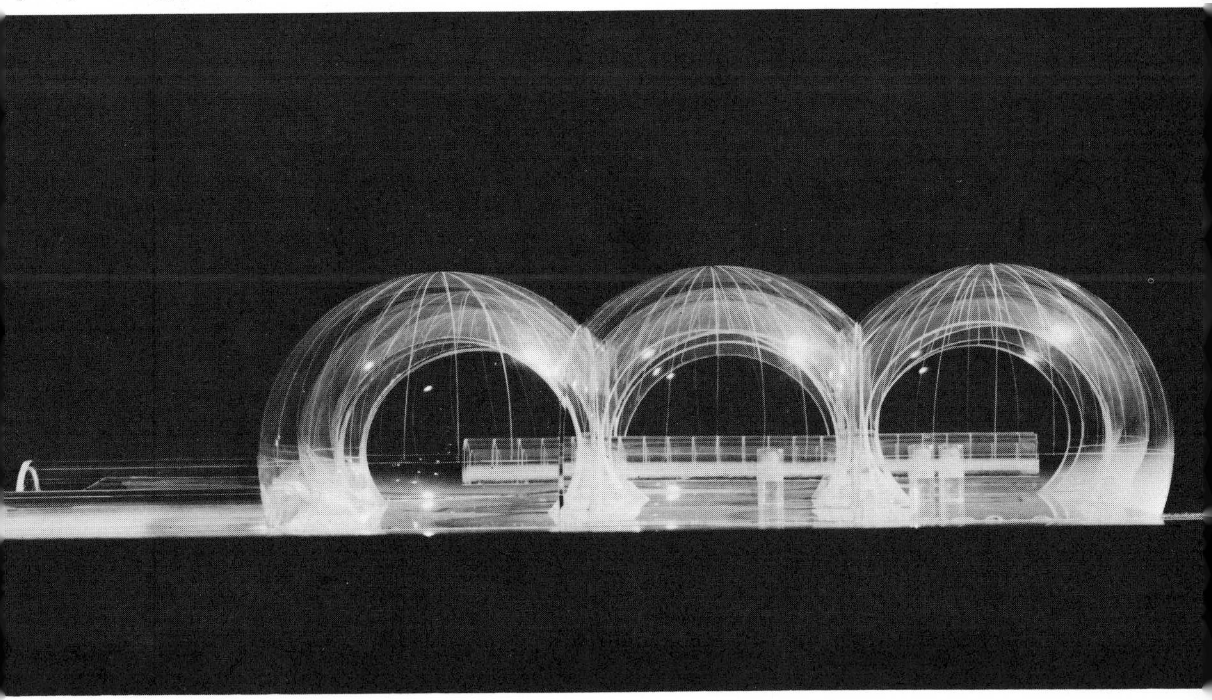

Ted Hallman uses fibers (sometimes synthetic) and dyed acrylic shapes that are woven into his wall hangings. (*Courtesy, Ted Hallman*)

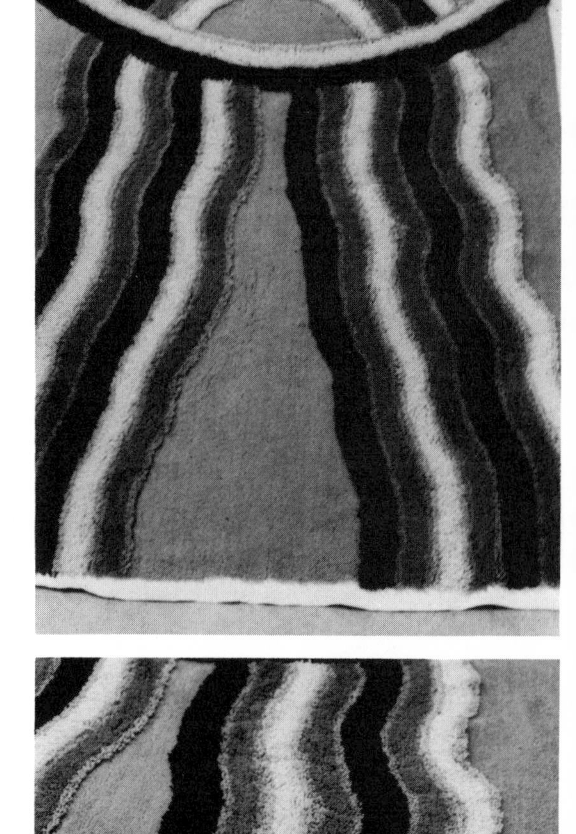

Nell Znamierowski's rug (4′ x 6′), called "Eternity," is made of Caprolan nylon. Washing has curled the pile heights down. The background color is orange and is the lowest pile height. (*Courtesy, Nell Znamierowski*)

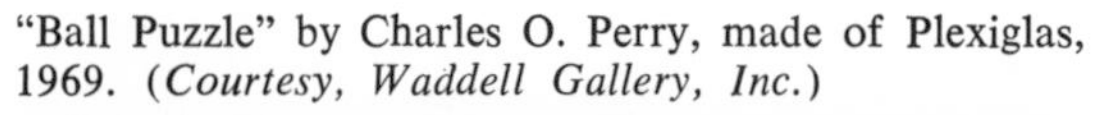

"Ball Puzzle" by Charles O. Perry, made of Plexiglas, 1969. (*Courtesy, Waddell Gallery, Inc.*)

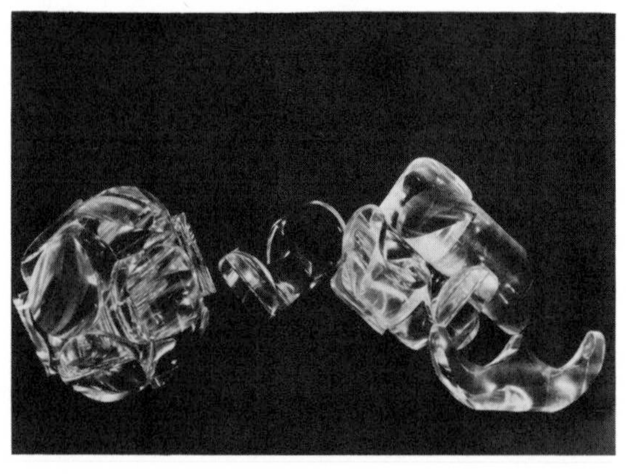

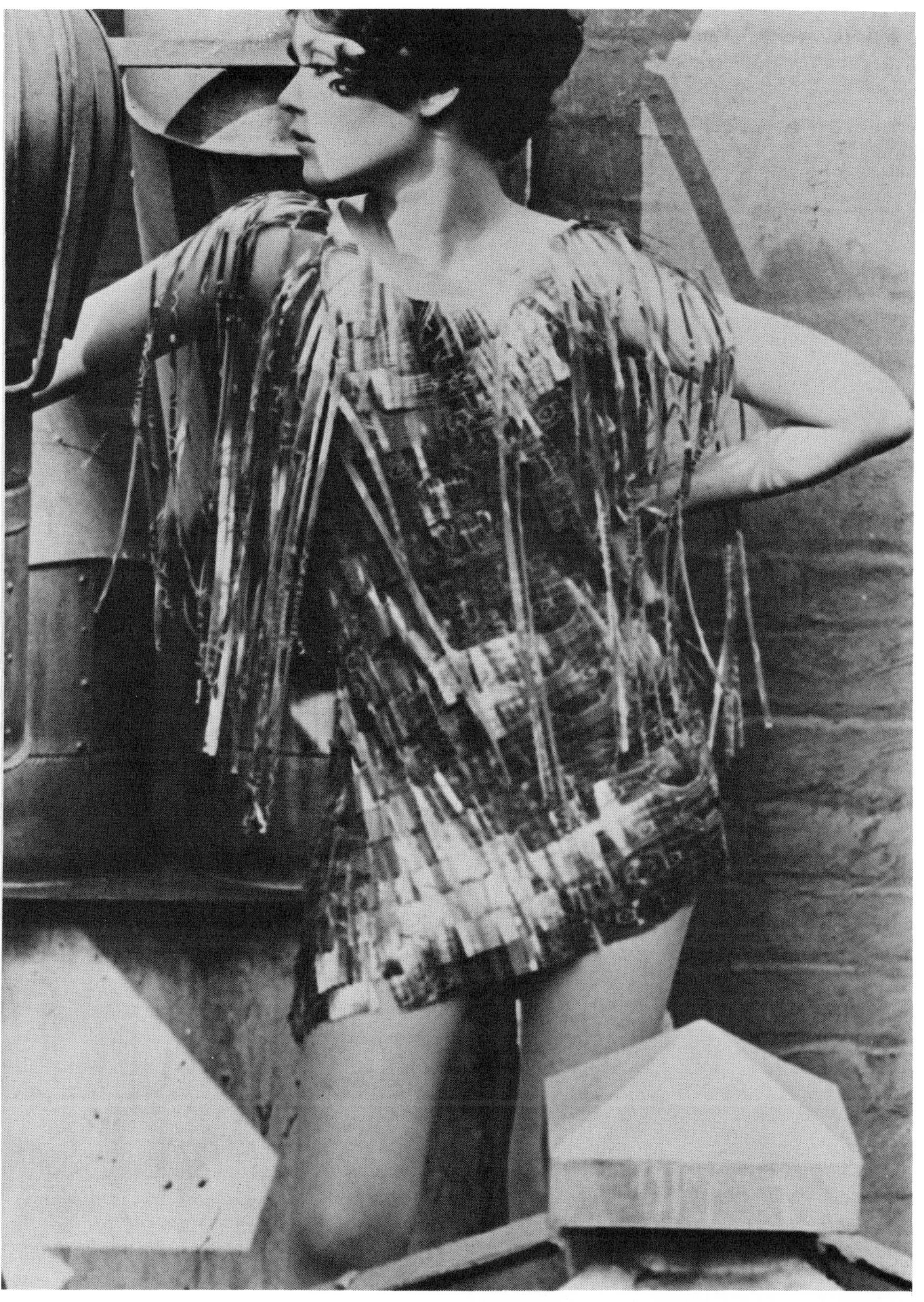

This minidress is constructed of flexible printed circuits, the same as those used in television sets. The circuits are etched from a laminate of polyester film and copper. Elisa Stone is the designer-engineer. (*Courtesy, Celanese Plastics Co.*)

2

plastic facts and generic extents

It is difficult to believe, as we look around at all the sophisticated plastic products upon which we are dependent, that the first man-made plastic was discovered over one hundred and fifteen years ago. From buttons and stockings to boats and buildings, man-made plastics have become a vital medium of today.

HISTORY OF PLASTICS IN A CAPSULE

Athough nature's plastics have been with us for hundreds of years—waxes, amber, clay, glass, resins, gums, and rubber, it was not until the beginning of the nineteenth century that these materials were pressed into heavy use as the Industrial Revolution began its exploitation of natural resources. Shellac and asbestos were made into insulators for the budding electrical industry; shellac and slate powder became the raw material of phonograph records. Latex rubber became a practical material when Charles Goodyear discovered vulcanization. And seven years later, in 1846, the first vital link to the making of synthetic plastic was forged when Dr. Friedrich Schönbein, a professor of chemistry at Basle University, discovered that he could convert the cellulose in wood and other plant products into a clear, horny, tough material by treating it with nitric acid. This became known as nitrocellulose (cellulose nitrate) and at first was used as an explosive gun cotton. At the point when Schönbein was patenting his nitrocellulose in England, Alexander Parkes of Birmingham, England, became interested in it as a lacquer. He noticed that, as the nitrocellulose hardened, it at first passed through a jelly-like stage and could at that point be molded like clay. But nitrocellulose contracted and sometimes cracked as it hardened. Parkes tried mixing it with other materials with no success, until he tried molten camphor. He called this first plastic "Parkesine" and exhibited it in 1862 but did very little with it commercially. (Some hair clasps, buttons, a paper knife, and other objects of Parkesine are displayed in the Science Museum in London.)

In 1868 an American newspaper advertised a $10,000 prize for the man who could provide a substitute for ivory billiard balls. John Wesley Hyatt rose to the occasion and made his billiard balls by grinding up nitrocellulose with camphor and heating the mixture. He called his discovery "celluloid," and patented it. At Hyatt's death in 1920 his company, the American Celluloid and Chemical Corporation was absorbed by the Celanese Corporation.

Two German chemists, Adolf Spitteler and W. Krische, joined forces in 1897 in experimenting with casein. When they combined casein with formaldehyde, the casein "cured" and became resistant to the softening action of water. Man learned to improve on nature's plastics—rubber (vulcanizing with sulphur), casein (curing with formaldehyde), shellac (combining with fillers—slate, asbestos)—and invented his first synthetic plastic, Parkesine or celluloid. At that time nobody could explain the peculiar behavior of these materials. There was no plausible scientific theory to account for the semisolid-semiliquid state. Plastics had not yet become part of theoretical science until Dr. Leo Baekeland published the results of his work in 1909. He found that phenol and formaldehyde when combined formed a resinous substance—phenolic plastic which he called "Bakelite." It was a *plastic*—it could be softened with heat and then molded into shape and could be set into a final form by continued heating under pressure while in the mold.

Baekeland's discovery triggered the imagination of organic chemists and research began. Within a very short time plastics has become a complex and specialized field with over forty different families of synthetic plastics.

THAT PLASTIC QUALITY

Moldability or plasticity is probably the reason why these materials have been named "plastics." Man's ability to alter the shape of a material and the property of its retaining its shape when heat or pressure is no longer exerted is what proves plastics so precious a material. These dual characteristics are commercially invaluable. As the technology of plastics grew, it became apparent that there were two basic kinds of plastics—*thermoplastics* and *thermosets.*

Thermoplastics are those materials that in the final state are hard but in processing can become soft and moldable. Like wax or ice, thermoplastics can be melted and re-formed many times. On the other hand, *thermosets* retain their final shape after earlier plasticity and cannot (theoretically) be melted and reformed, much like clay. Acrylics, vinyls, styrenes, polyethylenes are examples of thermoplastics; polyesters, epoxies, melamines, silicones are thermosetting plastics.

Some plastics are combinations of several plastics, much as metals that are alloys. ABS is a mix of acrylonitrile, butadiene, and styrene. Polyvinyls may be combined with acetate, chloride, acetal, et cetera.

Chemists have been able to combine molecules with heat and/or pressure to create new categories of plastics. They are only new in the sense that they are man-made under controlled conditions. Atmospheric nitrogen, milk, coal, limestone, fluorospar, petroleum, salt, sand, wood, and some vegetable matter become the raw materials with which the scientist works. He makes simple chemical compounds called "monomers," such as phenol formaldehyde, urea, melamine, methyl methacrylate, ethylene, vinyl acetate, adipic acid, et cetera, and from these basic materials whole ranges of plastics are made. When monomers are built up or linked to produce large molecules, the result is called *polymerization* and these huge molecules often consisting of thousands of atoms linked together are called *polymers.* "Poly" is placed in front of the name of a monomer to describe the polymer made from it; thus styrene polymerizes to form polystyrene; methyl methacrylate polymerizes to form polymethyl methacrylate (acrylic).

Polymers are mixtures comprised of molecules which have a wide variety of different absolute sizes; can be present in a variety of differing amounts of each size of molecule; can have a wide variety of molecular shapes and configurations; can have a homogeneous nature (identical repeating units) or they can be composed of segments with differing chemical compositions; can arrange themselves in various ways with respect to one another in space to form regular aggregates of molecules of crystals or in a random fashion to form amorphous polymers; can owe their structure, relative sizes, and distribution of sizes to the manner in which they are made; and can be mixed physically with other molecules, such as other polymers, fillers, or additives, to form secondary mixtures with properties different from the base polymer.

These differences in polymeric mixtures that give rise to the great variety of properties that potentially can be incorporated into plastics impart a tremendous potential, and this is only the beginning of the plastics era. As you can see, this variety also gives rise to the complexity in understanding polymer chemistry and engineering.

SPECIFIC FAMILIES

Some specific families are described in short form in the following charts—their formulation, pertinent characteristics, and uses. More detailed specifications for particular "brands" or generic groups can be obtained from the manufacturers. Each trade-named material has its full range and variety of polymers, with wide latitudes in characteristics and with specific differences tailored for prescribed applications.

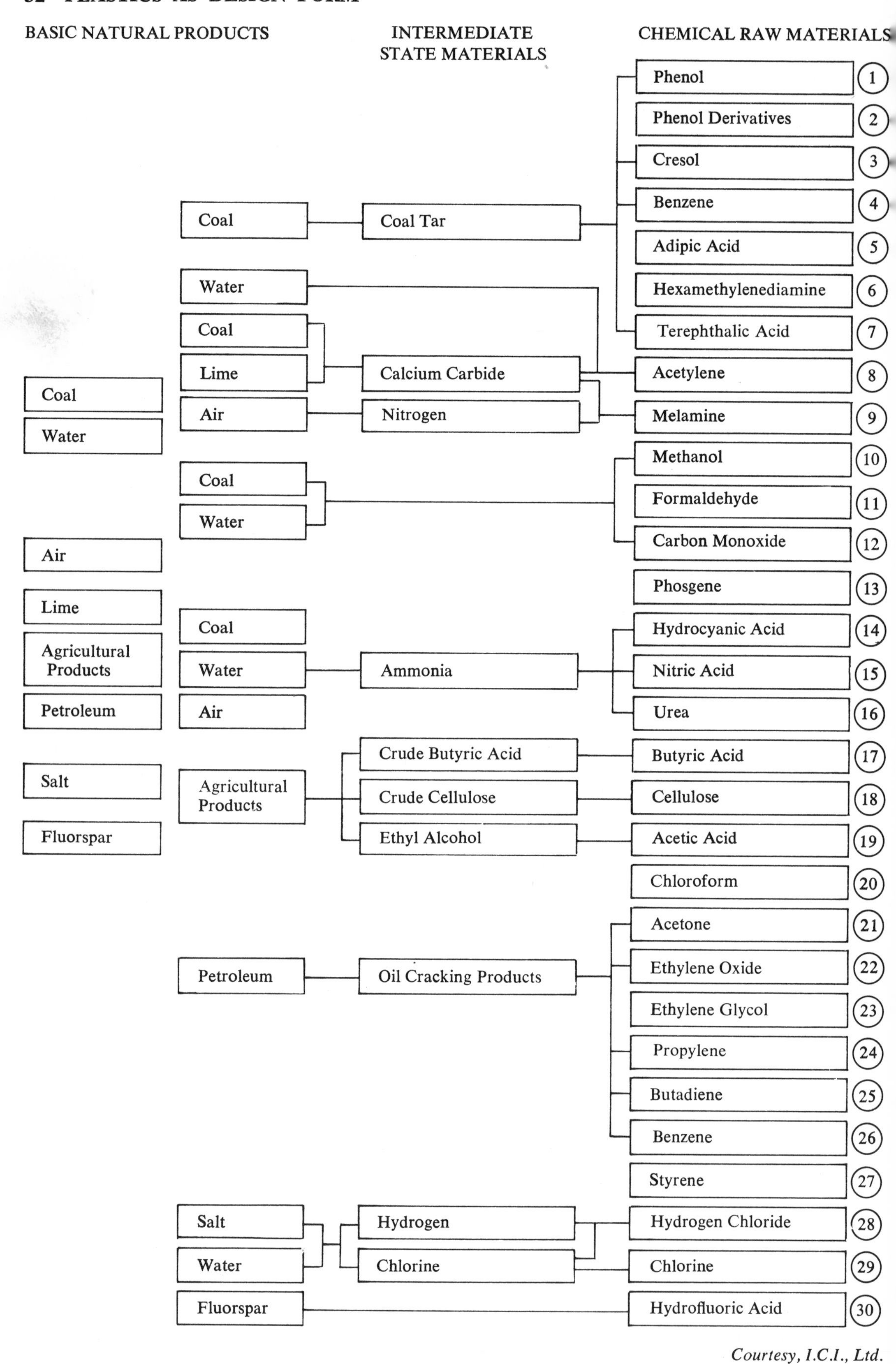

Courtesy, I.C.I., Ltd.

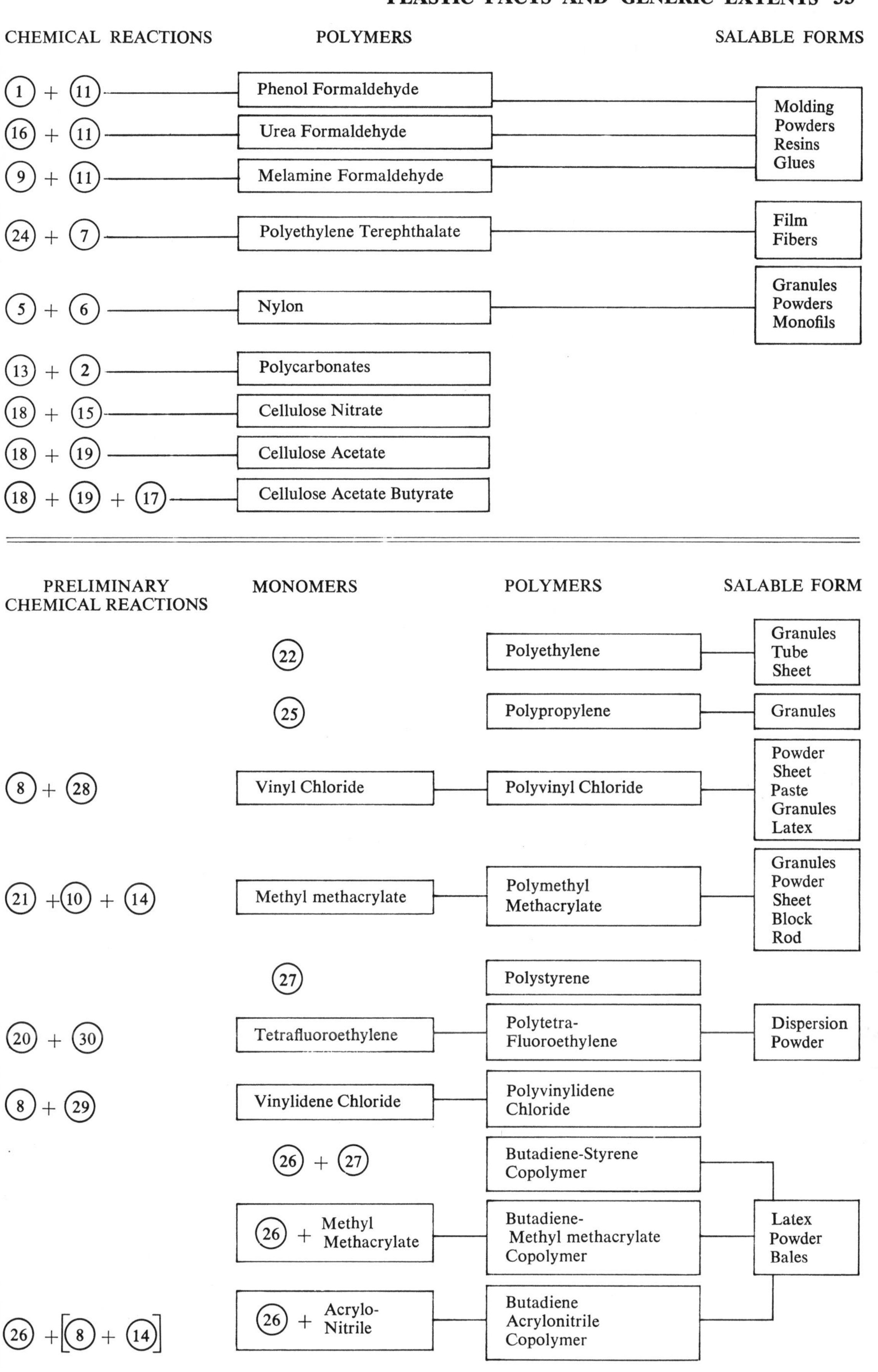
CHEMICAL REACTIONS
POLYMERS
SALABLE FORMS
1 + 11
Phenol Formaldehyde
16 + 11
Urea Formaldehyde
9 + 11
Melamine Formaldehyde
Molding
Powders
Resins
Glues
24 + 7
Polyethylene Terephthalate
Film
Fibers
5 + 6
Nylon
Granules
Powders
Monofils
13 + 2
Polycarbonates
18 + 15
Cellulose Nitrate
18 + 19
Cellulose Acetate
18 + 19 + 17
Cellulose Acetate Butyrate
PRELIMINARY
CHEMICAL REACTIONS
MONOMERS
POLYMERS
SALABLE FORM
22
Polyethylene
Granules
Tube
Sheet
25
Polypropylene
Granules
8 + 28
Vinyl Chloride
Polyvinyl Chloride
Powder
Sheet
Paste
Granules
Latex
21 + 10 + 14
Methyl methacrylate
Polymethyl
Methacrylate
Granules
Powder
Sheet
Block
Rod
27
Polystyrene
20 + 30
Tetrafluoroethylene
Polytetra-
Fluoroethylene
Dispersion
Powder
8 + 29
Vinylidene Chloride
Polyvinylidene
Chloride
26 + 27
Butadiene-Styrene
Copolymer
26 + Methyl
Methacrylate
Butadiene-
Methyl methacrylate
Copolymer
Latex
Powder
Bales
26 + [8 + 14]
26 + Acrylo-
Nitrile
Butadiene
Acrylonitrile
Copolymer

PROPERTIES OF THERMOSETTING RESINS

SOME THERMO-SETTING PLASTICS	FILLER TYPE	BURNING RATE	EFFECT OF SUNLIGHT	CLARITY	REMARKS ABOUT PROPERTIES	USES
Alkyd and Polyester (from molding powders to liquids)	No filler	Slow	Slight—after over twenty years turns brownish	Transparent to translucent	Good surface hardness, colors well in a wide range, can be formed without heat and pressure	Light switches, tuning devices, coatings, structural coverings when reinforced with fiberglass such as boat hulls, and car bodies, furniture shells, castings (embedments), decorative panels, sculptures, buttons
	Glass reinforced (woven cloth)	Slow to self-extinguishing	Slight	Translucent to opaque		
	Granular putty (mineral filled)	Slow to none	Virtually none	Opaque		
	Asbestos filled	Self-extinguishing	Virtually none	Opaque		
Diallylphthalate Molding Compounds (DAP)	Glass filled	Self-extinguishing to none	None	Opaque	Exceptional dimensional stability, surfaces are hard and tough, low moisture absorption, easy to mold at low pressure	Handles, knobs
	Mineral filled	Self-extinguishing	None	Opaque		
	Synthetic-fiber filled	Burns	None	Opaque		
Epoxy Molding Compounds	Glass filled	Self-extinguishing	Slight	Opaque	Exceptional adhesive qualities, low shrinkage, high strength when reinforced, transparent but light straw-colored	Same as for polyester with additional use as bonding agent. Used for encapsulating rather than embedment where light straw color would not be critical
	Mineral filled	Self-extinguishing	Slight	Opaque		
Epoxy Cast Resins	No filler	Slow	None	Transparent to opaque		
	Silica filler	Self-extinguishing	None	Opaque		

Melamine (from molding powders, foams, and liquids, amino-type resins)	No filler	Self-extinguishing	Colors fade	Translucent to opalescent	Low water absorption, tasteless, odorless, resists scratching and marring	Tableware, buttons, distributor heads, laminated surfaces such as table tops, industrial baking enamel for hard-finish coatings on stoves and refrigerators
	Asbestos filled	Self-extinguishing	Slight	Opaque		
Phenolics Compounds for molding, casting, and laminating	No filler	Slow	Darkens	Transparent	Has strength, hardness, and rigidity, low cost, colors limited to black or brown	Handles, knobs, switches, appliance parts
	Wood-flour filled	Slow	Darkens	Opaque		
Polyimide	Laminating, adhesive, filament winding, press cured, encapsulating	Self-extinguishing	Nil	Opaque	Resistant to ionization and radiation and most organic substances, better tensile strength than nylon	Aerospace parts, bearings, seals, piston rings, cable wrap, insulators for lunar cameras, cryogenic applications
Silicone Molding Compounds for laminating, foaming, casting, molding, coating	Asbestos filled	None to slow	Slight	Opaque	Resistant to heat, low moisture absorption, non-stick surface, can be cured without heat or pressure, physiologically inert	Insulation for motors, switch plates, laminations, molds for casting, and fiberglass layup
	Glass-fiber filled	None to slow	Slight	Opaque		
	Mineral filled	Self-extinguishing	Slight	Opaque		
Urethane Elastomer solids and foams	No filler	Slow	Browns but does not decompose	Opaque to transparent	Weather resistance even though changes color, tear resistance	Cushions, mattresses, pillows, gaskets, molds for casting

PROPERTIES OF THERMOPLASTICS

SOME THERMO-PLASTICS	TYPE	BURNING RATE	EFFECT OF SUNLIGHT	CLARITY	REMARKS ABOUT PROPERTIES	USES
ABS (Acrylonitrile Butadiene Styrene)	Molding, extruding, calendaring, vacuum forming	Slow	Chalks slightly	Translucent to opaque	Hard, rigid, exceptional strength and toughness, dimensionally stable	Pipes and pipe fittings, football helmets, telephones, trays, and tote boxes, furniture—tables, storage seating, shoe heels, refrigerator liners, automotive parts
Cellulosics Cellulose Propionate	Molding compound	Slow	Slight	Transparent to opaque	Ease of fabrication, accepts color well. Cellulose acetate butyrate is suitable for outdoor use. Cellulose propionate has high impact and tensile strength; therefore is dimensionally stable	Eyeglass frames, dice, drawing instruments, toothbrushes, packaging, light shades, lighting globes, tool handles radio cabinets, pen barrels, telephones, auto interior parts
Cellulose Acetate	Sheet	Slow	Very slight	Transparent to opaque		
Butyrate	Molding	Slow	Slight	Transparent to opaque		
Polypropylene	Molding and extruding	Slow	Must be protected	Transparent to opaque	Higher density than polyethylene and more rigid, good chemical resistance, very lightweight, good heat resistance and good mechanical properties	Pipes, heat-sterilizable bottles, wire coating, blownware, packaging, textiles, rope, electrical components, housewares
Polystyrene (granules, liquids, powder, sheets, foams, adhesives, coatings)	General purpose to high impact molding & coating extruding	Slow	Yellows	Transparent to opaque	Transparency, colors well, dimensionally stable. With the addition of rubber, greater impact resistance possible	Lamp globes and shades, housewares, storage boxes, decorative panels, coatings, adhesives
Vinyls and Copolymers	Molding, casting extruding	Slow to self-extinguishing	Slight	Transparent to opaque	Strong, abrasive resistance, low moisture absorption, good chemical resistance	Louvers, corrugated sheet, pipe, phonograph records, gaskets, textiles—upholstery, curtains, protective clothing, garden hose, wire jacketing, shoes, boots, floor tile, automobile roof covering
Polyallomer (propylene-ethylene)	Vacuum forming, injection molding, extrusion	Slow	Not advisable for outdoor use	Translucent to opaque	High impact strength, good melt strength, superior resistance to fatigue from flexing, surface hardness	Pipe fittings, self-hinged containers, bookbinders, toys, shoe lasts

Parylene (polyparaxylene)	Vacuum coating	Self-extinguishing	———	———	Dimensional stability outstanding, thermal efficiency increases in absence of air, chemical resistance excellent	Insulating and protective coatings, encapsulating of reactive materials
Polyphenylene Oxide (PPO)	Injection molding, extrusion	Self-extinguishing	Very slight	Opaque	Wide temperature range (—275° F. to +375° F.), constant mechanical properties over a wide range, unusual resistance to acids, bases, and steam, will bend but not break, excellent dimensional stability	Battery cases, coil forms, medical and surgical instruments, plumbing equipment, household appliances
Cellulose Acetate	Sheet and film	Slow to self-extinguishing	Very slight	Transparent to opaque		
	Molding	Slow to self-extinguishing	Slight	Transparent to opaque		
Fluorocarbons	Molding	None	None	Transparent to opaque	Inertness to most chemicals and resistance to high temperatures. Strong, hard, have good impact strength	High voltage insulation, gaskets, coatings, nonlubricated bearings, heating cable, transistor bases
Nylon	Molding and extruding	Self-extinguishing	Slight color change	Translucent to opaque	Resistance to temperature extremes, toughness, chemical resistance, mechanical durability. Not good for outdoor use	Tumblers, combs, gears, bearings, bristles, battery cases, washers, textiles—upholstery, clothing, stockings, draperies
Polysulfone	Injection molding, extruding, blow molding, thermoforming, adhesives	Self-extinguishing	Slight	Light amber to opaque	Stable over 300° F. with 30% loading of glass, chemically resistant to mineral acids, salts, alkalies, detergents, oils, and alcohol	Hand power tool housings, switches, computer parts, underhood auto applications, pipe and sheet, structural bonding, and laminating formulations
Polycarbonate	Molding	Self-extinguishing	Slight color change and embrittlement	Transparent to opaque	Exceptional impact strength, rigidity, and stability. If ultraviolet stabilized, has good weatherability. Good light transmission. Expensive	Aircraft parts, appliances, business machine parts, sight glasses, high temperature lenses, outdoor lighting globes

PROPERTIES OF THERMOPLASTICS

SOME THERMO-PLASTICS	TYPE	BURNING RATE	EFFECT OF SUNLIGHT	CLARITY	REMARKS ABOUT PROPERTIES	USES
Polyethylene	Molding & extruding vacuum forming (varying density) pellet, powder, sheet, film, rods, foam	Slow	Must be protected, not advisable for outdoor use	Transparent to opaque	Resistance to breakage. Can withstand low temperatures	Housewares, pipe, bags for food products, cable coating, bottles, light diffusers, blown globes for lighting when reinforced with fiberglass as rigid seating
Acrylic (Polymethyl Methacrylate)	Casting	Slow	None	Transparent to opaque	Excellent light transmission, good weatherability, rigid, hard, colors well. Heat distortion low—220° F.	Display signs, lenses, dials, window glazing, furniture—tables, shelves, seating, decorated panels, sculptures, parts in jewelry, decorative boxes, display cases, medical displays
	Molding	Slow	None	Transparent to opaque		
Styrene-Methyl Methacrylate	Molding	Slow	Very slight to nil	Transparent to opaque	Excellent light transmission, resistance to weathering, toughness, colors well, good heat resistance	Lighting lenses, automobile parts, signs, appliance parts, displays, furniture elements

POLYMER (COMMON NAME IN PARENTHESES)	FABRICATION METHODS
Acrylonitrile/Butadiene/Styrene Copolymer † (ABS)	E B M T
Butadiene/Methyl Methacrylate Copolymer †	
Butadiene/Styrene Copolymer †	
Ethylene/Vinyl Acetate Copolymer	E B M Fm Fo
Melamine-Formaldehyde	M
Nylon	E B M Fi
Phenol-Formaldehyde	M
Polycarbonates	E B M
Polyethylene	E B M T Fm
Polyethylene Terephthalate	Fi Fm
Polyformaldehyde (Acetal)	E B M
Polymethyl Methacrylate	E M T
Polymethylpentene	E B M T
Polypropylene	E B M T Fm Fo
Polystyrene	E B M T Fo
Polytetrafluoroethylene (PTFE) ‡	
Polyurethane Foams and Coatings	Fo
Polyvinyl Chloride (PVC)	E B M T Fm
Urea-Formaldehyde	M
Vinyl Chloride/Vinyl Acetate Copolymer	E B M T Fm

† = Used mainly in combination with other materials or as latices
‡ = Does not melt—processing involves sintering of particles

E = Extrusion
B = Blow molding
Fi = Fiber
Fm = Film
Fo = Foam
M = Injection and/or compression molding
T = Thermoforming

Information courtesy I.C.I., Ltd., England

A PRIMER OF PROCESSES

As molecules were aligned and connected to form new materials, new processes were being developed to complete the chain from material to product. Processing techniques are becoming more sophisticated every day. Although new applications bring forth more refinements, processing theory remains essentially consistent.

INJECTION MOLDING

Injection molding is a mass-production, high-speed operation. Plastics used for injection mold-

Cross-section of a typical injection molding machine.

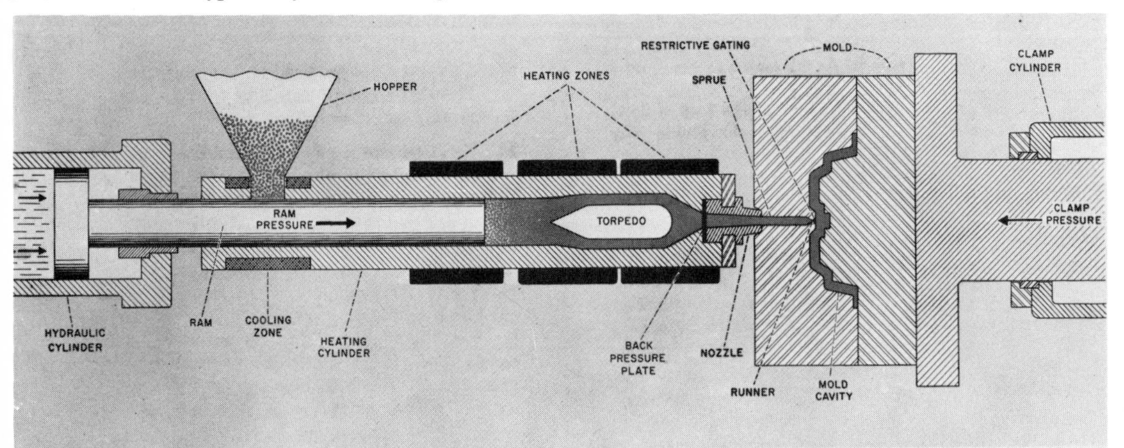

ing (usually thermoplastics but sometimes thermosets) must be held in a molten condition without degrading. Some thermoplastics such as cellulose nitrate are too sensitive to heat and cannot be used.

Raw material is fed into a hopper and then into a heating chamber, where it becomes thoroughly fluid and molten as a plunger pushes the plastic through a long heating chamber via a revolving screw. At the end of this chamber the fluid plastic is forced through a nozzle, thereby ejecting the plastic under pressure into a cool, closed mold. As the plastic cools, the mold opens and the finished piece is ejected. The process repeats in rapid succession so that no plastic remains too long in the heating chamber.

Compression Molding

Compression molding has been used in the rubber industry for over a hundred years. It is a forming method used for thermosetting plastics utilizing three factors—pressure, temperature, and time. Either cold or preheated compounds are put into an open mold cavity and then heated. The mold is closed by a ram that presses down on the plastic, forcing the material to flow and take on the shape of the mold. The part is removed after it has cured.

Extrusion

Extrusion resembles a meat-grinding operation. In application to plastic, dry thermoplastic material in the form of pellets is fed into a hopper and then into a long heating chamber

PROCESS SELECTION CHART

MOLDING PROCESS	PART SIZE RANGE	TIME TO PRODUCE ONE PART	EQUIPMENT REQUIRED	GENERAL PROPERTIES
Hand layup	1–50 ft.	1–10 hr.	Hand tools, curing oven	Low mechanical strength
Bag molding	4 in.–8 ft.	1–4 hr.	Curing oven, vacuum press	Good strength, complex parts
Autoclave molding	12 in.–10 ft.	3–6 hr.	Autoclave	Highest strength, critical parts
Vacuum injection molding	12 in.–8 ft.	1–3 hr.	Vacuum and pressure pumps	Void free, higher resin content, dimensional stability, difficult contour, complex inserts
Premix molding	3–24 in.	5–25 min.	Press and molds	Low mechanical strength, good surface and chemical resistance
Transfer molding	3–24 in.	10–30 min.	Press and molds	Same as premix
Mat and preform molding	6–60 in.	10–30 min.	Press and molds, preform machine	Medium strength, good chemical resistance
Prepreg molding	6–50 in.	10–30 min.	Press and molds	High strength
Wet fabric molding	6–50 in.	10–30 min.	Press and molds	High strength
Filament winding	Diameter 3 in.–26 ft.	1–24 hr.	Filament winding machine, curing oven	Very high strength, limited to surfaces of revolution

Stacking dishes injection molded of polycarbonate by Angelo Mangiarotti. (*Courtesy, Architect Angelo Mangiarotti*)

via a long, rotating, helix-finned screw. At the end of the chamber the fluid plastic is forced out by pressure through a specially shaped aperture (die) that gives the product its shape and then is fed onto a conveyor belt, where it is cooled by blowers or by immersion in water. Extrusions may be hair-thin monofilaments, various-sized rod, sheet, or pipe several feet in diameter.

CALENDERING

Using something similar to the rollers in an old-fashioned washing machine, calendering converts thermoplastics into sheet or film or applies coatings to textiles and other supporting materials by pressing them between a series of three or four large, hot, revolving rollers (film thickness up to 10 mils, sheet thicknesses over 10 mils). As the rollers turn, the thermoplastic is squeezed under heat and pressure through the adjusted thickness and assumes the surface texture of the rollers.

Injection molded parts of colorful transparent polystyrene in a color toy designed by Clement Meadmore for Trenton Toys, England. (*Courtesy, Clement Meadmore*)

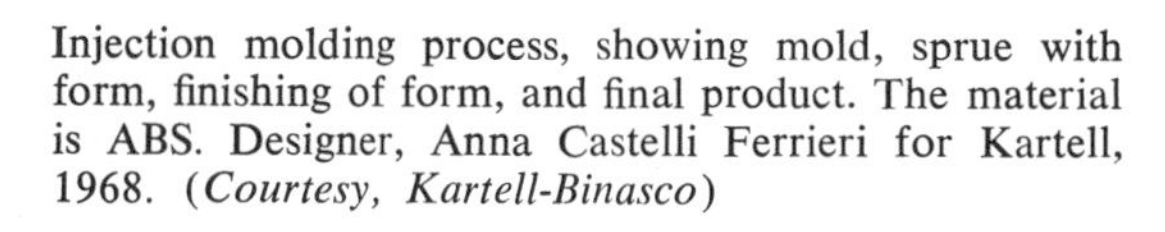

Injection molding process, showing mold, sprue with form, finishing of form, and final product. The material is ABS. Designer, Anna Castelli Ferrieri for Kartell, 1968. (*Courtesy, Kartell-Binasco*)

In anchoring plastic to a fabric, the plastic coating compound is passed through two top horizontal rollers, while the uncoated material is passed through two bottom rollers and emerges as film anchored to fabric.

Blow Molding

There are many variations of blow molding, through direct and indirect methods. Essentially blow molding is the stretching and then hardening of thermoplastic against a female mold. In the direct approach a gob of molten thermoplastic is at first formed into the rough shape of the final product. It is then inserted into a female mold; air is blown into it as into a balloon, which forces the hot plastic against the sides of the mold. When cooled, the form is removed from the mold.

In the indirect method thermoplastic sheet or a special shape is at first heated and then clamped between die and cover. Air pressure introduced between the plastic and cover forces the plastic to contact the die and assume its contour. After the plastic has cooled, the form is removed.

Casting

Casting is a simple method that requires no pressure and sometimes no heat. Either thermoplastics or thermosetting plastics are used. Thermoplastics are softened (thermosetting plastic liquids are used) and placed in an open-ended mold. By cooling or by curing with or without heat the material takes the shape of the mold. Film, sheet, rods, tubes, embedments, and special shapes can be made this way.

Coating

There are various methods of coating—knife or spread coating, spraying, roller coating, dipping and brushing, fluidized bed coating, and metallizing. (Calendering is also a form of coating.) Either thermoplastics or thermosets can be applied by machine or manual methods. Generally the appropriate amount of plastic is metered out and spread—usually by rollers, knives, or blades—onto whatever is being coated.

Rotational Molding

If a form could feel, it probably would feel as one does in an amusement-park ride that rotates a cubicle around a wheel and at the same time revolves on its own axis. Hollow, plastic, one-piece parts can be made in this manner. The machine consists of two or three arms extending outward like spokes from a central form. These arms move consecutively from a loading-unloading position into a heating chamber and then into a cooling chamber, while the molds attached to the arms are biaxially rotated. At the loading station cool, preweighed, micropulverized molding powders are placed inside a two-piece metal mold and bolted shut. The mold which is attached to a spider or holding device at the end of each arm moves to a heating chamber, where it is rotated. As the thermoplastic softens it adheres to the mold wall because of centrifugal action. After it has been moved to a cooling station, the part is removed. Since the plastic is not oriented, it is stress free. Street lamps, mannikins, dolls, and luggage are among the hollow forms made this way. Usually vinyl or polyethylene is used.

Filament Winding

Cylindrical or spherical parts are produced by this method of continuous glass filament or yarn fed through catalyzed thermosetting polyester or epoxy resin. Excess resin is removed by doctor blades. The glass filament is attached to a wood, metal, paper, elastomer, plastic mandrel shaped to the interior of the part to be molded. The saturated filament is pressure tensioned with an automatic device as it is tightly and evenly wound around the mandrel. After build-up, curing is accelerated by heat and the part is then removed from the mandrel.

Forming (Thermoforming)

Forming includes a wide variety of methods to shape thermoplastic sheet such as acrylic. Basically this process consists of heating thermoplastic sheet to a formable flexible state and then applying air and/or mechanical assists to shape it to the contours of the mold. Vacuum forming is one variety, and pressure forming is another variation of the principle. Drape forming, plug and ring forming, slip forming, snap-back forming, reverse-draw forming, plug assist and machine die forming constitute some other refinements.

Hand Layup

Hand layup is a room-temperature operation of impregnating fiberglass with thermosetting polyesters or epoxies in an open mold. There are also variations on this theme. Machinery

Injection molding process using ABS, showing top section of a two-part chair by Joe C. Colombo for Kartell (*Courtesy, Kartell-Binasco*)

Cross-section of an extruder.

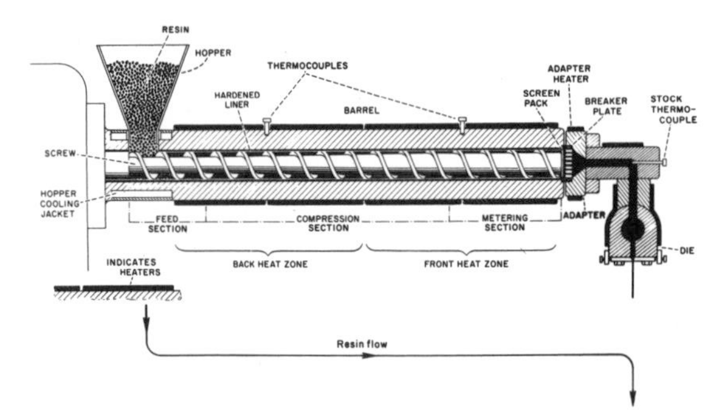

Plug assist, thermoforming, a vacuum forming operation, making a globe of patterned acrylic. In illustration (a) the acrylic sheet has been heated; a ring is placed on top of the softened sheet. (b) Clamps are tightening the sheet and ring into place. (c) The plug is drawn into the mold and the vacuum pulls the hot acrylic to the female mold while the male mold forces the acrylic downward. (d) Plug and clamps are released, ring is removed, and the acrylic form is taken off the mold. (e) A shaper with a veneer saw blade is trimming away the excessive acrylic. (*Courtesy, Dura Plastic of New York*)

The finished acrylic globe made by Dura Plastic of New York. (*Courtesy, Dura Plastic of New York*)

Two aspects of vacuum forming. The top design utilizes a male mold. Heated thermoplastic sheet is vacuum drawn over the mold. In the lower diagram a plug assists the vacuum that draws the heated sheet into a female mold. The Dura Plastic globe was made this way.

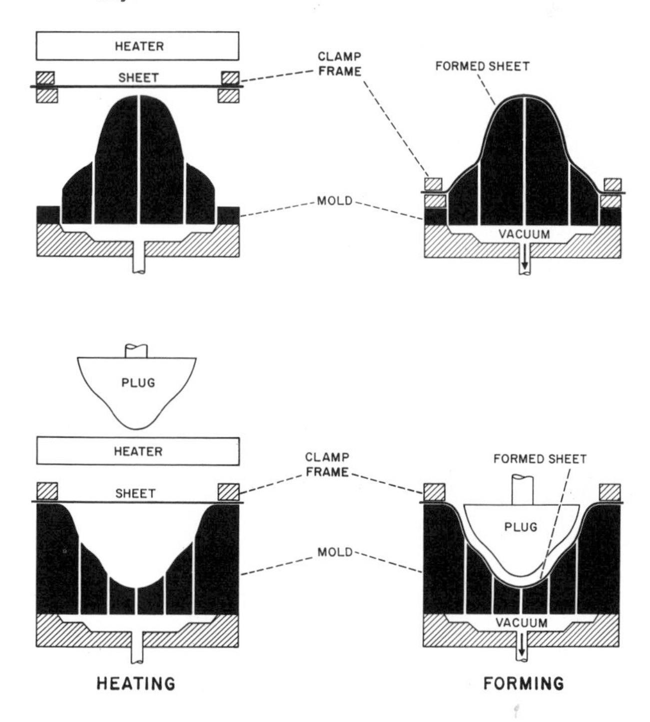

can be used to spray the glass-resin mixture into the mold. Then brush and roller daubing forces out entrapped air bubbles. As the resin soaks the reinforcing layers of glass, thicknesses are built up and the form is left to cure. Another variation of hand-layup techniques includes drape molding, bag molding, and low compression molding.

Laminating and High Pressure Laminating

Thermosets are generally used for laminating with or without application of heat or pressure. Generally, impregnated substrates are pressed into a sheet. Thin wood, cloth, paper, and fiberglass (reinforcements) are held together by impregnating plastic, sometimes melamine. In producing a flat surface, impregnated sheets are stacked between a sandwich of two highly polished steel plates and subjected to heat and high pressure in a hydraulic press which cures the plastic and presses the layers of material into a single sheet of predetermined thickness. In producing continuous lamination a web of reinforcing material passes through an impregnating bath of resin, between rollers or under wiper bars that pull off excess resin, and then

An acrylic blank already silk-screen printed with a Pepsi logo is clamped onto a moveable bar that is pushed into a vertical oven in order to soften the acrylic sheet. In the second photo the hot sheet is placed onto a press for pressure forming. (*Courtesy, Dura Plastic of New York*)

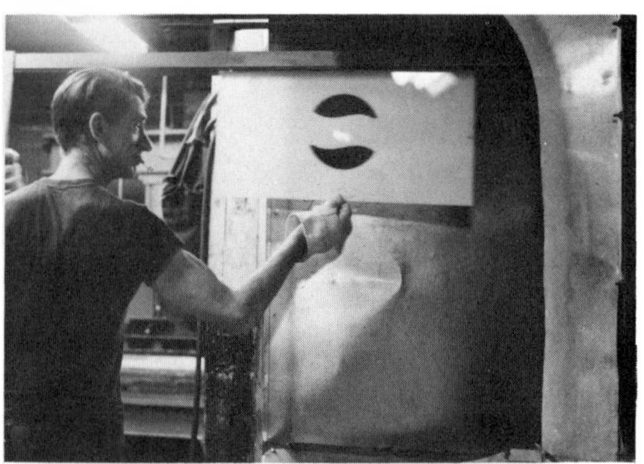

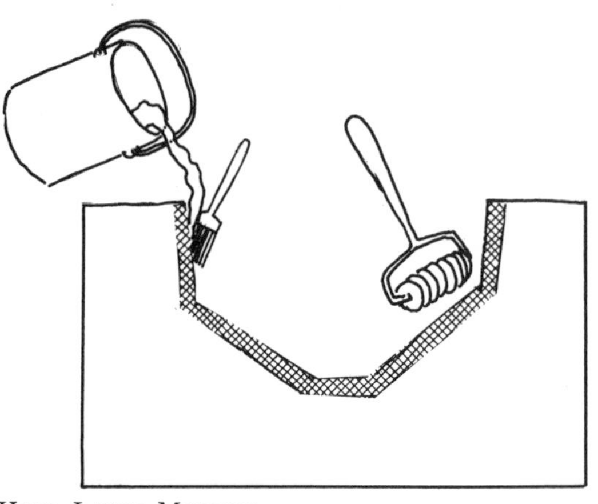

Hand Layup Molding
Reinforcement is cut and laid on the surface of either a male or female mold. Thickness of reinforcement can vary where extra support is needed. This reinforcement is then impregnated with catalyzed resin that is applied with a brush by working the resin into the glass fibers. (Catalyzed resin can be sprayed on as well.) The surface is then rolled to even out the saturation of reinforcement with resin and press away any air bubbles that may have been entrapped. Sometimes cellophane is applied during or after the rolling to improve the surface. This process is often used for making prototypes and sample runs because changes may be made as work progresses. Costly equipment and molds are not necessary here.

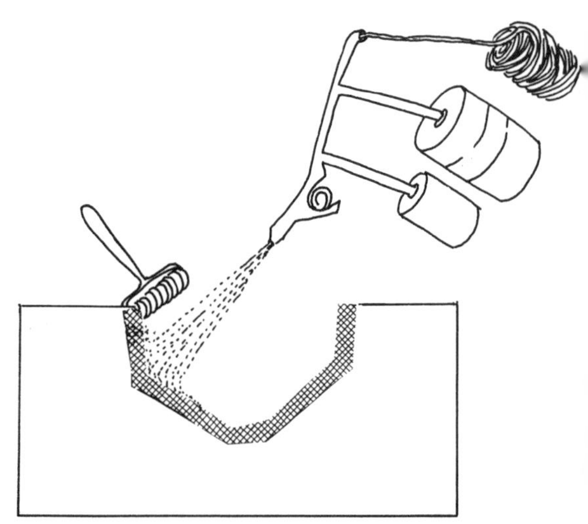

Spray-up Molding
This is mechanized extension of hand layup. Glass roving is fed through a chopper that cuts it to predetermined lengths and forces these chopped pieces into a stream of catalyzed resin (polyester or epoxy) that is deposited on the surface of the mold. The deposited mixture of impregnated resin-glass is rolled and cured then as in hand layup. This method can cover complex forms with little waste and the equipment is portable; this makes possible on-site fabrication.

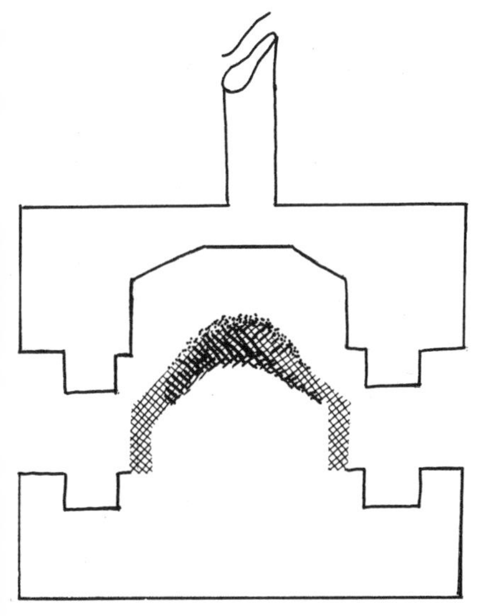

PREFORM MOLDING

Masses of reinforcement are built up as mats or felts on screens that are shaped closely to the contours of the final shape. This mass is held together with a small quantity of binder resin. The build-up of thickness in the preform is accomplished by either blowing or sucking chopped or continuous roving against the preform screen or by suspending a coater slurry and passing this suspension through the preform screen. The preform is dried by hot air until it is strong enough to handle or the binder resin has cured. It is then placed between the matched metal molds. An accurate measure of epoxy or polyester is poured over it in a special pattern, and the mold is closed. Heat and pressure cause the resin to flow and impregnate the reinforcement. This process can accommodate complex shapes economically on an assembly-line basis.

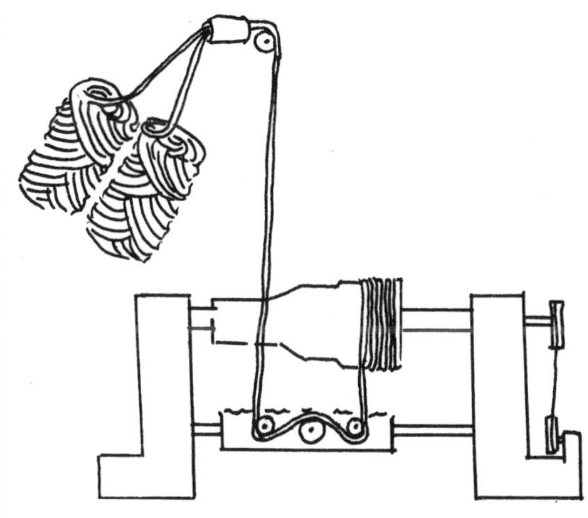

FILAMENT WINDING

Fibrous glass strands, rovings, or tapes are wound in continuous precise patterns with tension around a stationary or revolving mandrel or mold after it passes through an epoxy or polyester resin bath. (Preimpregnated resin is sometimes used.) Curing may be at room temperature, in an oven, by infrared or ultraviolet, or by vacuum or bag molding. Results have very high strength/weight ratios, sometimes higher than heat-treated steel.

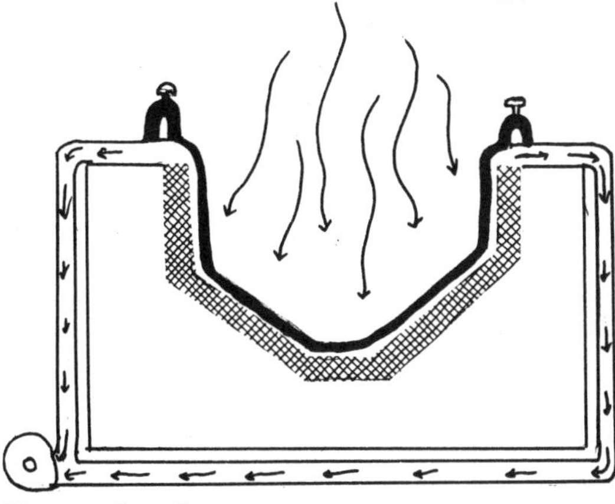

VACUUM BAG FORMING

A layup of either hand, spray-up, contour woven, or even filament wound reinforcement and resin is covered with a sheet of cellophane or polyvinyl acetate and the joints are sealed. When a vacuum is drawn under the sheet, atmospheric pressure on the layup forces entrapped air and excess resin out. It improves uniformity of the form and gives a good surface to the side not facing the mold.

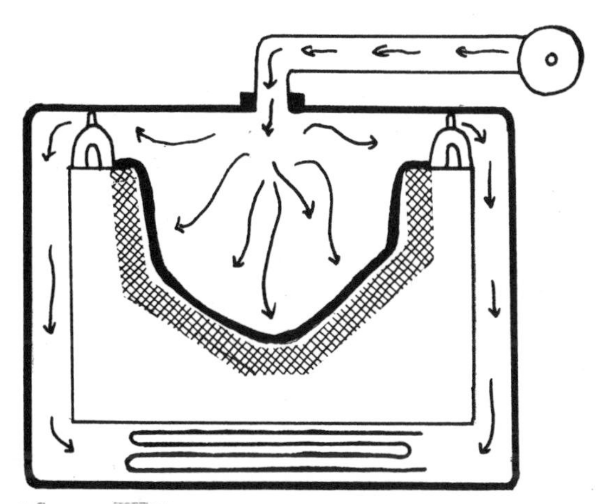

AUTOCLAVE MOLDING

This process differs from vacuum bag molding by placing the entire layup in a steam autoclave where pressure is later raised as high as 100 p.s.i. This method permits a higher glass reinforcement content with uniform resin impregnation.

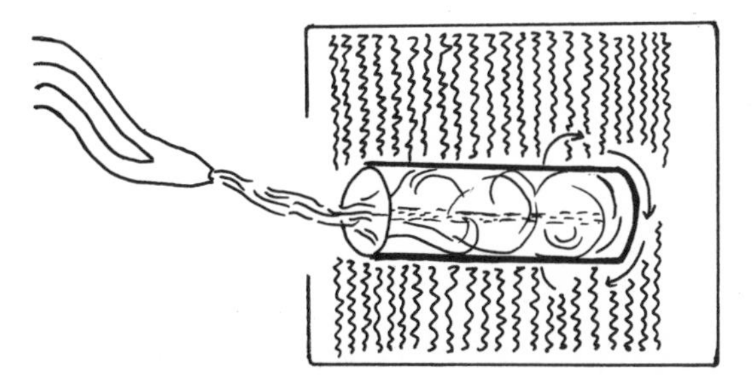

CENTRIFUGAL CASTING

This method produces parts of uniform thickness. Chopped reinforcement and resin are fed inside the hollow mandrel and are uniformly distributed in the mandrel as it is rotated inside an oven.

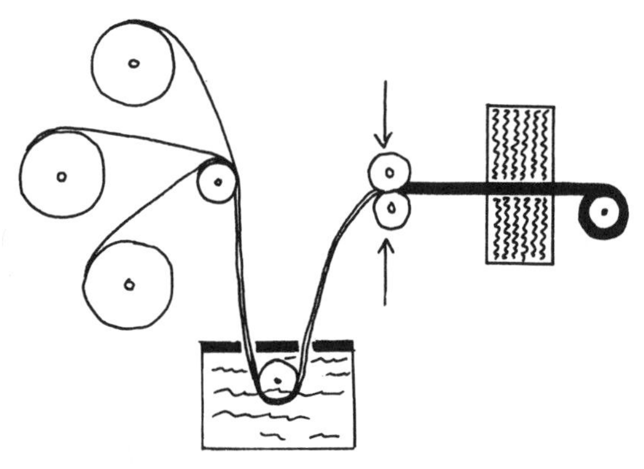

CONTINUOUS LAMINATING
Reinforcement fabric or mat is impregnated with acrylic or polyester resin by passing it through a dip tank. The saturated reinforcement is then laminated under pressure of rollers between a sandwich of cellophane covering sheets. Pressure of the rollers determines thickness and resin content. The laminate then is cured by passing through a heating zone.

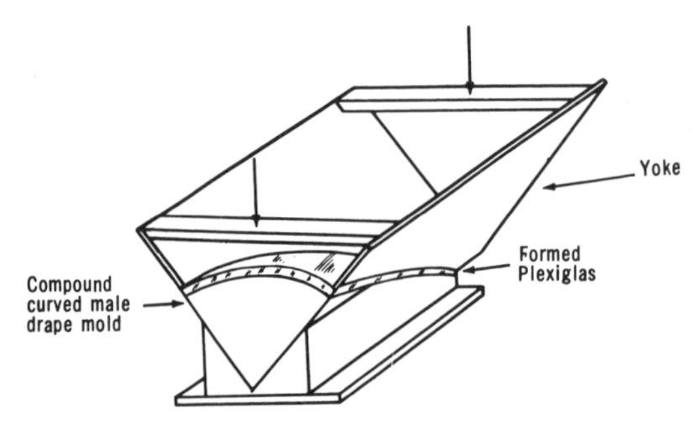

MANUAL STRETCH FORMING
Manual stretch forming is often used when the compound curvature is not great, when optical distortion is not objectionable, and when the number of parts to be made does not warrant setting up mechanical equipment. The sheet is heated to forming temperature and placed on the compound curved male drape mold. The yoke is placed over the acrylic sheet to stretch and form it to the compound curve and to hold it in position during the cooling process. (*Courtesy, Rohm & Haas Company*)

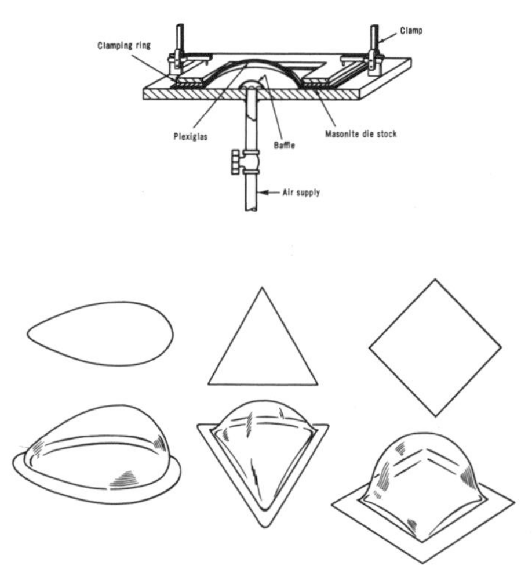

FREE BLOWING
Free blowing is used to form three-dimensional shapes by the use of positive air pressure—without the use of male or female forms. A heated sheet is clamped to a piece of plywood or Masonite. The finished part will assume the inside shape of the plywood or Masonite. After it has been clamped, air is introduced slowly until the heated acrylic is brought to its final shape. The air is maintained until the shape cools. (*Courtesy, Rohm & Haas Company*)

PLUG AND RING FORMING
A hot sheet of acrylic is clamped over an opening in a flat plate. This plate opening is designed to correspond with the outside shape of the formed part. A male plug, smaller than the clamping ring opening, is forced downward into the ring, which stretches the heated sheet to the desired cross-section. (*Courtesy, Rohm & Haas Company*)

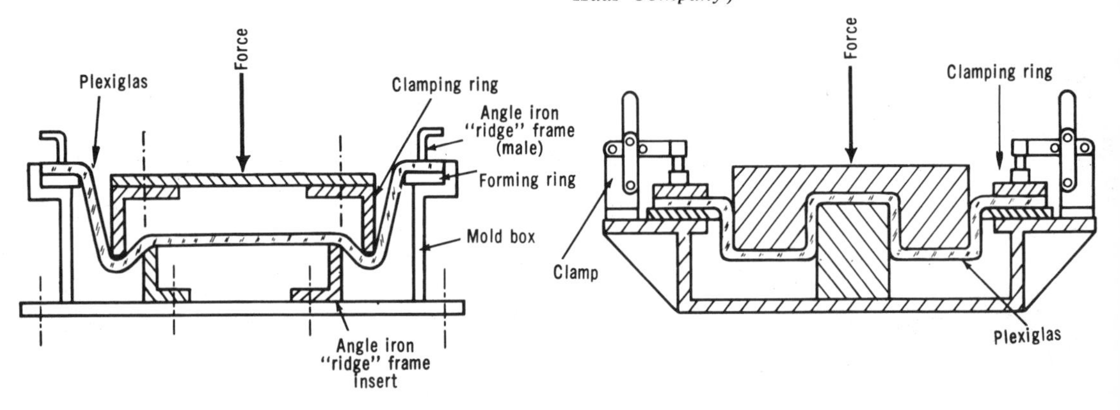

Drape forming large parts with undercuts. Heated thermoplastic is draped over male mold. It is then pressed onto the mold and, when cool, removed. (*Courtesy, Modern Plastics*)

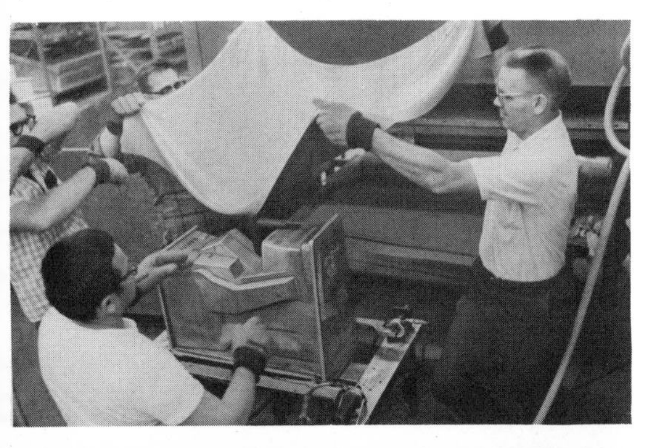

moves between sheets of cellophane, mylar, or coated paper and into a heat zone for curing. After it has cured, the cellophane is pulled off.

Pulp Molding

A porous form having a desired shape rests in a tank. Into it go thermosetting plastic, pulp, and water. As the water is drained off through the porous form by means of a vacuum, the pulp-resin mixture is drawn to the sides of the form and adheres to it. When a sufficient amount has been accumulated, it is removed and molded into a final shape, perhaps by compression molding.

Solvent Molding

Thermoplastic materials are the plastics used in this method. When a mold is immersed in a solution and withdrawn, or when it is filled with a liquid plastic and then emptied, a layer of plastic film adheres to the sides of the mold. Afterward the plastic can be removed or left on the form as a plastic coating.

Et Cetera

The preceding are some basic processes. Most techniques utilize age-old principles with such highly sophisticated refinements as combining with computers for metering, et cetera. On the other side very simple handcraft techniques exist that merely replace nature's materials for man-made plastics. For example, in hand-block printing textiles, polyethylene and polyurethane can be used to replace wood and linoleum blocks. Acrylic-modified inks and vinyl have replaced oil-based inks for silk screen, and silk in the silk screen has given way to nylon and Dacron screening. Polyethylene is used as the model in the lost wax process, vaporizing away with heat instead of wax. Polyester has replaced transparent enamel for *plique-à-jour* and *champlevé* techniques in enameling for jewelry. Synthetic fibers and acrylic shapes have joined with cotton, wool, and linen for woven materials and hangings. Polyester and epoxy set new directions as an approach to stained-glass windows, as well as acrylic and polystyrene for window-lighting effects. Latex and plaster molds have given way to room-temperature vulcanizing or setting silicones and urethanes and vinyls. Polyesters, urethanes, and acrylics have been used in many graphics techniques from etching to collography. Combinations and permutations expand potential processes in neoteric and creative directions.

BONDING AND JOINING TECHNIQUES

Plastics are capable of being bonded and joined in techniques not usual for other materials. Use of solvents, for instance, softens the surface of some thermoplastics such as acrylics and causes mating of surfaces when the solvent has completely evaporated. Thermal bonding is another method that can utilize ultrasonics with high frequency sound vibrations, hot tool and hot gas welding, where a welding rod is used to soften a rod of the same material by hot air or nitrogen while at the same time joining the parts. Or spin welding where parts to be bonded, spun at a high speed, develop friction at the bond area. When spinning stops, parts cool in a fixture under pressure. Dielectrics, another thermal bonding technique, uses high frequency voltage applied to film or sheet, which causes the material to melt at the bonding surfaces. Also, induction energizes with an electromagnetic field a metal insert or screen that has been placed between the parts to be welded and causes bonding as parts around it melt and then cool.

Use of adhesives and the development of new adhesives from plastics that have adhesive properties have also grown into a highly developed technology. Adhesives may be liquids, solvents, water-based materials or anaerobics, mastics, hot melts that flow when heat is applied, films, and pressure-sensitive tapes and sheets.

Mechanical fasteners in the form of staples and screws are used, and the innovative molded-in inserts, snap fits, and a whole line of proprietary fasteners are used to join dissimilar materials to plastics and to join plastics to plastics when other techniques will not work. Mechanical fasteners will certainly give strength immediately to a joint and can provide the possibility of a movable joint or a removable joint (for disassembly), which can be cheaper to apply and more attractive in certain instances.

Polyethylene wax of a bracelet and ring before lost wax casting. Designer, Alexander Kower. (*Courtesy, Alexander Kower*)

BONDING AND JOINING

TABLE 1—Bonding and Joining Techniques: What is Available and What They Offer

TECHNIQUE	DESCRIPTION	ADVANTAGES	LIMITATIONS	PROCESSING CONSIDERATIONS
Solvent Cementing & Dopes	Solvent softens the surface of an amorphous thermoplastic; mating takes place when the solvent has completely evaporated. Bodied cement with small percentage of parent material can give more workable cement, fill in voids in bond area. Cannot be used for polyolefins and acetal homopolymers	Strength, up to 100% of parent materials, easily and economically obtained with minimum equipment requirements	Long evaporation times required; solvent may be hazardous; may cause crazing in some resins	Equipment ranges from hypodermic needle or just a wiping media to tanks for dip and soak. Clamping devices are necessary and air drier is usually required. Solvent recovery apparatus may be necessary or required. Processing speeds are relatively slow because of drying times. Equipment costs are low to medium. The high range might be $1,000-$2,000
THERMAL BONDING				
1. Ultrasonics	High-frequency sound vibrations transmitted by a metal horn generate friction at the bond area of a thermoplastic part, melting plastics just enough to permit a bond. Materials most readily weldable are acetal, ABS, acrylic, nylon, PC, polyimide, PS, SAN, phenoxy	Strong bonds for most thermoplastics; fast—often less than one sec. Strong bonds obtainable in most thermal techniques if complete fusion is obtained	Size and shape limited. Limited applications to PVC's, polyolefins	Converter to change 20 KHz electrical into 20 KHz mechanical energy is required along with stand and horn to transmit energy to part. Rotary tables and high-speed feeder can be incorporated. Cost: from $2,400 up for stand, converter; $65-$1,000 for horn
2. Hot Plate & Hot Tool Welding	Mating surfaces are heated against a hot surface, allowed to soften sufficiently to produce a good bond, then clamped together while bond sets. Applicable to rigid thermoplastics	Can be very fast, e.g., 4–10 sec. in some cases; strong bonds	Stresses may occur in bond area	Use simple soldering guns and hot irons, relatively simple hot plates attached to heating elements up to semiautomatic hot plate equipment. Clamps needed in all cases. Costs run gamut from $4.99 gun to $2,000-$20,000 range for commercial hot plate units
3. Hot Gas Welding	Welding rod of the same material being joined (largest application is vinyl) is softened by hot air or nitrogen as it is fed through a gun that is softening part surface simultaneously. Rod fills in joint area and cools to effect a bond	Strong bonds, especially for large structural shapes	Relatively slow; not an "appearance" weld	Requires a hand gun, special welding tips, an air source, and welding rod. Regular hand gun speeds run 6 in./min; high speed hand-held tool boosts this to 48–60 in./min. Cost begins at under $100

TECHNIQUE	DESCRIPTION	ADVANTAGES	LIMITATIONS	PROCESSING CONSIDERATIONS
4. Spin Welding	Parts to be bonded are spun at high speed, developing friction at the bond area; when spinning stops, parts cool in fixture under pressure to set bond. Applicable to most rigid thermoplastics	Very fast (as low as 1–2 sec.); strong bonds	Bond area must be circular	Basic apparatus is a spinning device but sophisticated feeding and handling devices are generally incorporated to take advantage of high-speed operation. Costs go up to the $30,000-$40,000 range
5. Dielectrics	High-frequency voltage applied to film or sheet causes material to melt at bonding surfaces. Material cools rapidly to effect a bond. Most widely used with vinyls	Fast seal with minimum heat applied	Only for film and sheet	Requires rf generator, dies, and press. Operation can range from hand-fed to semiautomatic, with speeds depending on thickness and type of product being handled. 3–25 kw units are most common, with costs running $500/kw and up
6. Induction	A metal insert or screen is placed between the parts to be welded, and energized with an electromagnetic field. As the insert heats up, the parts around it melt and, when cooled, form a bond. For most thermoplastics	Provides rapid heating of solid sections to reduce chance of degradation	Since metal is embedded in plastic, stress may be caused at bond	High-frequency generator, heating coil, and inserts (generally 0.02–0.04 in. thick). Hooked up to automated devices, speeds are high. Cost: in the $1,000/kw range. If 1–5 kw's used, work coils, water cooling for electronics, automatic timers, multiple-position stations may also be required
ADHESIVES [a] 1. Liquids: Solvent, Water Base, Anaerobics	Solvent-and-water-based liquid adhesives, available in a wide number of bases—e.g., polyester, vinyl—in one- or two-part form fill bonding needs ranging from high-speed lamination to one-of-a-kind joining of dissimilar plastic parts. Solvents provide more bite, but cost much more than similar base water-type adhesive. Anaerobics are a group of adhesives that cure in the absence of air, with a minimum amount of pressure required to effect the initial bond. Adhesives are used for almost every type of plastic	Easy to apply; adhesives available to fit most applications	Shelf and pot life often limited. Solvents may cause pollution problems; waterbase not as strong; anaerobics toxic	Application techniques range from simply brushing on to spraying and roller-coating lamination for very high production. Adhesive application techniques, often similar to decorating equipment, from hundreds (e.g., a glue pump for $400) to thousands of dollars with sophisticated laminating equipment costing in the tens of thousands of dollars. Anaerobics are generally applied a drop at a time from a special bottle or dispenser

2. Mastics	Highly viscous single- or two-component materials which cure to a very hard or flexible joint, depending on adhesive type	Does not run when applied	Shelf and pot often limited	Often applied via a trowel, knife, or gun-type dispenser; one-component systems can be applied directly from a tube. Various types of roller coaters are also used. Metering-type dispensing equipment in the $2,500 range has been used to some extent
3. Hot Melts	100% solids adhesives that become flowable when heat is applied. Often used to bond continuous flat surfaces	Fast application, clean operation	Virtually no structural hot melts for plastics	Hot melts are applied at high speed via heating the adhesive then extruding (actually squirting) it onto a substrate, roller coating, using a special dispenser or roll to apply dots or simply dipping
4. Film	Available in several forms including hot melts, these are sheets of solid adhesive. Mostly used to bond film or sheet to a substrate	Clean, efficient	High cost	Film adhesive is reactivated by a heat source; production costs are in the medium-high range, depending on heat source used
5. Pressure Sensitive	Tacky adhesives used in a variety of commercial applications (e.g., cellophane). Often used with polyolefins	Flexible	Bonds not very strong	Generally applied by spray with bond effected by light pressure
Mechanical Fasteners (Staples, screws, molded-in inserts, snap fits, and a variety of proprietary fasteners)	Typical mechanical fasteners are listed on the left. Devices are made of metal or plastic. Type selected will depend on how strong the end product must be and on appearance factors. Often used to join dissimilar plastics or plastics to non-plastics	Adaptable to many materials; low-to-medium costs; can be used for parts that must be disassembled	Some have limited pull-out strength; molded-in inserts may result in stresses	Nails and staples are applied by simply hammering or stapling. Other fasteners may be inserted by drill press, ultrasonics, air or electric gun, hand tool. Special molding (i.e., molded-in-hole) may be required

[a] Because of the thousands of formulations available within the various adhesive categories, it is not practical to attempt an analysis here of which adhesives will satisfy a particular application's need. However, typical adhesives in each class are: Liquids: 1. Solvent—Polyester, vinyl, phenolics, acrylics, rubbers, epoxies, polyamide; 2. Water Base—Acrylics, rubber, casein; 3. Anaerobics—Cyanacrylate; Mastics—Rubbers, epoxies; Hot Melts—Polyamides, PE, PS, PVA; Film—Epoxies, polyamide, phenolics; Pressure Sensitive—Rubbers.

Courtesy Plastics Technology

ADHESIVES FOR BONDING PLASTICS

	ABS	Acetal	Acrylic	Cellulosics	Fluorocarbons	Nylon	PPO	PVC	Polycarbonate	Polyethylene	Polypropylene	Polystyrene	Polysulfone	Alkyds	Epoxy	Melamines & Ureas	Phenolics	Polyesters	Polyurethanes
Metals	23	3,23	2,3 43,44	2,3	22,23	2,23	2,4,23	2,3,15 23,36 43,44	23 43	2,31 41	1	31,43 44	4,23		4	3,43	2	4,43 44	3,4
Paper		3,23	42	42	22,23	3,41	6,23	42	36	41	1,41	4,31 36			3	41,42	42	41	4,36
Wood	23	23	2,3 42	3	23	2,3	2,4 23	3,23 36,42	23 36	2,41	1,41	31,36			23,31	2,3	2,42	2,36	36
Rubber	23	3	1-4	1-4	23	2	2,4 23	3,4 15	4,36	2,41	1,41	5			3	2,3 43	2,3	1-4	4,36
Ceramics		23	2,3	3	23	3,23	4,23	3,4	23,36	2,41	1,41	41,42 43,44			23,31	2	2	2	3
ABS	23,43 44	2,4 11,23				21,23		23	4,23	23									
Acetal	2,4 11,23	3,23	23		23	23		23									23		
Acrylic	2,4 11,23	2,4 11,23	S,13 43,44		23	2,3 15,22		3,4 11,23	43,44								4,21 23	23,36 43,44	4,23
Cellulosics				3,4 14,36	23	2,3 15,22		23									23		
Fluorocarbons		23			22,23			23		23		23							23
Nylon	21,23	21,23	2,3 15,22	2,3 15,22		2,22 23,36	21,23							21,23	21,23	21,23	21,23 43	21,23	
PPO							4,23 43							4,23 43,44	4,23 43,44	4,23 43,44	4,23 43,44	4,23 43,44	4
PVC	23	23	3,4 11		23			3,4 11,36 42					23				4,15	23	
Polycarbonate	23							3,4 11,36 42	S,43 44									43,44	
Polyethylene	23				23					23,31 41			4						

Polypropylene											23,31 41,43 44								
Polystyrene			43,44		23							4,5,13 23,31 36,43,44					4,23 43	36,43 44	
Polysulfone										4			4,23						
Alkyds							4,23 43	23										23,36	
Epoxy			23				4,23 43,44	23							2,3 23,31 36			23,36 43,44	
Melamines & Ureas							4,23 43,44									2,3 23,31 36			
Phenolics	23,43	43	4,21 23	4,21 23	23		4,23 43,44	4,15				4,23				3,21 23,24	2-4 23,31 43,44		
Polyesters			23,36 43,44	23			4,23 43,44	23										3,23 31,36 43,44	
Polyurethanes			4,23		23		4												3,4 23 36

KEY

Elastomeric

1. Natural rubber
2. Neoprene
3. Nitrile
4. Urethane
5. Styrene-Butadiene
6. Silicones

Thermoplastic

11. Polyvinyl acetate
12. Polyvinyl alcohol
13. Acrylic
14. Cellulose
15. Polyamide

Thermosetting

21. Phenol Formaldehyde (Phenolic)
22. Rescorcinol, Phenol-Rescorcinol
23. Epoxy
24. Urea-Formaldehyde

Resin

31. Phenolic-Polyvinyl Butyral
32. Phenolic-Polyvinyl Formal
33. Phenolic-Nylon
36. Polyester
37. Acrylic

Other

41. Rubber Latices
42. Resin Emulsions
43. Cyanoacrylate
44. Alpha Ethylcyanoacrylate

S. Solvent only recommended

NOTE

This information contains a compilation of suggestions and guidelines offered by various adhesive and materials manufacturers and plastics molders. It is intended only to show typical adhesives used in various applications. Lack of a suggested adhesive for materials does not mean that these materials cannot be bonded, only that suppliers do not commonly indicate which adhesive to use. Before making a final selection, consult both materials supplier and materials manufacturer. Key for table and part of the data was supplied by USM Corp., Chemical Div.

Courtesy Plastics Technology; *revised by author*

GENERAL APPLICATION CONSIDERATIONS

There are two application methods—manual or machine. Manual techniques can utilize brush, roller, knives, trowels, or spatula and can employ dipping. Machine methods can be automatic or semiautomatic, utilizing spray guns, caulking guns, air pressure devices (such as the syringe that is coupled with air pressure), and automatic devices hooked into assembly for mass production systems.

For successful bonding, humidity and temperature need to be controlled; the room, tools, and surfaces to be adhered must be clean and free of dust. Some techniques require the temporary application of uniform pressure for a uniformly thin adhesive layer. Amounts of pressure vary with materials and systems. If there is too much humidity, solvent types will turn surfaces white.

Some materials are difficult to bond and require special treatment, notably polyolefins (including polyethylene), fluorocarbons, and acetals. Treating methods include electronic or corona discharge, flame treating (particularly for large and irregular-shaped forms), acid treating by dipping articles in a solution of potassium dichromate and sulfuric acid, and solvent treating. Fluorocarbons, for instance, must be cleaned with a solvent such as acetone, then treated with a special etching solution.

Materials such as polystyrene and vinyls can cause problems when solvent-based adhesives are used because solvents may cause surfaces to craze or cause extraction of additives used in the original material formulation.

Most plastics require surface preparation such as chemical cleaning with a solvent such as acetone; or abrasive cleaning by sandpaper, wire brush, sandblasting or vapor honing; degreasing to remove residual contaminants such as mold release agents.

SELECTION FACTORS

When getting ready to select an adhesive, analyze critical product requirements—for example, is the material transparent, will the bond show, what are environmental factors, will the bonding be subject to stress, is the joining in an inaccessible location, what are cost factors? Consideration of requirements will immediately eliminate many adhesive types and techniques of application. As environmental factors are listed, it is necessary to know what heat and humidity conditions will be present, and whether flammability is a key issue.

The charts details process and materials to aid in selection. More specific techniques are covred in chapters about specific plastics.

TYPICAL SOLVENTS FOR SOLVENT CEMENTING

ABS	Methyl ethyl ketone, methyl isobutyl ketone, tetrahydrofuran, methylene chloride
Acetate	Methylene chloride, acetone, chloroform, methyl ethyl ketone, ethyl acetate
Acrylic	Methylene chloride, ethylene dichloride
Cellulosics	Methyl ethyl ketone, acetone
Nylon	Aqueous phenol, solutions of resorcinol in alcohol, solutions of calcium chloride in alcohol
PPO	Trichloroethylene, ethylene dichloride, chloroform, methylene chloride
PVC	Cyclohexane, tetrahydrofuran, dichlorobenzene
Polycarbonate	Methylene chloride, ethylene dichloride
Polystyrene	Methylene chloride, ethylene ketone, ethylene dichloride, trichloroethylene, toluene, xylene
Polysulfone	Methylene chloride

These are solvents recommended by the various resin suppliers. A key to the selection of solvents is how fast they evaporate: a fast evaporating product may not last long enough for some assemblies; too slow evaporation could hold up production.

Courtesy Plastics Technology

FILLERS

All fillers reinforce thermoplastics and thermosetting plastics to some extent. They also help to extend, by dint of their bulk, the quantity of polymer used, but extending is not their primary function. Using certain fillers with appropriate polymers can change the properties of plastics, even though some fillers are inert.

The newer nonwovens can create desirable textures but also impart breathability to polymers such as vinyls (particularly good for upholstery). Nonwovens may make a polymer easy to drape or give it more bounce and recovery. They can increase dimensional stability, provide stitch and tear strength, afford an adhesive anchor or a reinforcement for fragile forms. Nonwovens are grain free. Some are made of modacrylic fibers; others from polyester, glass, polypropylene, nylon, cotton, or rayon.

Fibrous Materials

Fibers of straw were used thirty centuries ago by Israelites to reinforce their bricks. Animal hair has also been used for centuries to reinforce plaster. The use of cellulose fibers, nylon and other synthetic fibers, carbon and graphite fibers, filaments, flock and whiskers (single crystal fibers) provide varying degrees of strength and internal meshing to polymers. Perhaps the most important reinforcement is fiberglass mat, roving, yarns, and woven fabrics because of their durability, high strength, light weight, rigidity, economy, and design flexibility.

Spheres

Glass in solid spherical form can be used in conjunction with other fillers. Characteristically it provides for a smooth surface, keeps resin viscosity down, and permits high loading. With a specific gravity of just 2.48, spherical glass fillers reduce the price of any resin system. Besides promoting good resin flow in compression and injection molding when used in conjunction with fiberglass, reinforced epoxy or polyester laminates increase their flexural strengths without loss of other physical properties.

Other spheres may be "Saran" and phenolic, each with its own special benefits. Use of "Saran" spheres in a polyester form may make it easier to drive in a nail and also withdraw it with a "healing" property; the hole practically disappears.

Inorganic Fillers

Inorganic fillers are detailed in the following chart.

INORGANIC FILLERS

FILLERS	TYPE OF PARTICLE	DESCRIPTION	COLOR	WETTABILITY	OTHER COMMENT
Aluminum Oxide	Plate-shaped crystal	Aluminum oxide specially calcined to have hard and sharply defined alumina crystals. 90% plus alpha alumina	White	Good	Low shrinkage
Antimony Oxide	White powder	Antimony oxide is processed from high quality antimony ore	White	Good	
Antimony Trioxide	White powder	Antimony oxide is processed for use where the finest particle is desired	White		
Mineral Filler (Asbestos)	Fibrous	Fibrous mineral filler is somewhat between a talc and an asbestos	Gray	Good	
Barium Carbonate	Precipitated white powder	Barium carbonate is chemically precipitated	White		

FILLERS	TYPE OF PARTICLE	DESCRIPTION	COLOR	WETTABILITY	OTHER COMMENT
Barites	Granular crystals	Barites are water-ground from highest quality ore	White		
Blanc Fixe	Rhombic microcrystalline	Blanc Fixe is chemically manufactured by a direct process precipitation	Bluish-white		
Low-Micron Barium Sulfate	Granular crystals	Low-micron barium sulfate is jet-attrition-milled to give an exceptionally low particle size	White		
Low-Micron Barites	Granular crystals	Low-micron barites is water-ground, bleached, and classified to a fineness below 12.5 microns	White		Good dispersion
Volclay Bentonite	Massive, claylike	Bentonite is of the domestic swelling type having high suspension	Buff	Good	High gel
Low-Micron White Bentonite	Massive, claylike	White bentonite has high, good suspension properties without swelling when wet	Cream	Good	Nongelling Swelling power—29 cc
Calcium Carbonate	Calcite crystalline	Calcium carbonate is produced from white sugar calcite; it has low water-soluble salts and low reactivity	White	Good	
Whiting (Rambo)	Crystalline	Whiting is dry-ground from highest quality calcite ore	White	Good	
Marble Dust	Crystalline	Marble dust is a custom-ground calcium carbonate	White	Good	
Calcium Sulfate (Terra alba)	Cryptocrystalline	Calcium sulfate is chemically manufactured	White		
Precipitatic Chalk	White powder	Precipitated calcium carbonate is an extra-light grade of U.S.P. quality and is nonabrasive in nature	White		
English Kaolin (Clay)	Soft kaolinite	Kaolin is a colloidal clay produced by an elutriation process	White	Good	
Clay—Ultra Fine, Pulverized	Predominately thin hexagonal plates	This clay has controlled, low-soluble salts. It is compatible with both aqueous and nonaqueous vehicles	White	Good	
Diatomaceous Earth (Natural)	Diatom	Diatomaceous earth is a dry, chemically inert, natural silica powder formed by diatoms	Cream		

FILLERS	TYPE OF PARTICLE	DESCRIPTION	COLOR	WETTABILITY	OTHER COM-MENT
Diatoma-ceous Earth (Calcined)	Diatom	Diatomaceous earth is a dry, chemically inert, calcined diatomaceous silica powder	Pink	Good	
Feldspar	Fine crystal	Feldspar has a distinct crystal structure even in its finest particles	White	Good	
Fuller's Earth	Claylike	Fuller's earth is in the clay family but has an inherently larger particle	Cream	Good	
Graphite	Flaky and massive mixed	Graphite is approxi-mately half coal carbon; the remainder is graphitic carbon. This is not amorphous ore	Black	Poor	
Ground Glass	Fine powder	Ground glass is made by milling glass cullet	White	Good	
Lime (Chemically Hydrated)	Fine powder	Lime is made by calcining calcium carbonate and re-hydrating	White	Good	
Magnesite (Caustic Burned)	Fine powder	Magnesite is made by highly calcining natural magnesite		Good	
Magnesium Carbonate, Light	Fine powder	Magnesium carbonate is made from sea water by the Marine Magnesium Division of Merck & Company	White	Good	Index of refrac-tion: 1.15
Magnesium Hydroxide	Fine powder	Magnesium hydroxide is precipitated from sea water by the Marine Magnesium Division of Merck & Company	White	Good	Index of refrac-tion: 1.58
Magnesium Oxide-K	Fine powder	MgO-K is processed from sea water by the Marine Magnesium Division of Merck & Company	White	Good	Index of refrac-tion: 1.67
Magnetite	Massive	Magnetite is a native ferrous-ferric oxide also known as lodestone. It possesses magnetic properties	Black		
Mica	Thin plate	This mica is water-ground from high quality muscorite	Off-white	Poor	Index of refrac-tion: 1.56 to 1.59
Silica Sand	Angular	Silica sand is a silica quartz that has been thoroughly washed, kiln-dried, pulverized, and air-floated to give a silica flour	Off-white	Good	

FILLERS	TYPE OF PARTICLE	DESCRIPTION	COLOR	WETTABILITY	OTHER COM- MENT
Flint Quartz	Sharp quartzite	Flint quartz is a hard, crushed quartzite	Gray	Good	
Aluminum Silicate	Cellular	Aluminum silicate is a superfine expanded volcanic glass	Off-white	Good	
Mullite	Long needles	Mullite is formed by calcining aluminum oxide and pure kaolin at a high temperature	Off-white	Good	
Rottenstone	Indefinite	Rottenstone is one of the oldest known mineral fillers	Brown	Good	
Slate Flour	Cryptocrystalline	Slate flour is a popular filler because of its low cost	Dark gray	Good	
Sericite	Micaceous plate	Sericite is a selected type of micaceous mineral. It has low calcium and chloride salts	Gray	Poor	Index of refrac- tion: 1.59
Talc	Semiplaty	This talc is outstanding because of the softness of its particle as the result of a flotation treatment during processing	Off-white	Poor	
Magnesium Silicate S.F.	Semimassive	Magnesium silicate is a low-micro talc processed in a jet-attrition mill to be very white in color and of extreme softness	White	Fair	
Tripoli	Crystalline	Tripoli is a porous, siliceous rock ground to graded meshes	Cream	Good	Index of refrac- tion: 1.54
Vermiculite (A4)	Expanded plates	Vermiculite has been expanded from a micaceouslike ore by heat	Light brown	Good	
Volcanic ash	Pumice	Volcanic ash is a high surface area filler due to its cellular structure	Gray	Good	

In Conclusion

Some fillers, because of their low cost, function primarily as extenders; others act as reinforcement enhancing polymer's properties in various ways. Selection is determined by a wide variety of attributes and factors such as availability, function of a form, utility, cost, and physical chemical properties. Even though proportions of filler to resin vary, it is customary to use at least 60% resin in most composites.

PROCESSING AIDS

Like many of nature's materials, such as clay, metal, and fibers, plastics can be altered to emphasize particular properties not necessarily inherent in the base material. Through additives, resins can be made resistant to oxidation, ultraviolet radiation, flame, other environment degradation; can be made flexible or rigid, brightly colored, shiny, or matte.

Mold release agents permit parting of resin from its mold; slip agents modify plastics to provide surface lubrication; antioxidants stop discoloration and surface cracks; flame retardants are vital if we are to comply with federal and state codes. Other additives, such as plastisols, increase viscosity, change gel properties, and in the final state can impart flexibility and softness to a polymer. Stabilizers vary from protection against light and heat to creating better adhesion properties. For ultraviolet protection a stabilizer may absorb UV light and dissipate energy, or it may remove energy that the polymer has absorbed before degradation sets in.

COLORANTS

Coloring of plastics is the marriage of two technologies—pigment and polymer. Pigment technology is concerned with hue, value, chroma, dyes, organic and inorganic pigments, and its form in application to polymers. Polymer technology relates to the general performance of the colorant in the resin.

Dyes are organic pigments that are soluble in ordinary solvents. They are bright and transparent. Nearly all have limited light stability and poor resistance to processing temperatures, particularly above 370° F. They migrate and bleed easily. Chemical resistance is very poor.

Organic pigments are dyes that have been modified. They can be opaque or transparent, bright in hue, have light and heat stability far better than dyes, but are not always so good as inorganic pigments.

Inorganic pigments are dense, heavy colorants that are insoluble in solvents. Tinting strengths are good but mass tones are not so brilliant as those of organic pigments. Heat and light stability is excellent.

Special effect pigments include metallic flakes based on aluminum or copper, pearlescent pigments based on bismuth and lead compounds, and fluorescent pigments that have a limited period of excitation.

Colorants are available as dry colors, pellets, pastes, or liquids. In selecting colorants, end use has to figure as a vital consideration: the need for transparency, opacity, permanence, chemical reactions of some colorants on catalysts, for instance, processing temperatures, and temperature degradation of the colorant. Metallic flake pigments may reduce the shelflife of polyester resin; colorants with free copper, zinc, manganese, and magnesium should not be used with PVC. Zinc, tin and lead compounds cause improper curing of silicones.

SURFACE COLORING

Although colorants usually are integrated into polymers during processing, sometimes it is desirable to apply surface coloring, usually by hot stamping or by spraying.

Hot stamping consists of transferring a pattern/color from a thin cellophane or polyester film backing onto a form by the momentary application of heat and pressure. Design or pattern is achieved by the pattern or the metal die or silicone pad. It is a dry process that requires no cure time. Stampings can be permanent and inexpensive if rendered in large enough quantities. Heat is applied in a temperature range of 175° F. to 500° F., depending on the plastic melt point.

Acrylics, urethanes, epoxies, polyesters, vinyls have been used as coatings applied by brushing, rolling, or spraying. Some have achieved a high degree of sophistication, such as Durethane 600, an elastomeric urethane lacquer, and Durethane 100, an elastomeric thermosetting enamel that has stretchability particularly over flexible expanded urethane foam surfaces. They have been used over high-density polyurethane foam car bumpers that can "heal" themselves of minor dents by expanding back to normal contours after an accident.

FLAMMABILITY

The entry of plastics into new areas of consumption and the field's greater depth and exposure have established plastics as a major concern of consumers and the Federal government. Car interiors, some outside parts and motor parts, furniture and fabrics, clothing are just a few areas that are being subject to Federal regulations. In 1968, 4,000 persons died because of flammable fabrics (some of these synthetic). "The Flammable Fabrics Act" (Public Law 90-189, December 14, 1967) states that all items of apparel and interior fabric furnishings be fire-resistant. New restrictions are bound to be placed on materials used in construction, including restrictions limiting the generation of smoke and toxic gases.

Just as no one plastic can be used across the board for every application, so no one method or no single flame-retardant plastic can have universal application.

Flame retardance can be achieved in several ways: the basic polymer can be constructed so that exposure to heat and oxygen will not produce combustion; a polymer can be chemically altered to produce resistance to combustion; a coating can be placed on the material to exclude oxygen. Before beginning work with any polymer it is best to inquire about its flame resistance. In the near future it is expected that at least 50% of polymers will meet Federal regulations.

TOXICITY

All plastics should be used in well-ventilated areas. Skin should be protected with clean clothing and/or creams. Many plastics are innocuous, affecting only highly allergic people. It is the plasticizers, catalysts, and solvents that are potentially hazardous. Thousands of workers are exposed daily to various plastics with no ill effects and yet medical literature is accumulating case histories of organ malfunctioning, blood, nerve, and skin disorders.

Some applications require extreme precautions, such as good ventilation and use of miners' or oxygen masks. Spraying of polymers and the fine mist that can be inhaled, ingested, and deposited on your skin is not good for you. A mask should be used as well as a spray booth and adequate covering. Foaming of urethane and the use of Freon as a foaming agent are very dangerous, because the Freon can "freeze" your lungs. The amine and anhydride catalysts for epoxies are toxic. There are less potent catalysts that can do the job of catalyzing epoxies. The styrene odor in polyester is not harmful, but spraying with polyester can be.

Any polymer brought to its pyrolizing temperature will decompose, some into toxic chemicals. Although every gas released from the combustion of plastic is not dangerous, burning plastic should be considered harmful because some plastics release carbon monoxide, hydrogen cyanide, hydrogen chloride, and low-molecular weight hydrocarbons. Carbon dioxide, the major component from combustion, is not toxic, but it will suffocate.

SOME JOINT CONFIGURATIONS

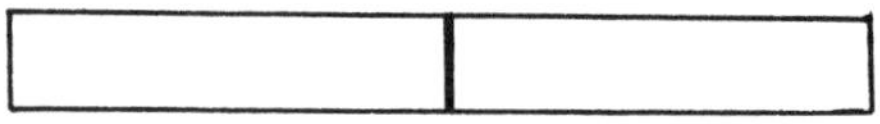

Butt—Unsatisfactory

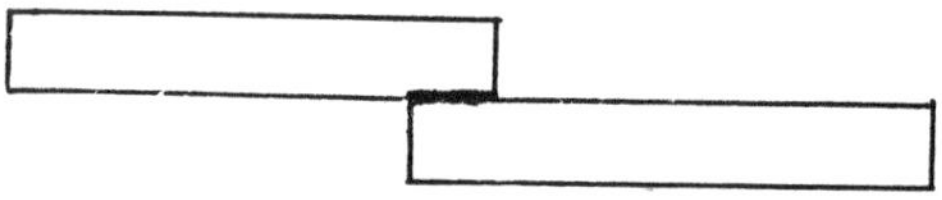

Straight overlap—Good, practical

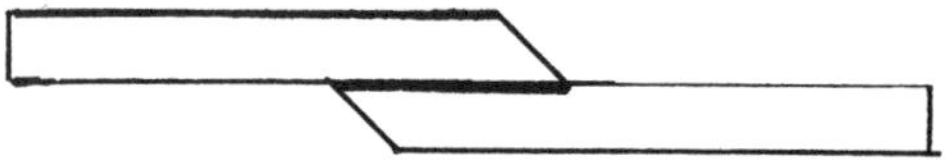

Beveled lap—Good, but impractical, difficult to mate

Scarf—Good, requires machining

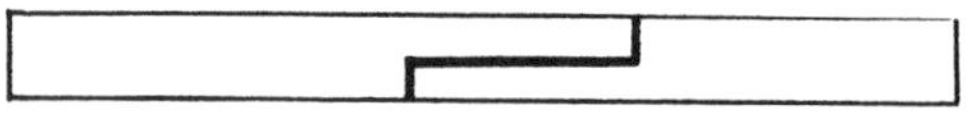

Double butt lap—Good, requires machining

Toggle lap—Good, practical

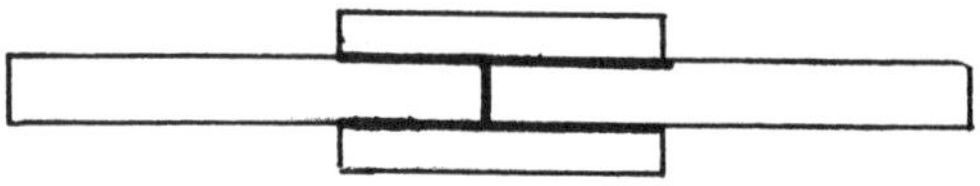

Double strap—Good, practical

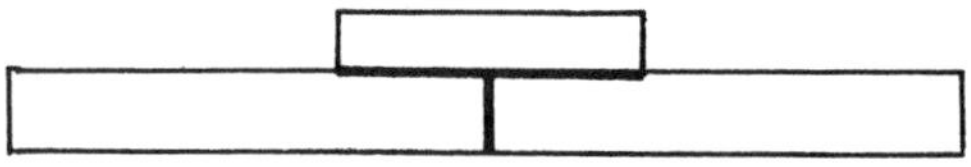

Strap—Fair, but seldom used

Recessed double strap—Good, but expensive to machine

Tongue and groove—Good, but expensive to machine

Double scarf lap—Good, but requires precision machining

Generally, large molecules are eliminated from the human system within a matter of hours, days, or weeks. But monomers, small molecules, are not easily eliminated. Poor physical condition and some other (nonplastic-caused) organ disorders can trigger abnormalities spurred by some potentially toxic polymers.

Caution and care are key words. Unfortunately the study of the degradation of polymers in environments, including biological environments, has not been studied adequately to permit determination of clear cause-effect relationships. Many direct causes go undetected because doctors are not well informed here.

Yet we wear plastics with no apparent ill effects; we eat plastics—some low-cost, low-calorie "ice-creams" are made of plastic; synthetic "whipped cream" and some instant desserts are made of plastics too—with no observable ill effects. Plastics are used in packaging of food, cooking of foods (bags), and coating of pots and pans. We eat and drink from plastic dishes, glasses, spoons, knives and forks. We mix our foods in plastic bowls and store foods in plastic containers in plastic refrigerators. We sleep on plastic mattresses covered with plastic sheets, sit on plastic chairs, and look out of plastic windows. We paint our rooms and houses with plastic paint. Dentists have been using plastics to fill our teeth and doctors have been implanting plastic blood vessels, bones, and tissues in our bodies. Some recoveries would not be possible if it were not for plastic "internal" parts.

What does all of this mean? When you work with plastic make certain you have a large fan to ventilate and exchange the air; that you use a mask when potentially toxic materials are used; that you coat your hands with barrier cream and wear clean clothes (wash or throw away plastic-coated work clothes); that you vacuum plastic dust; and that you spray in a ventilated spray booth.

PERMANENCE

Once upon a time plastics were considered inexpensive, temporary materials and grew to fill the disposables market, replacing paper cups, plates, et cetera. Our waste piles have proved otherwise. It takes a very long time for these plastics to decompose and the plastics industry is researching solutions to this problem. Most plastics made of organic materials, as well as nature's organic material, will decompose in time and return to their carbon-based state. Even inorganic materials wear away in time or break down into smaller parts.

Decomposing or degrading are a matter of degree. Degrading can manifest itself in fading of color, discoloration and/or pitting of surfaces, and cracking. Decomposing, at the other end of the scale, is the absolute breakdown of form. Plastics that "live" indoors can be considered permanent materials. Outdoor weatherability imposes other variables: changes in temperature, immersion in water, humidity, cloaking with a variety of chemicals, and baking in sunlight. Some plastics fare better than others. Polystyrenes, related copolymers, and terpolymers such as ABS do not weather well. Epoxies and fiberglass reinforced polyesters hold up reasonably well—particularly well if coated (wood needs coating too). Fluorocarbons and acrylics, as well as urethanes can be used out of doors. Urethanes will brown but will not decompose easily. If plastic is coated, it will withstand a brutalizing environment as well as wood and some metals. It is best to check your colors for permanence and data sheets on weatherability of any specific plastic.

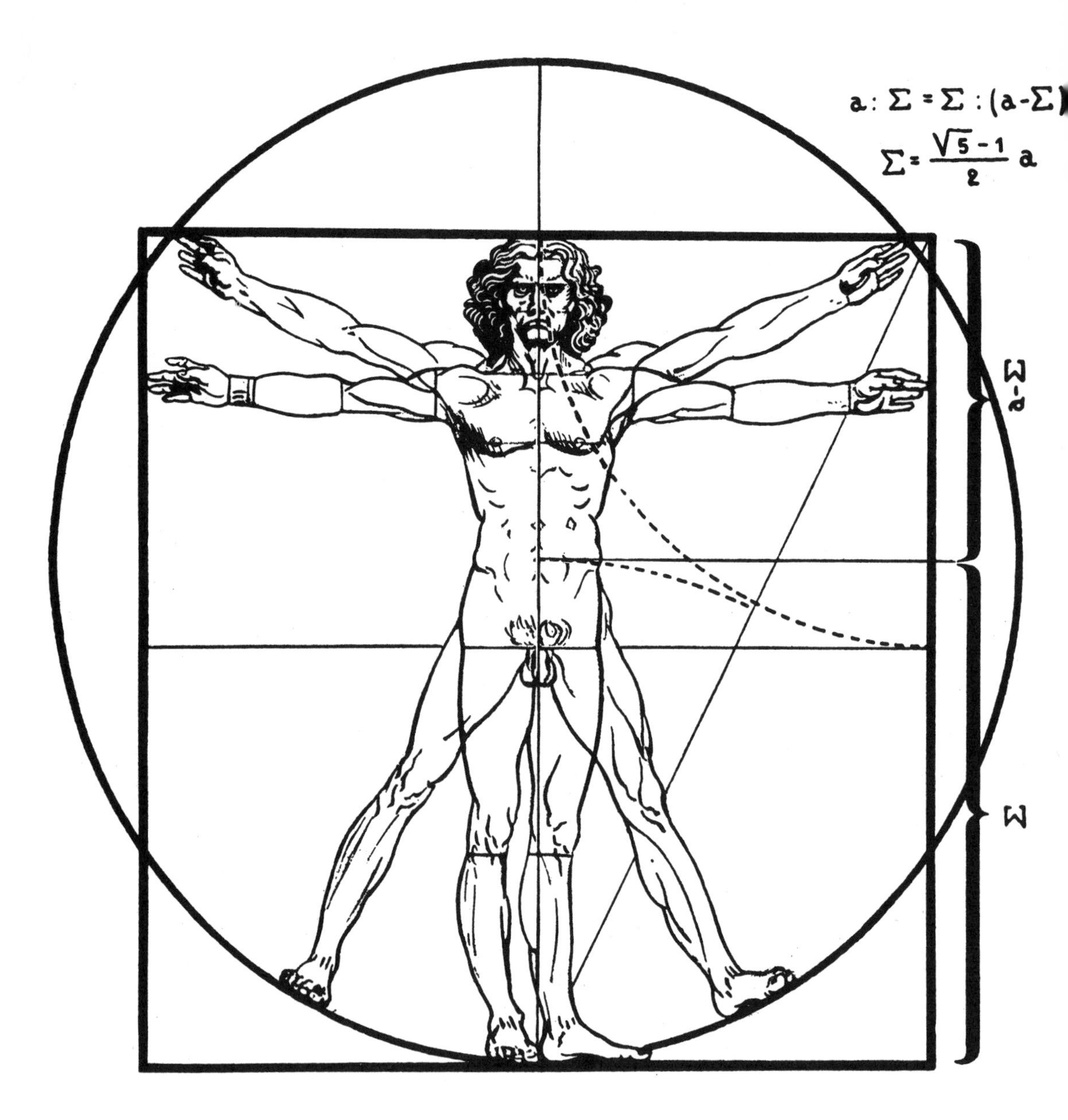

πάντων χρημάτων
μέτρον ἄνθρωπος

By Protagoras of Abdera. Translation: "Man is the measure of all things." Euclid restated this in the mathematical formula in the upper right-hand corner. And in the fifteenth century Leonardo da Vinci illustrated Euclid's formula (The Golden Section or Golden Rule) with this drawing.

3

design parameters and the sixth sense

DESIGN STANDARDS

Good design, beautiful design—what does that mean? Opinion may be so emotional and subjective that, as with religion and politics, expressing it can mean the loss of friends. But there have been standards that have prevailed through the ages and are as vital today as they had been at their conception. So much bad design exists, though, that one might be inclined to believe that we have no reference points upon which to base our judgments. Designers say that the decisions that go into designing involve the sixth sense because one draws upon sensory resources plus powers beyond, such as intuition. Regardless how one feels about this statement, creating a wonderful shape certainly is a productive act. For our purposes in this book, *design is conceiving and giving form to objects used in daily life.*

Design as it is related to current human activities—engineering, the sciences and the arts, politics, economics—should be endowed with human considerations. A design can be mechanically beautiful and possibly humanly useless. Man-made objects, however, are for use by human beings and are a part of man's environment; therefore designs need to be endowed with mechanical as well as human efficiency.

St. Thomas Aquinas listed *integrity, clarity,* and *harmony* as requisites to beauty. If we relate this to contemporary values, designs should fulfill the requirements of the form's function; its materials and processes should be suited both to the functions ascribed to the product and to its own qualities and potential. For example, designs that mimic French Provincial hand carving and are translated into plastic are totally bad. Not only does the style mimic a period and way of life long gone; co-ordinately it shows lack of faith in our own values, and, compounding the injury, simulates another material—wood—and another production technique—wood handicraft. If design is to have *integrity,* then it should be what it is, not what "should have been" or "had been." If a

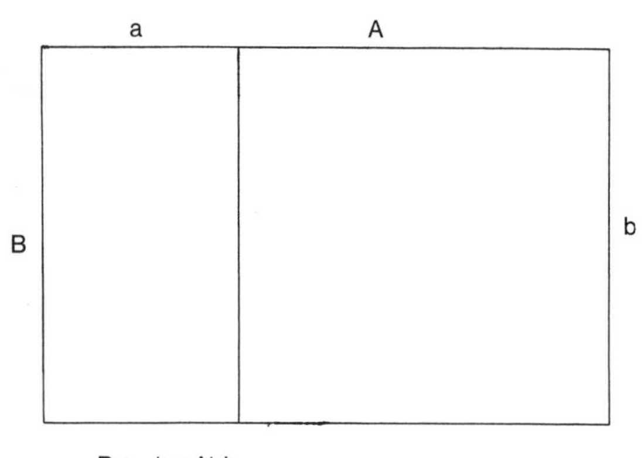

If you construct a square on the longer side of ϕ (Ba), then, taken together with the original Golden Section rectangle, you will have a new Golden Section.

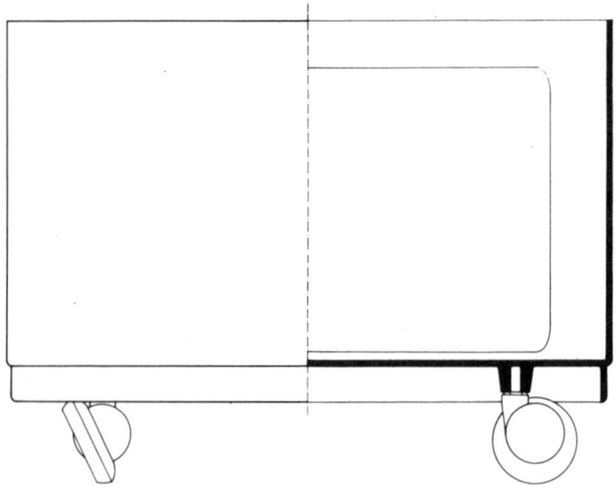

"Composable" table by Anna Castelli Ferrieri for Kartell is based on the Golden Section. (*Courtesy, Kartell-Binasco*)

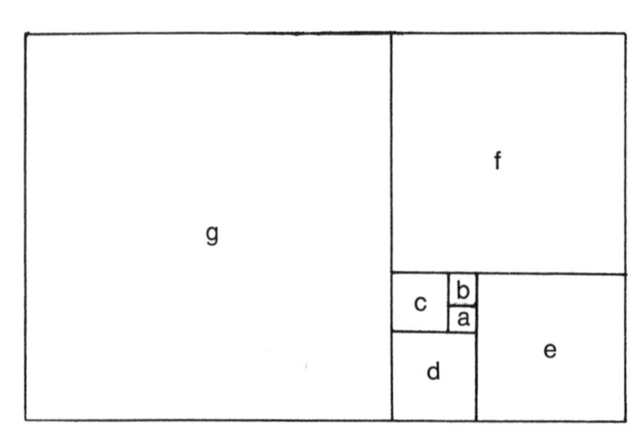

The smallest rectangle is ϕ. *b* is a square that has been added to *a*. Then square *c* was added to *a* and *b*. Together they form a new Golden Section. Square *d* was added to ϕ (*a,b,c,*) again to form a new Golden Section. You can continue to do this and form new Golden Sections of *e, f,* and *g*. This whole unit can become a design for a building front, divisions for a cabinet wall, et cetera.

design is to have *clarity,* then it should be a simple statement, its structure evident through its appearance without extraneous embellishment. And if a design is to have *harmony,* then it should serve the current needs of man, express the spirit of our times, meet his pocketbook, and blend materials, process, and use into a visually satisfying whole.

This does not mean that we should scrap our heritage. Objects that serve should stand. Some forms may never need to change; the furniture of eating, for instance, flatware, and china, except for style variations, have not drastically altered their form over more than two hundred years. Streamlining for the sake of change and its concomitant, forced obsolescence, though have to go. We cannot afford to support artificial stimulations to trade when our garbage heaps threaten to bury us.

These standards also do *not* imply that the "way-out" idea is a luxury. On the contrary, the avant-garde design predicts, leads, and inspires. Time and mind sharpened by practice will modify and adjust design solutions to useful purpose.

In the past, trial and error played a very important role in the discovery of what a material would do. Today, however, scientific and mathematical knowledge enables us to predict with precision and economy how a material will behave in a certain form. When this mechanical efficiency encompasses the human element, the results are usually excellent. Plastic, by the way, was the first group of materials to skip the trial and error route of the artist and craftsman, and to be designed by engineers. They borrowed standards from such other materials as metal and wood.

This may be a reason—one of many—for our problem in finding aesthetic as well as mechanical validity in useful forms. Most design programs gloss over aesthetic aspects and spend a great deal of time on engineering process and material considerations. The purpose of this chapter is to redress this grievance by stressing aesthetic-functional considerations and leaving the engineering factors to the engineers—with just a few exceptions.

MEASUREMENTS

Through history there are some mathematical models that keep popping up bearing different names. They are worth considering because of their application to nature, man, and man's creations. One of these is called the Golden Mean,

Golden Section, the Golden Number, or the Golden Ratio. In the Renaissance it was dubbed the Divine Proportion and the Greeks gave it a symbol ϕ (pronounced "Phi"). In the ϕ rectangle, side A is to side B as side B is to the sum of A and B. The ratio between the two sides is in the range of 1.618:1. All kinds of mathematical and arithmetic relationships evolve. For instance, if you start with a ϕ rectangle and attach a square on its longer side, the square taken together with the original ϕ rectangle will create a new larger ϕ rectangle. The pentagon used by the Greeks includes a number of ϕ relationships.

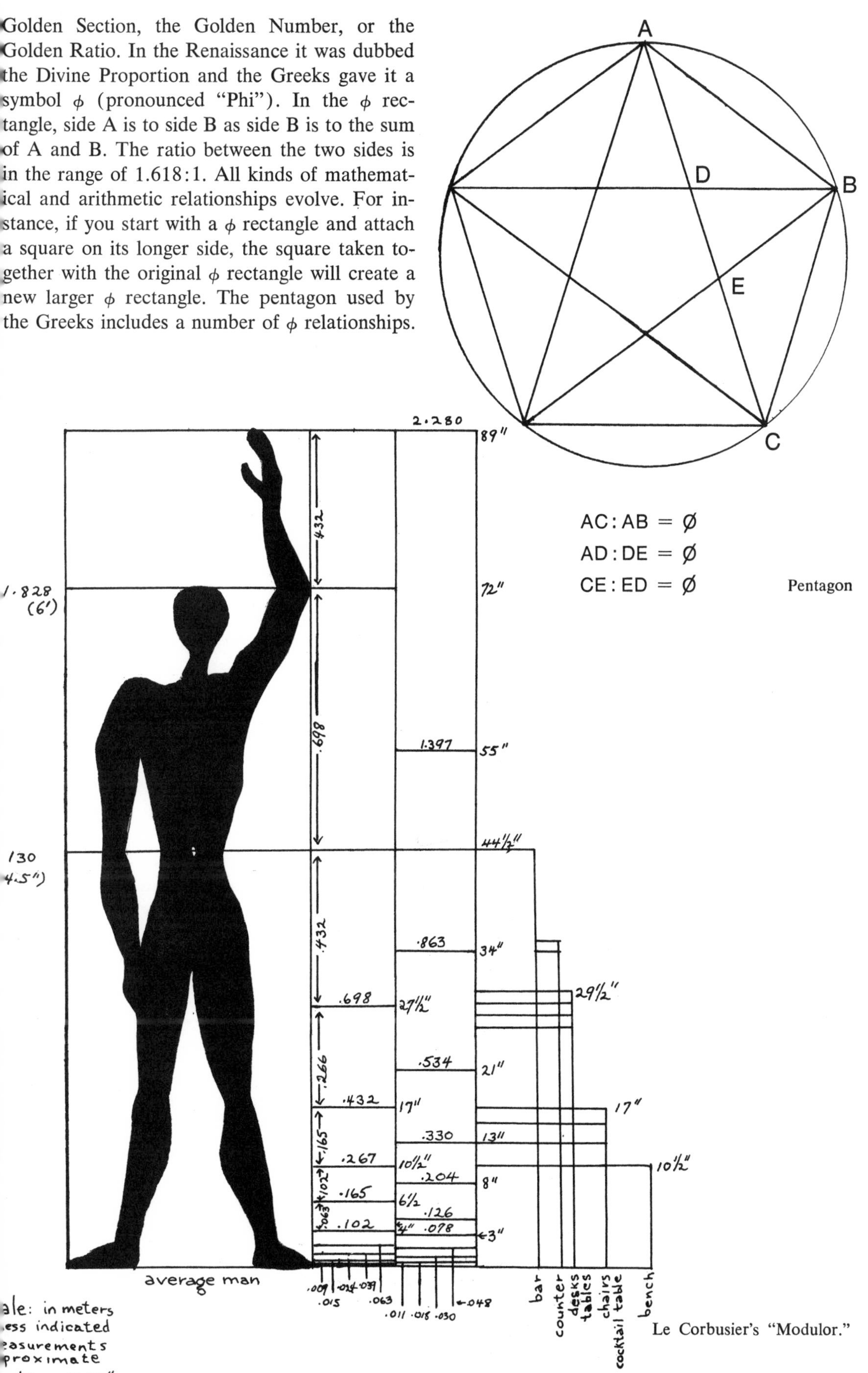

Pentagon

Le Corbusier's "Modulor."

There is a Golden Section relationship between any diagonal and any side of the pentagon. AC:AB $= \phi$. Diagonals also intersect each other according to the Golden Section, and so on. The Greeks thought the pentagon such a perfect shape that they attributed sacred significance to it. Architects of the past, perhaps deriving inspiration from Greek architecture, were preoccupied with these ϕ ratios. Windows and doors related to outside walls, et cetera. The ϕ proportion also appears in the human body and in other living things.

Another mathematical series that finds application to design, man, and nature is the Fibonacci series, discovered by Fibonacci in the thirteenth century. Each number is the sum of the two which precede it, such as 1, 1, 2, 3, 5, 8, 13, 21, 34, 55, 89, 144 . . . If you divide any number by the preceding one, strangely enough the result approaches the Golden Section ϕ. For example: $\begin{array}{r}1.6179\\ 89\,\overline{)144}\end{array}$ If you look closely at the formation of a pine cone or a dahlia, for example, and start counting parts and their relationships, you find they are part of the Fibonacci series. Nature is a logical system that is revealed by mathematics and so the human figure has often been related in its proportions to the Golden Section.

Le Corbusier designed a system of measurements and proportions which he called the "Modulor." This system is based upon the height of the average man and if his height is divided according to the Golden Section, we will find another Golden Section ϕ. Le Corbusier constructed a Fibonacci series based on man and used this as a design model. When buildings and furniture are proportioned to these measurements, they can be both practical and harmonious because the Modulor, with its basis the Golden Section and Fibonacci series, creates a harmonious link between man-made designs and nature. Le Corbusier gave a mathematical function to visual language. Proper application of the Golden Section as Le Corbusier or as Leonardo da Vinci illustrated it from Euclid's formulas should enable a designer to design almost anything from a helmet to a space station.

There are some design parameters that govern size and possibly proportion, if they are going to interact or function with man. In seating, for instance, it is impossible to design a chair without knowing something about the human fundament. All seating should maintain hip angle, knee angle, and ankle angle at 90° or greater if it is to be comfortable. Seating postures vary according to whether the seat is for a work posture, relaxing posture, or reclining posture. Seat height also varies from 17″ for a man's work posture and 16″ for a woman's work posture to varying heights up or down, depending upon the purpose of the seat. If seats are too hard, the flesh under the ischial tuberosities (bony protuberances) of the body suffer from too much compression; on the other hand, cushioning that is too soft increases flesh pressures around the peripheral edges. Pressure under the thigh should be relieved somewhat by having the feet support some weight. Optimum size for rectangular seats should be 16″ wide by 16″ deep—never less than 13″ for an adult. Round seats should be about 18″ in diameter.

Chair and table proportions for seated woman.

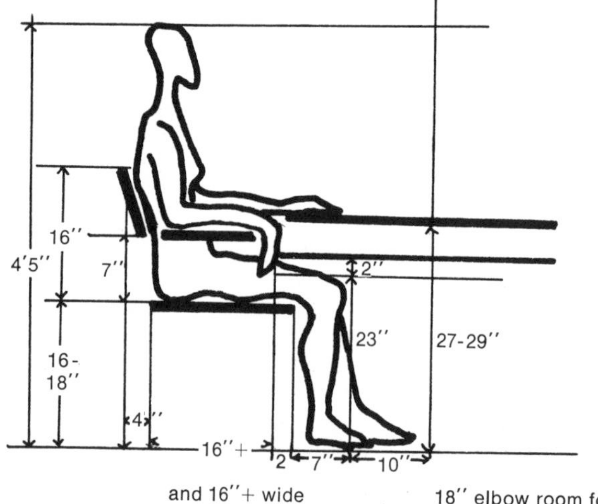

GENERAL DESIGN CONSIDERATIONS

Whether it is a chair, goblet, bowl, panel, box, or bead, every product has to be designed. Someone must go through the process of creating the form that is to be hand-crafted or manufactured. Materials as well as tools are affected by space (volume), time, and movement, and they can be modified with these factors in mind. The design process must be seen as a *Gestalt*. The whole process has to be considered in order to create an effective visual form that will function acceptably as part of the environment, for the human use for which it was intended. The process involves stating the need and defining

A stackable chair prototype in fiberglass to be produced by injection molding of ABS. Designers are Giorgina Castiglioni, Giorgio Gaviraghi, and Aldo Lanza for Kartell. (*Courtesy, Kartell-Binasco*)

HEIGHT-WEIGHT-CHART

YEARS	2	4	8	12	20
SEX	M F	M F	M F	M F	M F
HEIGHT	2'10"	3'4" 3'3"	4'2" 4'2"	4'10" 4'10"	6' 5'7"
WEIGHT	27 26	35 34	73 74	85 86	170 135

the preliminary design specifications. Information pertinent to the form should be collected. This should include specifics as to use, history of use, dimensions, a search for potential causes of failure, and all other relevant information. Then the most pertinent and essential factors have to be sifted from the collected data. Decisions then follow about material selection, methods of manufacture, form of the whole, and component parts. These decisions lay the groundwork for determining aesthetic, technical, structural, economic, and production aspects.

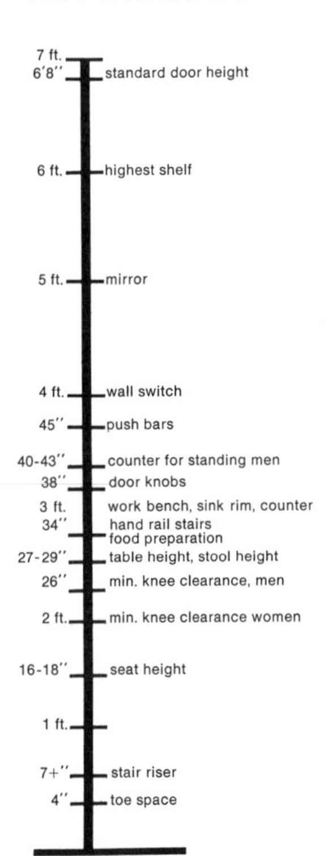

VARIATIONS IN FURNITURE SIZES

Upholstered Furniture and Bedding

Love Seats 32″ x 50-55-60″
Love Seats 34″ x 50-55-60″
Sofas 32″ x 72-78-84-90-96-102″
Sofas 34″ x 72-78-84-90-96-102″
Lounge Chair 33″ x 34″, 34″ x 31″, 32″ x 32″, 30″ x 30″
Barrel Chair 30″ x 30″, 29″ x 25″, 33″ x 31″
Reclining Chair 30″ x 29″ opens to 66″
Armchair 29″ x 27″, 27″ x 27″
Occasional Chair 26″ x 23″, 25″ x 20″, 24″ x 22″, 21″ x 18″
Twin Bed Mattress 75″ x 39″
Double Bed Mattress 75″ x 54″
Queen-Sized Mattress 80″ x 60″
King-Sized Mattress 80″ x 78″

Dining Room Tables and Chairs

Dinette Table 36″ x 48″. Chair 16″ x 16″
Round Table 44″. Chair 18″ x 18″ (4 chairs)
Oval Table 42″ x 60″. Chair 18″ x 18″ (6 chairs)
Oval Table 54″ x 74″. Chair 18″ x 18″
Rectangular Table 42″ x 64″. Chair 18″ x 18″ (6 chairs)
Extension Table 38″ x 60″ extends to 86″. Chair 18″ x 18″ (6 to 8 chairs)

Occasional Tables and Special Pieces

Square Table varies from 36″ x 36″ to 16″ x 16″
Cocktail Tables 22″ x 54″, 22″ x 60″, 22″ x 70″, 24″ x 58″, 28″ x 66″, 32″ x 66″
Stool 18″ diameter
Desks 30″ x 60″, 25″ x 50″, 21″ x 50″, 18″ x 40″

Interchangeable Storage Pieces and Special Pieces

Bookcases 18″ deep, 14″ deep, 11″ deep to any length
Dressers, serving carts, buffets, cabinets, consoles, bars, record cabinets, other storage units:
21″ deep to varying lengths from 30″ to 84″
19″ deep to varying lengths from 19″ to 72″
17″ deep to varying lengths from 25″ to 72″
13″ deep to varying lengths from 36″ to 52″

Exceptions are the Rule!

From an aesthetic viewpoint plastics *as* materials are subject to the same variables, the same physical laws, the same rules for good design as other substances—size, shape, position, direction, number, interval, density, surface textures and patterns, tensions and motions as well as the inner and outer structure. The designer has to be knowledgeable about what a material will do. For instance, we know that we can cut paper, score it, bend, fold, slit, stretch, curl, twist, tear, and crumple it. What can we do with polyester resin, or acrylic sheet, for example, if that is the best material to meet preliminary product specifications? All other determining aspects have to be studied—size, strength, environmental exposure, chemical reactions to the surface, the material's reaction to tools, fatigue, moisture, toxicity, life of the material, odor absorption, dimensional changes over a long period of time, stress, cracking and crazing, heat distortion. Failure can be caused by excessive load, fatigue of the material,

Design and model for a mayonnaise jar spoon. (*Courtesy, Architect Archilli Castiglioni*)

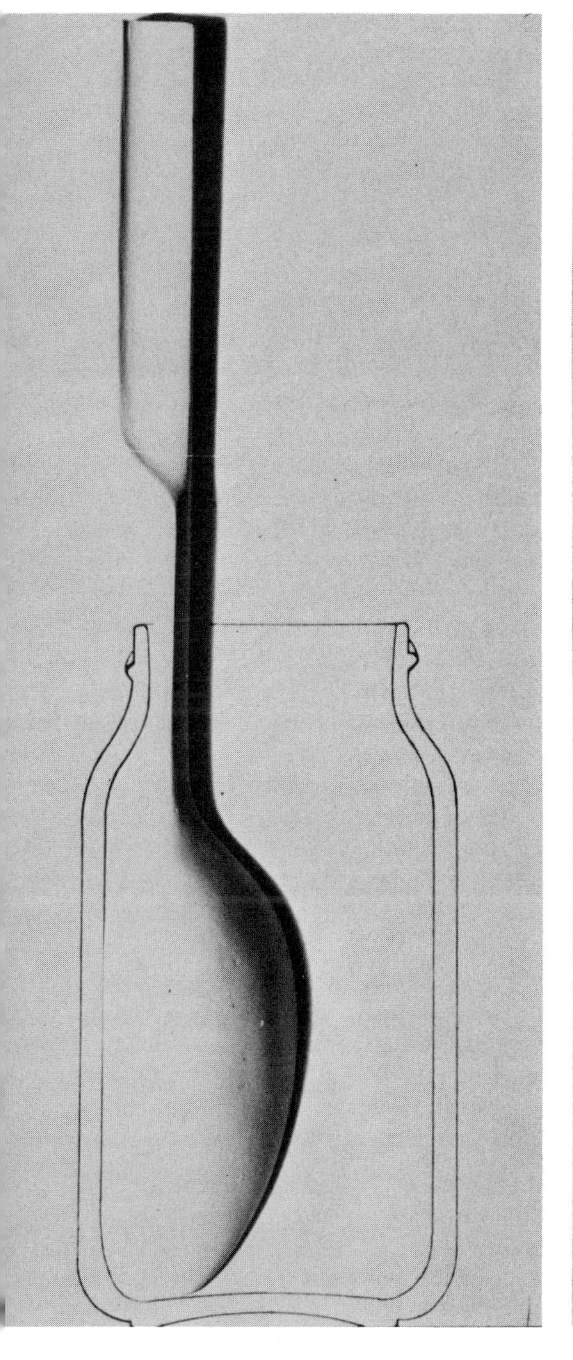

Architectural model for prefabricated building units. (*Courtesy, Sam Davis*)

wear, or chemical attack. Normally, only one of these factors can snafu a product, and only one of these considerations can head a priority list to determine design from the standpoint of performance.

Then process considerations have to be made —what forming methods will be used and what are the specifics about the design in relation to process requirements? For instance, in designing for the molding process, draft, radii, dimensional tolerance, wall thickness, bosses, inserts, methods of injection, sink marks, undercuts, gate location, flow pattern, shrinkage, and postmolding shrinkage, all are concerns.

After product requirements have been set, and after material and process decisions are made, finishing aspects finalize the product's development—and adhering and fastening parts, coloring and decorating, cleaning and polishing.

All along the way there may be experimentation and testing, particularly if it is a new design form. Until the production of the prototype, economic factors should take a second seat. At the prototype stage, critical evaluation will determine a design's acceptability from visual, beautiful new idea can be triggered by a very divergent experience. Rolling logs can inspire a

tube chair; a large balloon that someone sits on could present the idea for blow-up furniture, et cetera. Some of our most outstanding innovations were born in that way, and these serendipitous solutions inspired countless variations.
functional, and economic vantage points. Modifications are then made before production methods and numbers are determined. The best way to keep costs to a minimum is good design at every phase.

In the final analysis we start with a given set of factors and proceed. Proper use of proper materials will transform a raw material into a visual experience appealing to our minds and sensibilities. Preliminary work may be shortened by creative insight. As with Archimedes' bath,[1] a beautiful new idea can be triggered by a very divergent experience. Rolling logs can inspire a tube chair; a large balloon that someone sits on could present the idea for blow-up furniture, et cetera. Some of our most outstanding innovations were born in that way, and these serendipitous solutions inspired countless variations.

[1] His bathtime discovery was that a body immersed in fluid is buoyed up by a force equal to the weight of the fluid displaced by the body—and so he could weigh his gold.

CHECK LIST FOR PRELIMINARY DESIGN CONSIDERATIONS

General

What is the product's function?
Who will use it?
Why is it required?
When will it be used?
What space limitations exist?
What service life is required?
Is light weight desirable?
Are there government codes and specifications that set limitations?
Do analogous products exist?

Environment

Where will it be used?
What temperatures, chemicals, are present?
How much humidity is there?
How much use will it get?

Appearance

What are general ideas about style, shape, color, surface, size?
What are the creative possibilities?

Economic Factors

What is the cost/selling price of similar products?
What are preliminary estimates?
Will estimates justify production numbers and the possible market?
If not, re-evaluate all these questions and start again. Look for alternate solutions, no matter how attached you were to your preliminary product.

THE DEVELOPMENT OF PRINCIPLE

It may be that a design concept is known to us, but that we cannot carry it out effectively because material or process technology has not provided the means. A prototype may exist (cement was used to translate this idea),[2] but sometimes only a few experimental forms have been made. For example, the hyperbolic paraboloid concept has been known to us, but until recently creative use of the mathematical idea and the technology were not married into a practical solution. Professor John Zerning, working at the University of Surrey, Department of Civil Engineering, with a grant from Shell Chemical International Corporation, carried this mathematical idea to fruition. He started with a structural system consisting of an hyperbolic paraboloid shell element, with a material (fiberglass reinforced plastic), and with a process (cocooning) and proceeded to unite form (the idea) and construction technique (physical reality) into an integrality enclosing various *types of* space with structural forms that will be characteristic of the three beginning factors. A purpose of the empirical investigation that follows was to stimulate the imagination of architects and engineers.

[2] The hyperbolic paraboloid was used in Mexico by Felix Candela and by Corbusier (the Philips pavilion at the Brussels Exhibition in 1958), to name two translators.

Definition of Surface

A hyperbolic paraboloid surface is a translation surface because it is generated by translating a principal parabola (GOC) with an upward curvature upon a downward parabola (AOE) (3). It can also be called a doubly ruled surface. Two systems of straight lines intersect the nonparallel ABC-GFE and AHG-CDE, called directrixes. The directrixes are parallel to the plane XOZ and YOZ (director plane). Plane sections such as contour lines parallel to the Z axis will be parabolic, curving positively or negatively depending on the direction of the section. By pulling the "corners" and twisting the form, the surface can be transformed in four basic ways—slope (4), angle (5), tilting the Z axis (6), and cutting various segments from a standard hyperbolic paraboloid surface so as to obtain new HP segments with different perimeters (7).

These surfaces can be further transformed through organization. Any geometrical configuration can be organized according to the synthetic concept, which is a process of adding and putting together separate elements to form a whole (8), or according to the analytic concept. A given whole is resolved into small elements, in order to "fit" them into HP elements (9). Sometimes both these concepts of analysis and synthesis are combined. Repetitive HP units can be identical or quite different from one another (10). They can be combined in a single linear dimension, a two-dimension planar representation or a three-dimensional spatial form (11). There is almost no end to the infinite possibilities of arranging these forms. Some shapes are exceedingly difficult to analyze because of their complexities.

Studies of two of many possibilities of models explore ways in which the perimeter of an HP unit changes if various segments are cut out from a standard HP surface. The only factors that are varied are locations and perimeters of the seg-

3, 4, 5, 6, 7, 8, 9, 10, 11, 12, 12a, 13, 14, 15, 16, 17, 18, 19, 20, 21, 22, 23, 24, 25, 26, 27, 28, 29, 30, 31, 32, 33, 34, 35, 36, 37, 38, 39. Photos of John Zerning's hyperbolic paraboloid designs.

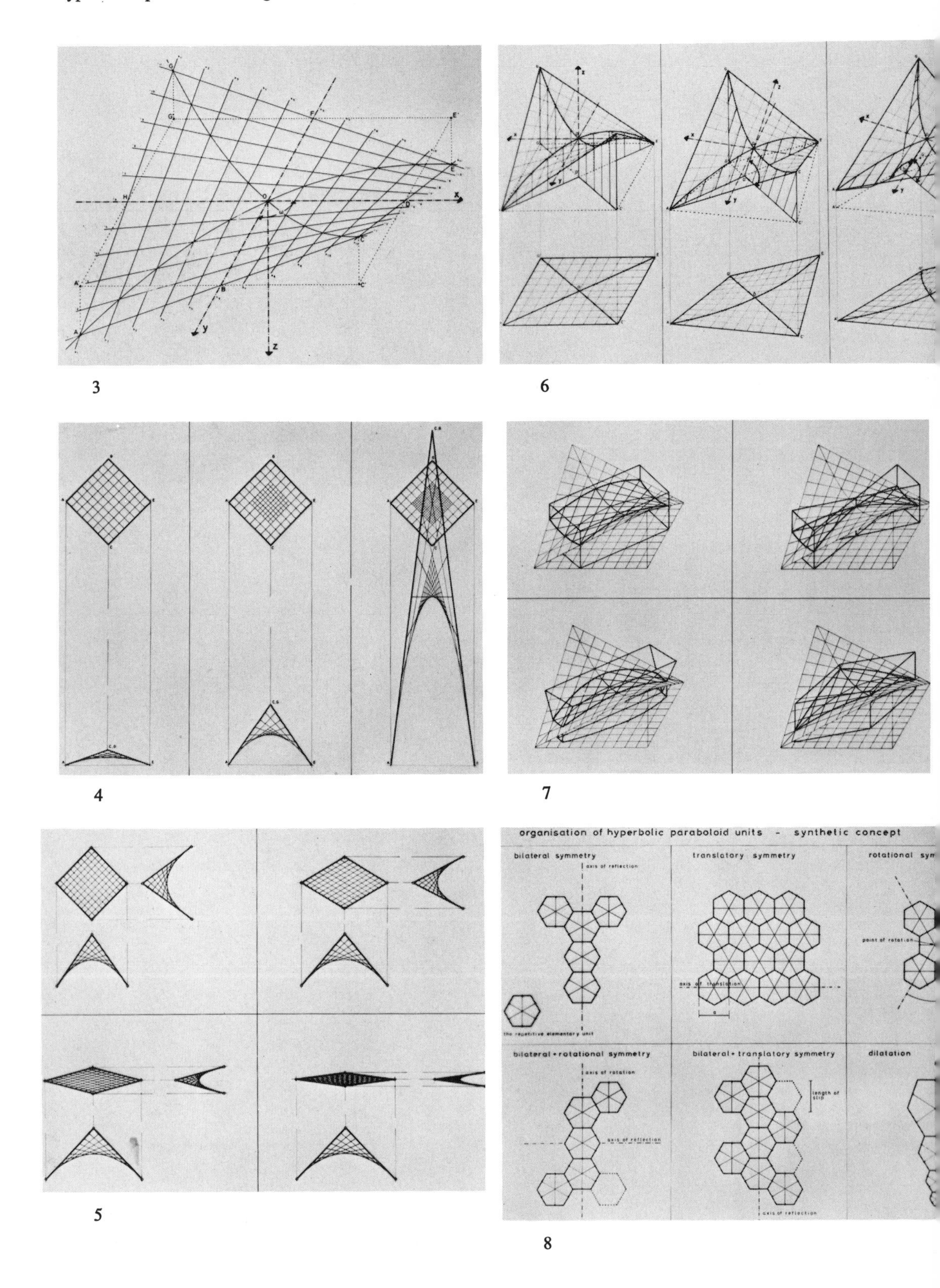

3

6

4

7

5

8

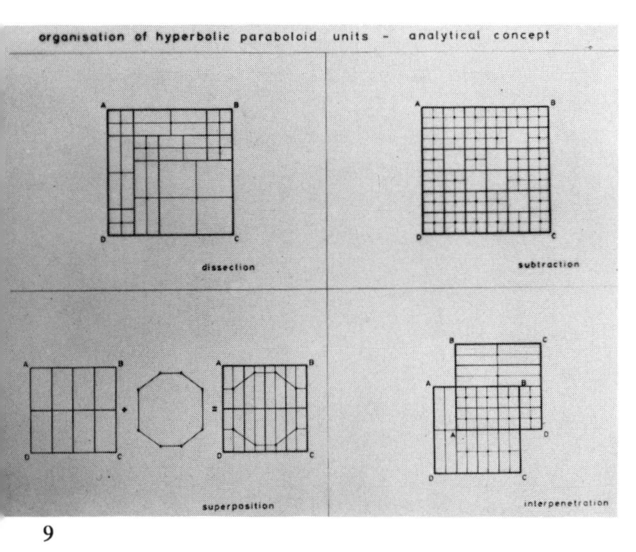

9

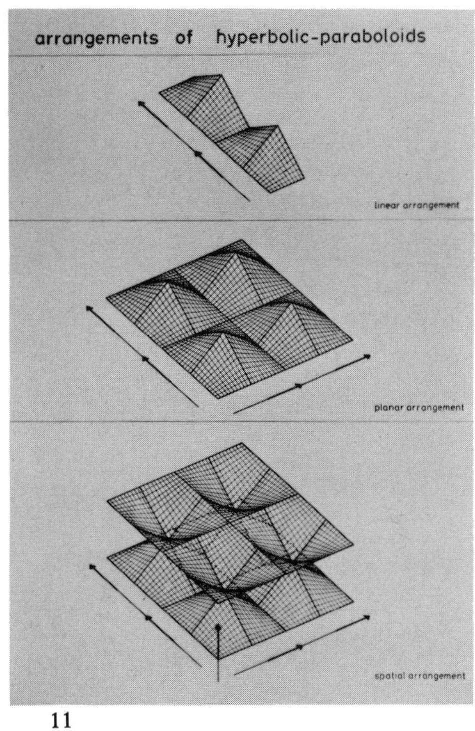

11

organisation of hyperbolic-paraboloids (hypars)
the "repetitive elementary" unit (with a higher degree of symmetry)
an elementary unit with 1 'hypar' surface
two 'hypar' surfaces
three 'hypar' surfaces
four 'hypar' surfaces
five 'hypar' surfaces
six 'hypar' surfaces
seven 'hypar' surfaces
eight 'hypar' surfaces
etc. etc.

10

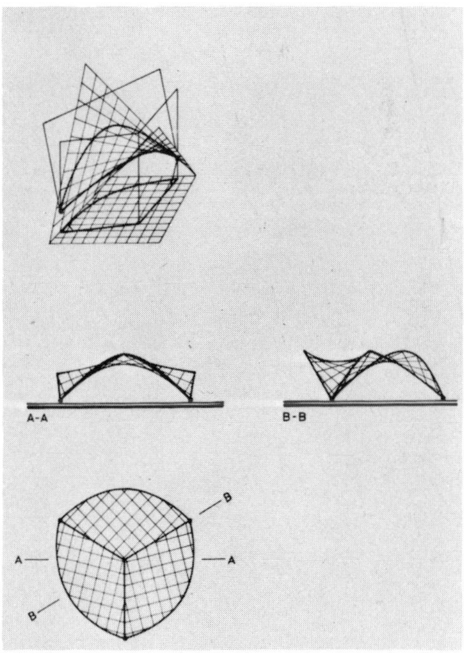

12

mental cut-outs. The boundaries and shapes change when segments are cut with triangular or rectangular perimeters. In the top left corner of (12) the invariant standard surface is depicted. The structural form consists of an arrangement of eight of these segments over an octagonal plan, radiating from a central low point. The free boundaries are undulated.

The structural form in the plan (13) and model (14) consists of two intersecting HP segments over an almost elliptical plan. Here the slanted free boundaries are hyperbolic and an

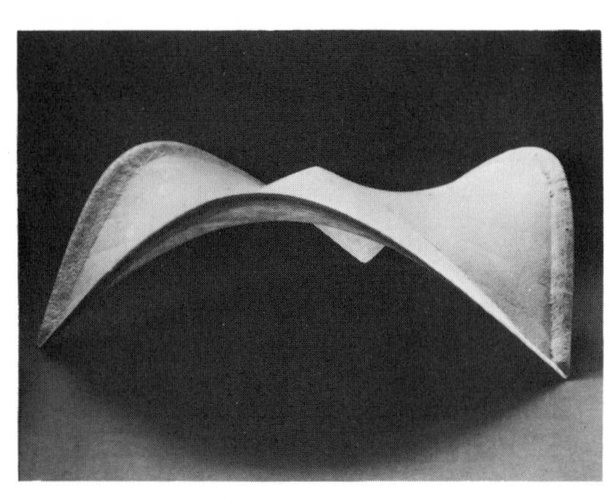

12a,

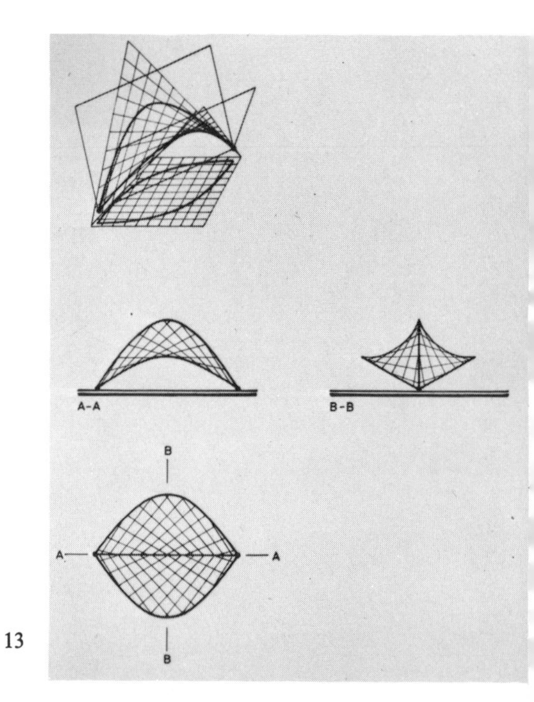

13

14

ıyperbolic vertical arch spans between the two ow points. This form explores the organization ɔf HP units according to the analytical concept. The common shell forms into a vault, cone, el-iptical—paraboloid, paraboloid of revolution, HP, et cetera. These are macro-modular units.

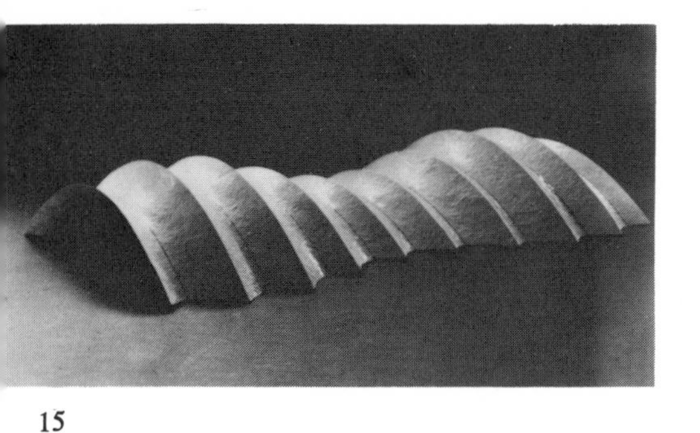

15

The infra-modular unit, in this particular project, was always an archlike HP segment taken entirely from the middle of an invariant HP surface.

Four different segments are needed to make up the macro-modular unit (15) and they all have to be taken from the same standard surface. An invariant standard surface, an HP over a rhombic plan, is seen in (16), from which the largest and smallest sections were taken.

In the next experiment the invariant factor was the plan of the segmental cut-out—a square and an isosceles triangle, and the variant was the standard surface. This permitted variations of the edges of the segment (17) and shows an HP section with two upward, one downward curving edge plus one straight edge over a square plan. The relationship of the segmental cut-out to the variant surfaces is shown in

16

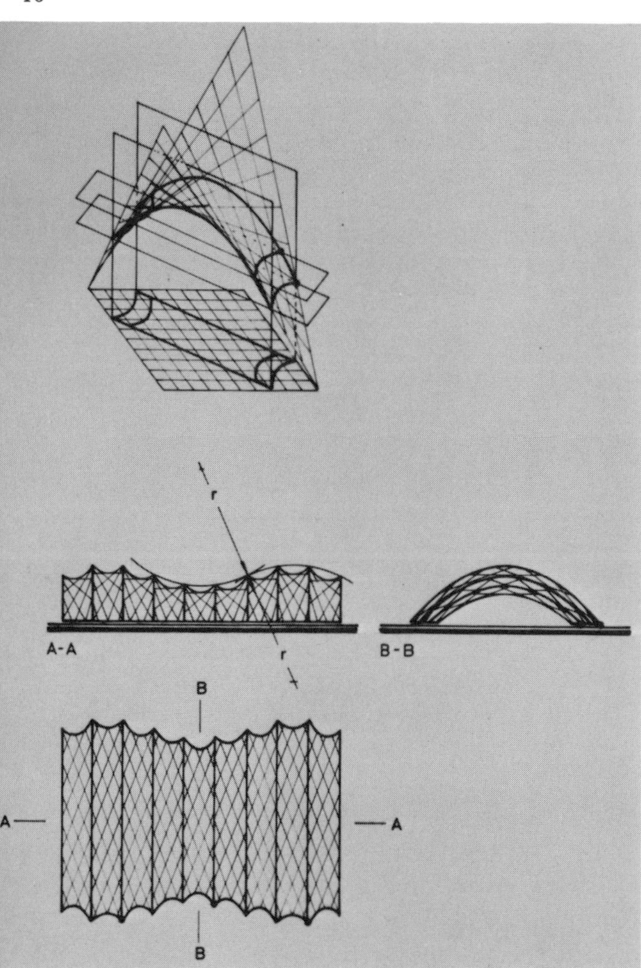

17

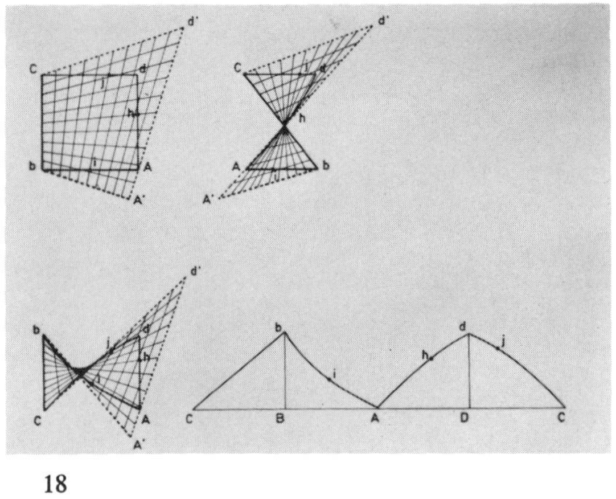

18

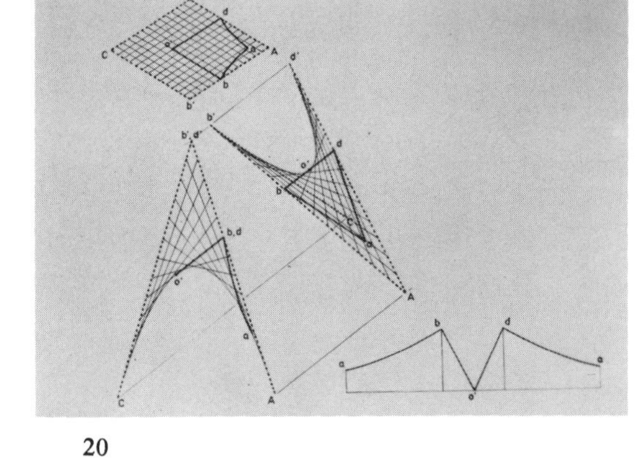

20

19

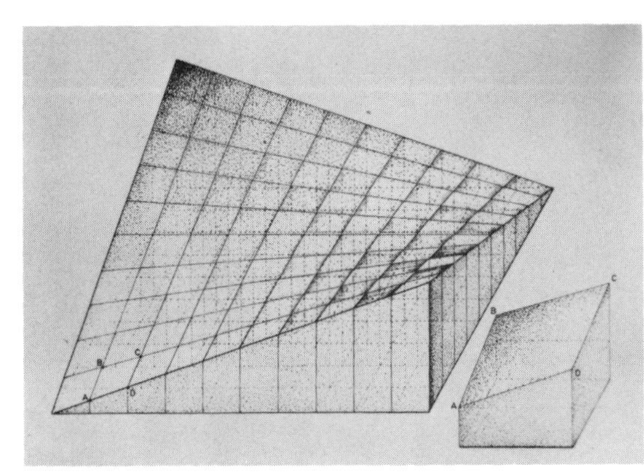

21

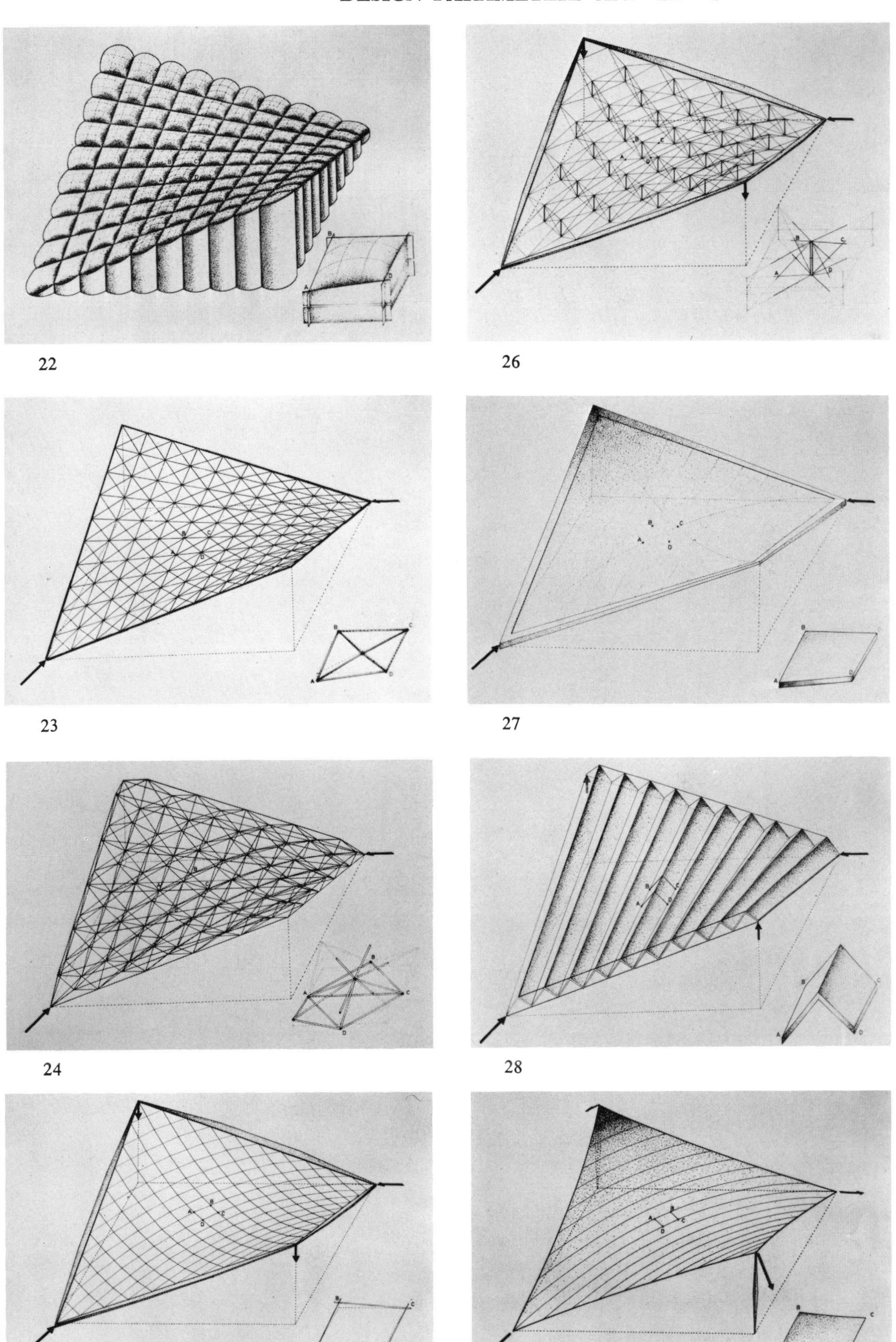

22

26

23

27

24

28

25

29

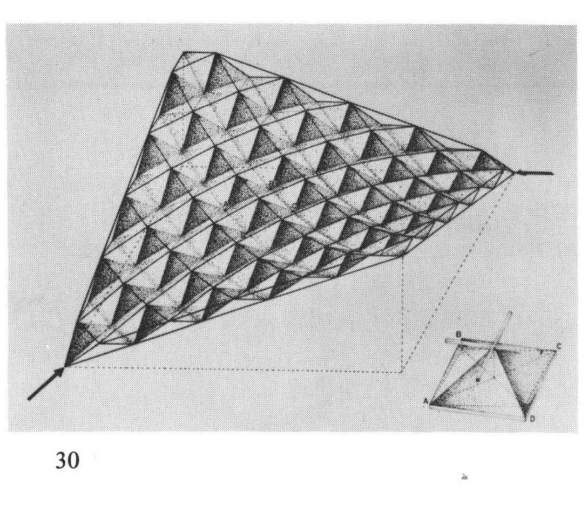

30

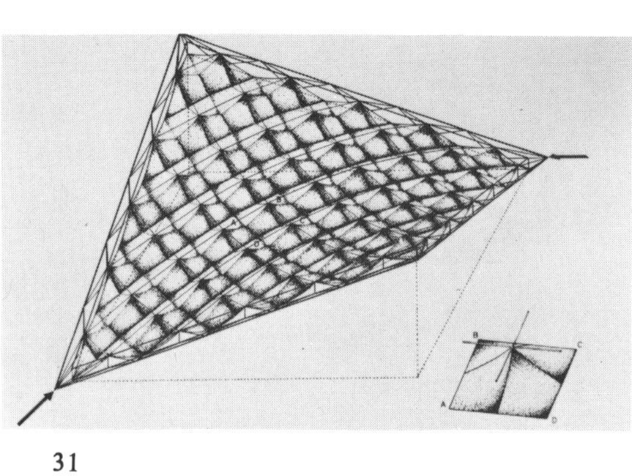

31

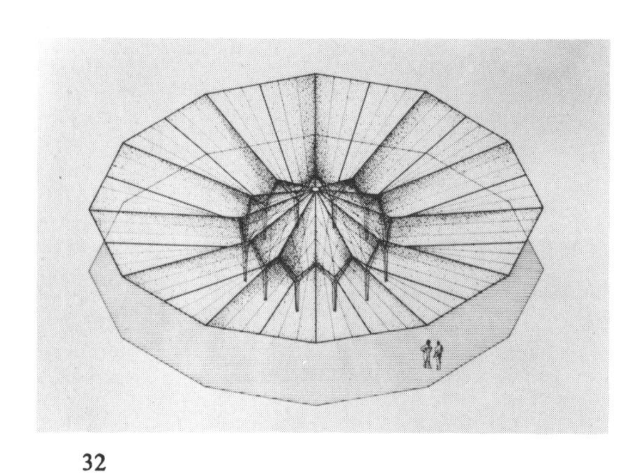

32

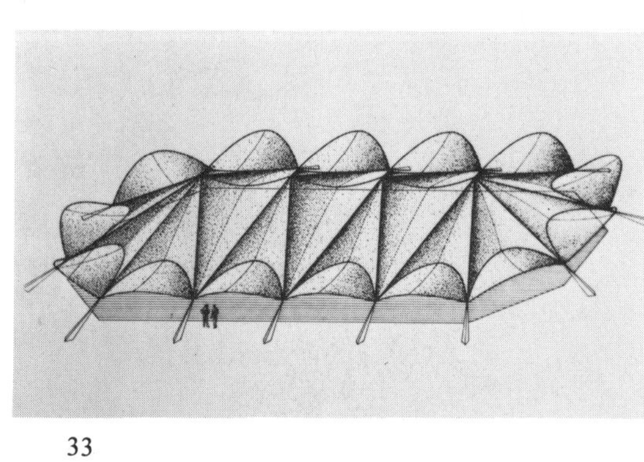

33

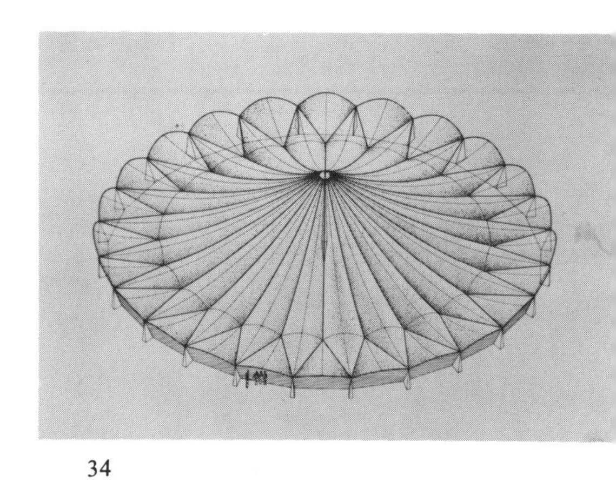

34

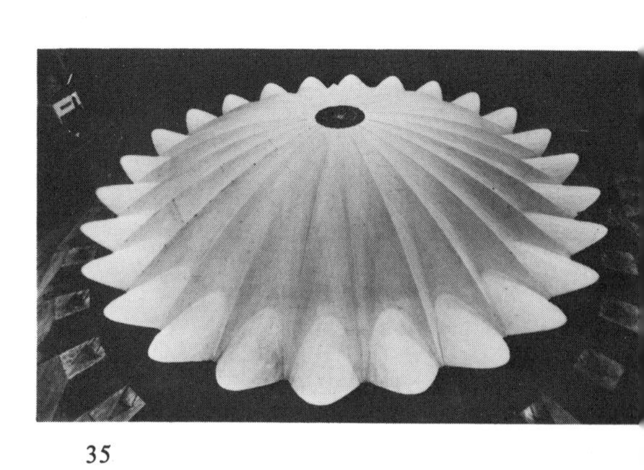

35

36

(18). Another variation is an HP segment over an isosceles triangular plan, as illustrated in (19). The relationships of the segmental cut-out to the variable surface is shown in (20).

Fundamentally, in the construction of hyperbolic paraboloids, all structures (as well as HP structures) can be traced back to solid construction as in (21) and (22); to reticulated construction as in (23), (24), (25), and (26); or in surface construction as diagrammed in (27), (28), and (29). This definition depends upon the relative dimensional relation between length "l," breadth "b," and height "h." In a solid construction, then, "l," "b," and "h" are of similar magnitude; a reticulated construction is made up of linear elements where "l" is always greater than "b" and "h"; in a surface construction "l" and "b" exceed "h."

Reticulated and surface construction can be integrated into a combination that forms a stress skin reticulated grid. As a single layer system it has been used quite frequently, but at a double layer system it is new (30, 31). None of these factors is independent of the other. There are definite relationships among them, sometimes complementary and sometimes conflicting, and it becomes the task of the creative designer to sort out so that these factors can be synthesized to form a whole.

Application of HP Structures

When we consider application of these structural forms, we discover that in principle they have not changed very much through the ages. Macro-modular units such as vaults, domes, and umbrellas have been with us for ages. But new technology and new materials have provided the means to build larger spans with less weight.

New materials have also made possible "new" structures such as cantilever and saddle-shape because these structures require materials that can resist high tension, compression, and shear. One of today's needs is for prefabrication. Up to now it has not been possible to use the HP form with its strong individuality. There is a way, however, if the HP surface is used as an elementary repetitive unit, in which it can be made to fit almost any enclosure. With a vertical "Y"-column supporting system, a scalloped cone-shaped roof, intersected by a cantilever circular fan-type roof structure, can span very large areas (32). Another spanning approach is with a single-folded scalloped system with half cone-shaped ends having an undulated edge at its perimeter and support through inclined buttresses (33). And a third is a scalloped tent system with an undulated edge at its perimeter and support through an interior column (34) and/or tension wires around the perimeter (35), (36), (37).

These HP's can be built of reinforced concrete, aluminum, and timber. Potentially it should be possible to use fiberglass reinforced plastic (polyester or epoxy) much more economically because of moldability, lighteners, corrosion resistance, variability of surface color and texture, and durability. There are disadvantages; the material has to be made fire resistant. Its low modulus of elasticity, high creep and high coeffi-

37

cient of expansion impose other limitations when it is used to span large areas.

COCOON CONSTRUCTION SYSTEM

John Zerning, engineer of this HP design development, also adapted an economical, practical building technique that avoids the use of costly molds necessary for fiberglass reinforced plastic construction. He constructed a network (of wires or cord) that became a temporary structure for spanning large areas (38), and then, by his using a sprayable vinyl cocoon system, sizable areas could be bridged over the network. Henry Miller used this system for his balloon lamps in the 1940's and the U. S. Government has used it for mothballing naval vessels.

A styrene webbing agent is sprayed over the network, by using a 10- to 20-lb. quart pressure cup with an air nozzle (66p for webbing, Binks Manufacturing Co., 3114–44 Carroll Avenue, Chicago) and a model-62 gun at 23 cubic feet of air per minute at 80 lbs. of pressure. With proper atomization a spiderlike web is "spun" that can span 12 to 18 inches. The webbing has its own structural strength. When the webbing is uniformly "knitted" opaque and firmly in place, a pigmented vinyl (Rynohyde 90) is sprayed with a high-velocity air nozzle cap (63 PB-Binks). (One gallon of webbing covers 20 square feet at 20 mils thickness.) When this has cured, a seamless leatherlike skin covers the sprayed area. Upon curing, the vinyl shrinks and forms a prestressed membrane that does not support combustion. This, then, becomes an economical mold for fiberglass reinforced plastic that has its own built-in release agent (although for epoxy an additional release spray is necessary—also a good idea for polyester resin because it helps to extend the life of the mold) (39).

Fiberglass reinforced plastic can be used as the prime structural element, with all stresses carried by this form, or it can become a cladding material over a base, as a framework for a concrete shell, or as a composite for sandwich construction.

38

Analysis of This Design Approach

Design is a visual language that combines human, physical, and functional properties with man's scientific, technologic, and aesthetic prowess into an expressive form. At every age design evolves as a means of expression that describes the thinking and feeling of that period in history. This is a complex age that finds many modes of expression with great differences in approach, style, and effect. We have an accumulated treasure from history as well as new potential solutions because of new materials and new processes. Sometimes new materials are put into the procrustean bed of old processes and design solutions. At other times innovation comes through a designer's way of seeing—his approach.

Most designers develop their products from defined problems. After problem analysis, the designer searches for the proper form. That is but one way of working. The "Development of a Design Principle," the hyperbolic paraboloid, illustrates that structural forms can emerge from a principle with applications to be found later. The formal vocabulary that evolves becomes the raw material to be modified later into a design form. John Zerning finds that this approach can free a designer's imagination. It certainly can grow toward a wider in-depth fund of design solutions, leading to more creative forms.

39

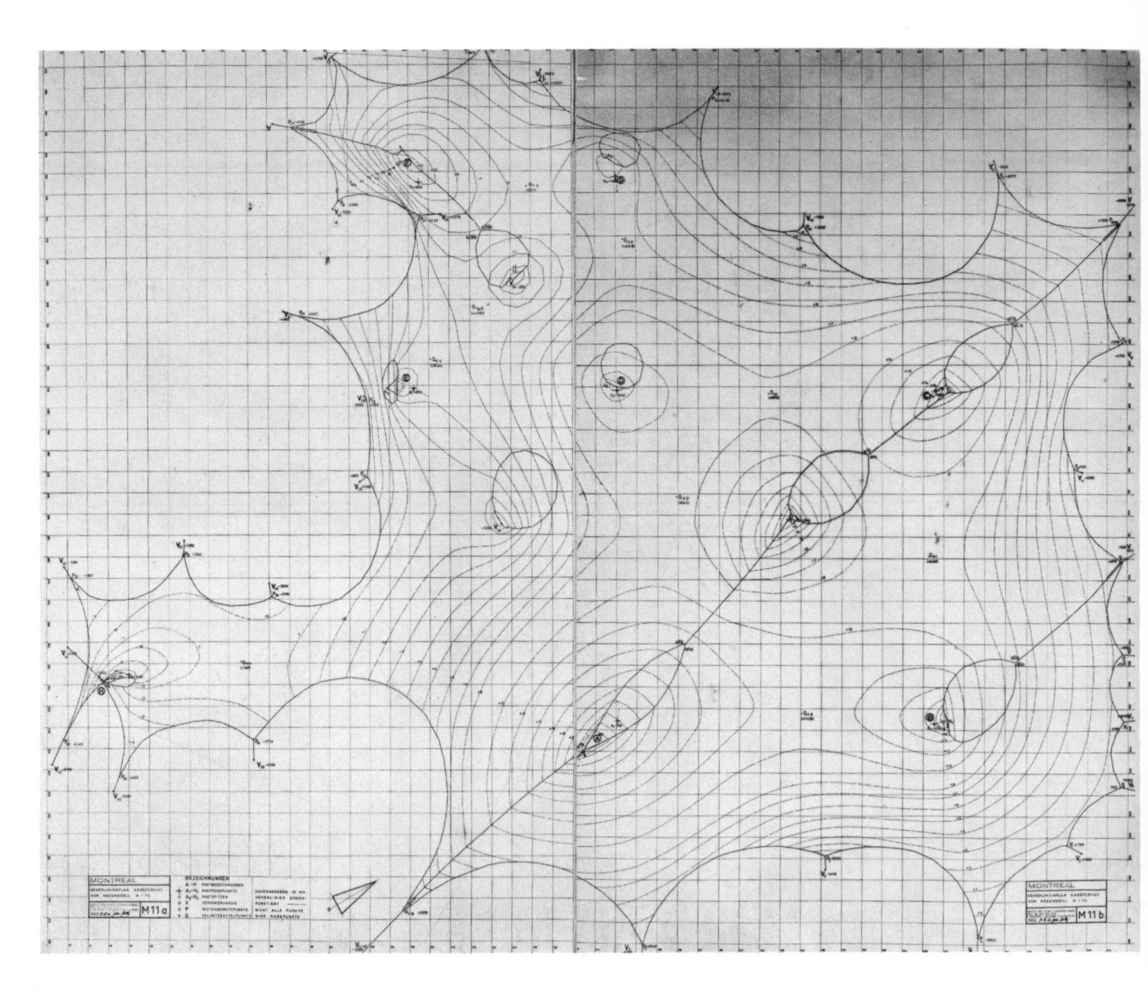
MONTREAL
BEZEICHNUNGEN
M11a
MONTREAL
M11b

Stages in the construction of the "roof." (*Courtesy, Institut für Leichte Flächentragwerke*)

Hyperbolic paraboloid at the Bangkok International Fair was translated to translucent fiberglass reinforced plastic, approximately $\frac{1}{10}''$ thick, fastened to steel edge ribs. (*Courtesy, Professor Albert Dietz*)

DESIGN PARAMETERS

Choosing the right plastic to use in a particular application is very difficult when we have to consider over forty different families of plastics and sometimes hundreds of individual types within each family. The choice becomes easier when the designer defines his material needs for a specific application. Some of the differences among plastics that need to be considered are: useful temperature range, processing equipment; degree of flexibility or stiffness; tensile, flexural, and impact strengths; stress and deformation; color; environmental effects of weather, sunlight, moisture; odor and taste; long-term creep properties under critical loads; design—material limitations; economics. Evaluations of how plastic behaves under standard conditions are usually available from manufacturers in their specification sheets. From this data the designer can eliminate some plastics and then refine his list with more critical analysis. For example, for "X" product, tensile strength would be essential at above 6,000 p.s.i., desirable above 8,000 p.s.i., but not important to be above 12,000 p.s.i. One of the most significant considerations is that of stress and deformation.

Stress and Deformation as Design Considerations

A beautiful shape that does not function or breaks too soon is a failure. (So is a smoothly operating piece a failure if it is ugly.) When an object breaks, it can fail because of stress or deforming. *Stress* is developed at some point when the load exceeds the strength of the material. It is usually expressed as load in pounds per square inch. *Deformation,* or *strain,* happens when a form cannot function as required under load. A unit of deformation or strain is measured by percentage or in inches per inch. The material is tested on a test bar. As the tensile stress is increased, the strain or elongation increases proportionately. When the bar elongates without the application of an additional load, the material has reached the *yield point.* Stress-strain data tell the strength of the material and the extent to which deformation takes place at any given stress and at the rate of loading used. For practical design work, materials should not be used that will have to function at the yield point. Lower stresses should be considered. *Working stress* is what should be used as a lesser figure. It is determined by dividing the yield point by a number from 4 to 10, known as the safety factor. If maintenance costs resulting from failure are very high, then the safety factor should also be high—a number such as 8. Resistance to deformation or stiffness is the *modulus of elasticity.* Effects of environment—temperature, water, and sunlight—are very important if the product has this exposure. Properties of all materials vary with temperature but not all materials are affected by moisture adversely. These aspects also figure into the safety factor that is set. The safety factor considers the future of the product —safety, reliability, and economics of the material.

Karl Gerstner's "Tension Picture" Number 3 illustrates stresses in plastics. Size, 22" x 22" (1963–64). (*Courtesy, Stampfil Gallery, New York*)

Another stress factor is called *creep* or *cold flow,* which is more peculiar to plastics than to metals; for example, a form that is subjected to stress over a long period of time must be designed so that deformation with time will not be excessive and so that fracturing or breakage does not occur. If a material has an initial modulus of 200,000 lb./sq. in. and in 100 hours the material behaves as though the modulus were 114,000 lb./sq. in., the deformation expected in 100 hours at a stress of 1,800 lb./sq. in. is $\frac{1,800}{114,000}$ or 0.016 in./in. The maximum deformation that can be tolerated should be compared with the above calculations.

Fatigue and impact are two more considerations relating to stress. Any material subjected to varying load would fail at a lower stress than if the load were continuous. Reversed stress—tension and compression—is the level where failure will not occur. This is the *fatigue endurance limit.* The appropriate fatigue value has to be determined. For designing working parts the working stress would be two thirds of the appropriate fatigue value.

Ability of a form to absorb impact energy has a great deal to do with the shape of the design. (The energy of the impact must be absorbed by a force within the part times the distance that the part can deform.) Thin-walled *flexible* open-ended forms will tolerate greater impact

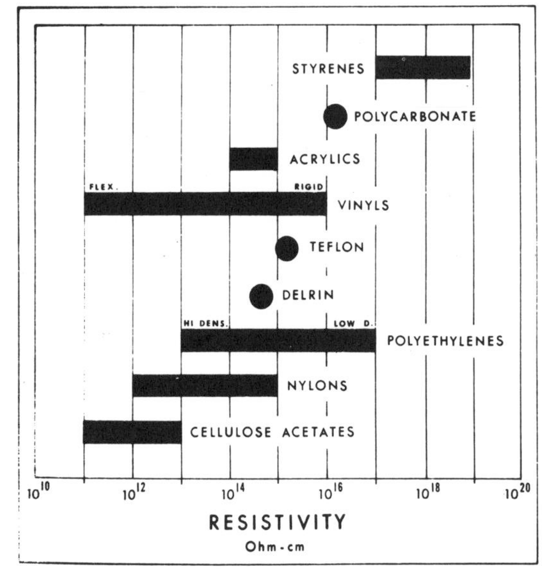

Design properties of plastics. (*Courtesy, Cadillac Plastics & Chemical Co.*)

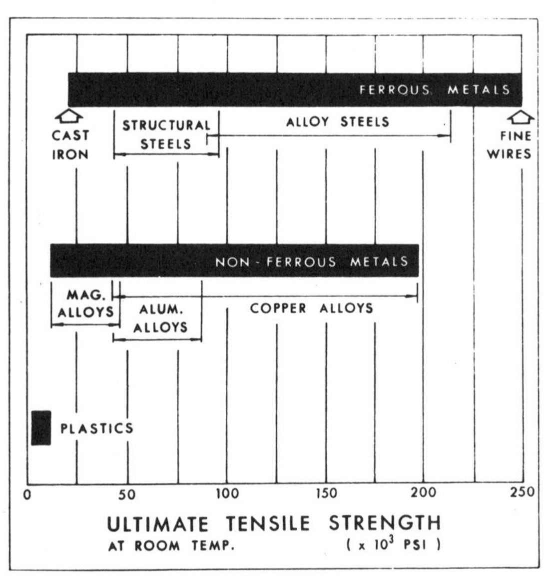

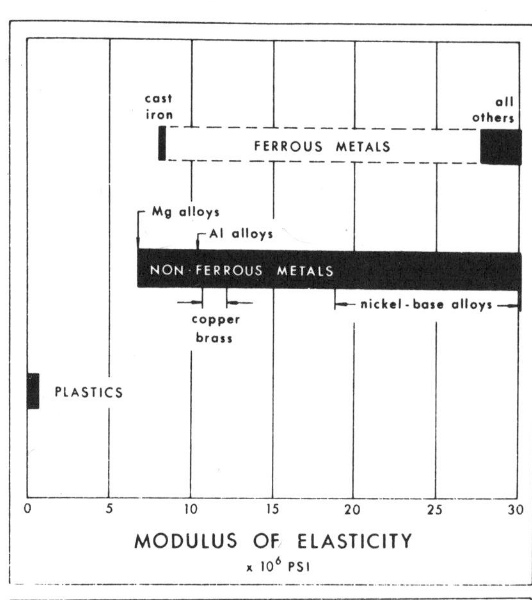

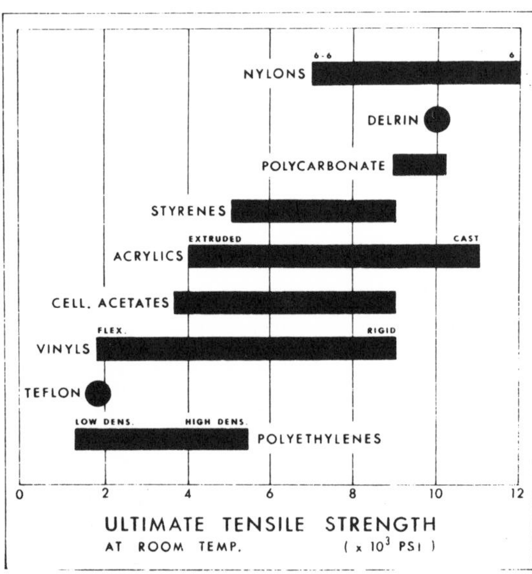

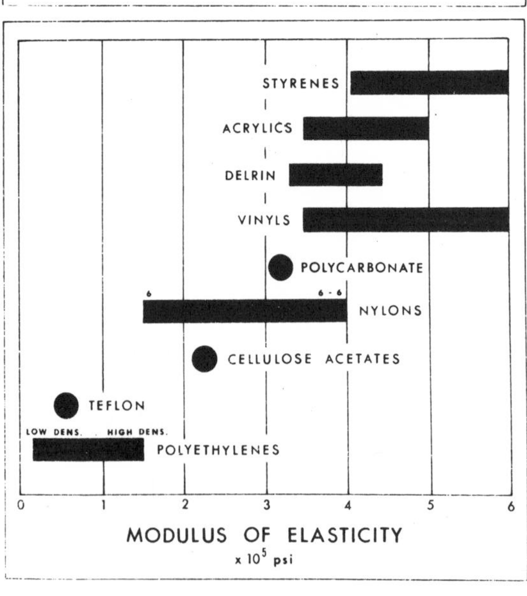

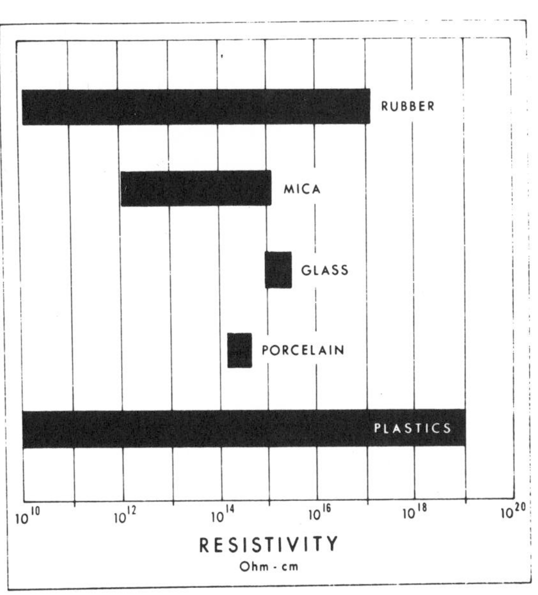
RUBBER
MICA
GLASS
PORCELAIN
PLASTICS
10^{10}
10^{12}
10^{14}
10^{16}
10^{18}
10^{20}
RESISTIVITY
Ohm - cm

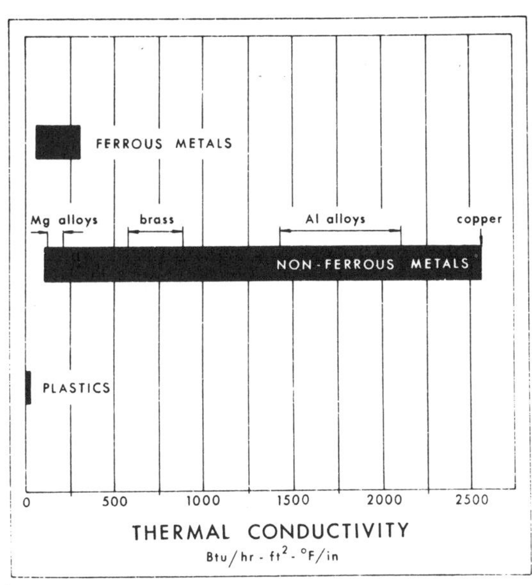
FERROUS METALS
Mg alloys
brass
Al alloys
copper
NON-FERROUS METALS
PLASTICS
0
500
1000
1500
2000
2500
THERMAL CONDUCTIVITY
Btu/hr - ft^2 - °F/in

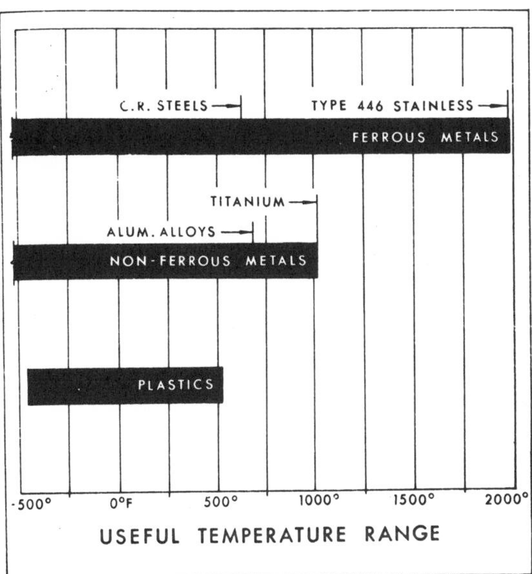
C.R. STEELS
TYPE 446 STAINLESS
FERROUS METALS
TITANIUM
ALUM. ALLOYS
NON-FERROUS METALS
PLASTICS
-500°
0°F
500°
1000°
1500°
2000°
USEFUL TEMPERATURE RANGE

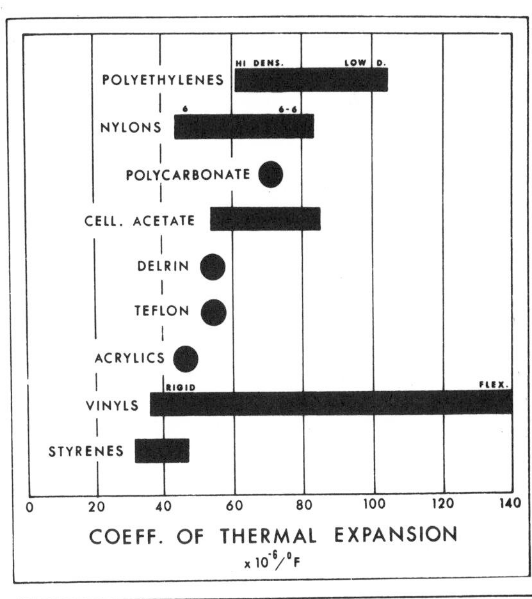
HI DENS.
LOW D.
POLYETHYLENES
6
6-6
NYLONS
POLYCARBONATE
CELL. ACETATE
DELRIN
TEFLON
ACRYLICS
RIGID
FLEX.
VINYLS
STYRENES
0
20
40
60
80
100
120
140
COEFF. OF THERMAL EXPANSION
x 10^{-6}/°F

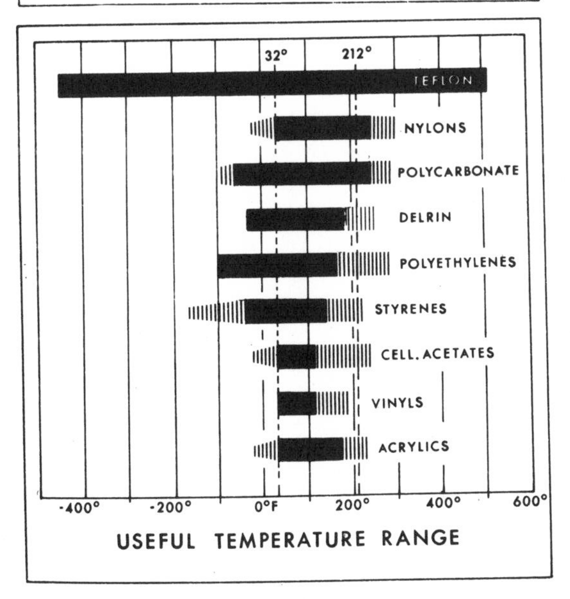
32°
212°
TEFLON
NYLONS
POLYCARBONATE
DELRIN
POLYETHYLENES
STYRENES
CELL. ACETATES
VINYLS
ACRYLICS
-400°
-200°
0°F
200°
400°
600°
USEFUL TEMPERATURE RANGE

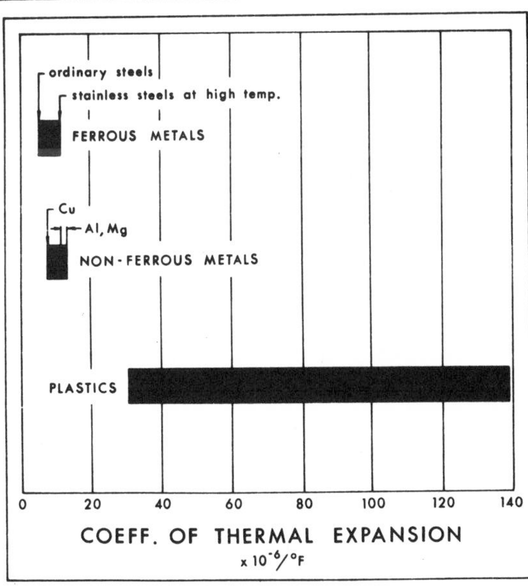
ordinary steels
stainless steels at high temp.
FERROUS METALS
Cu
Al, Mg
NON-FERROUS METALS
PLASTICS
0
20
40
60
80
100
120
140
COEFF. OF THERMAL EXPANSION
x 10^{-6}/°F

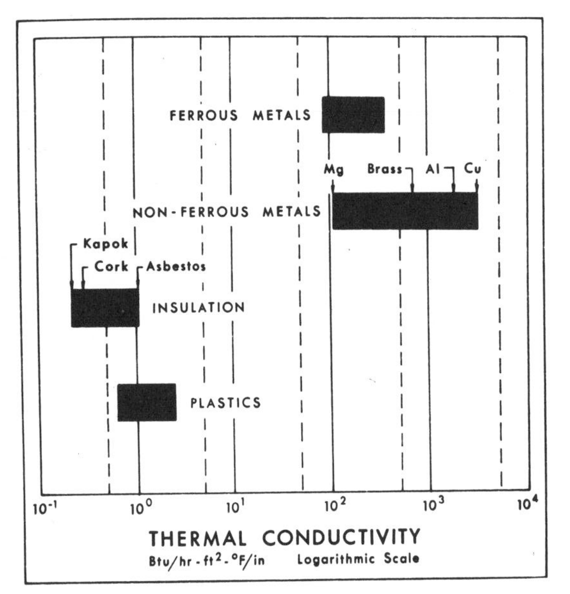

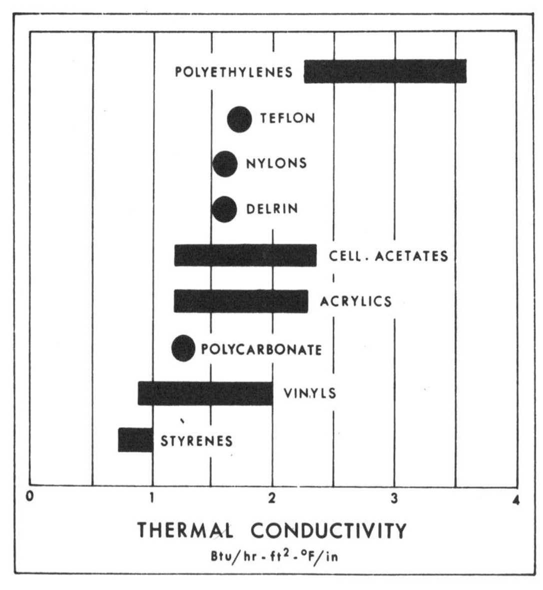

Elongation

Some plastics will stretch like taffy before breaking and others will pull apart like an egg shell. If the taffy is originally 4″ long and stretches to a total length of 12″ before breaking, its elongation would be expressed in percentages —in this case, 200% elongation.

Tensile Strength

Tensile strength is the resistance of a piece of plastic to being pulled apart. It is expressed in pounds per square inch. A puff of soft cotton would take very little force or total pounds to pull apart, whereas plastics have greater strength, usually from 1,000 to 50,000 per square inch (p.s.i.). Structural alloys such as steel have tensile strengths that run well into six figures.

Injection molding of ABS cuts down costs by providing the possibility for quick mass production. ABS itself is a very strong, durable, colorful material. Limitations are in designing for a mold; undercuts and tolerances, supports and ribs, gates and runners, flow patterns, warpage allowance, seams for multipart molds and parting lines, et cetera, all have to be considered. Naturally this can affect the original design.
For this chair "Seggio," designed by Joe C. Colombo for Kartell of Italy, ABS was chosen because of its structural strength, ability to be injection molded, and production economy. The chair will both stack and gang; is available in both 18″ and 15″ seat heights for adults and children, as you can see by the bottom leg insert; was made in bright orange, white, or black (potentially ABS can accommodate a wide opaque color range). The hole at the back serves two purposes: it provides ventilation where the body gets hot, and it can be used as a handle for lifting the chair. Curves were used because the straight line would parallel a break, whereas curves follow stress. Horizontals and verticals work better with wood; plastics work better with curves because form flexes and deforms with the body. Some fillets are used underneath for reinforcement. (*Courtesy Joe C. Colombo*)

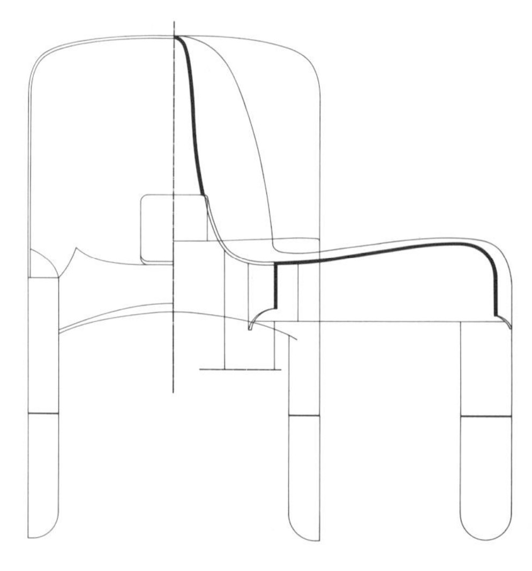

than the same flexible material joined with thicker walls by rigid corners, because the walls cannot deform enough on impact to reduce the generated forces.

Plastic Memory

Thermoplastics have an elastic memory. An area that is deformed or distorted by a pushing force will return to its former shape after the force of the tool is removed, much like a rubber ball which deforms while the pressure of a finger is held there, but returns to its former shape when pressure is released. Plastic memory, therefore will effect close tolerances where clearance is insufficient.

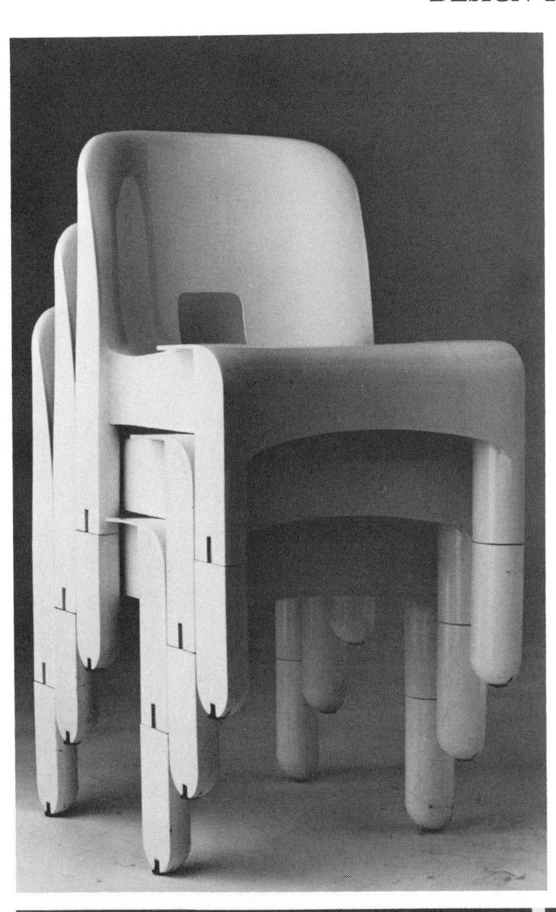

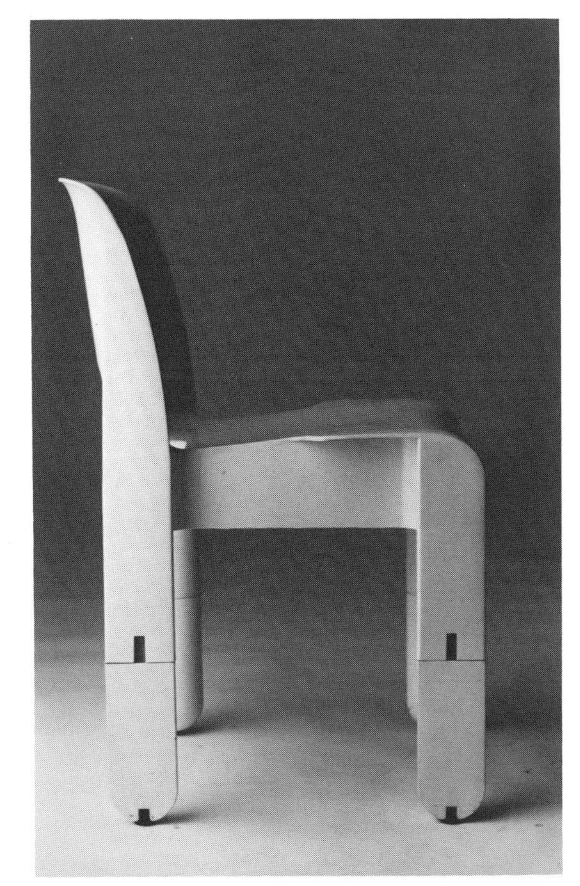

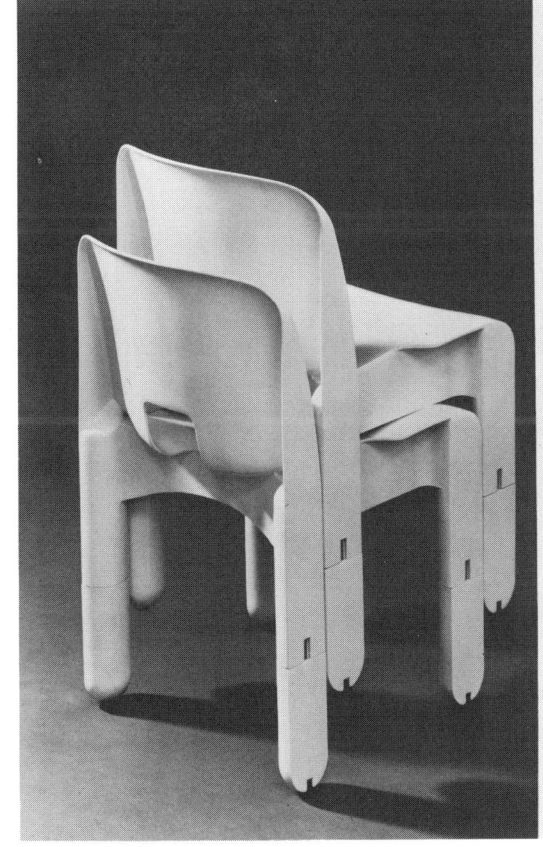

This programmer designed for Olivetti by Mario Bellini is also injection molded of ABS. It acts as a container, but also supports heavy mechanical equipment and is designed to "take a beating." The surface is pebbly to minimize surface scratching and to eliminate glare. (*Courtesy, Mario Bellini*)

Flexural Strength

Flexural strength is also expressed in pounds per square inch. When hung over the edge of a table, a sheet of paper will curve downward slightly, whereas a cotton handkerchief will hang almost perpendicular. This means that the flexural strength of paper is higher than that of the cotton handkerchief because it resists bending under its own weight.

Coefficient of Linear Thermal Expansion

The amount of growth that occurs in a material when it is heated is called "Coefficient of Linear Thermal Expansion." It is expressed in inches/inches/°F. Mercury in a glass thermometer when heated will rise considerably, but the glass container at that same temperature may remain the same.

UNPREDICTABLE VARIABLES

Some materials will change their properties under certain processing conditions. High impact thermosets containing glass fibers can lose almost 50% of their value when processed in

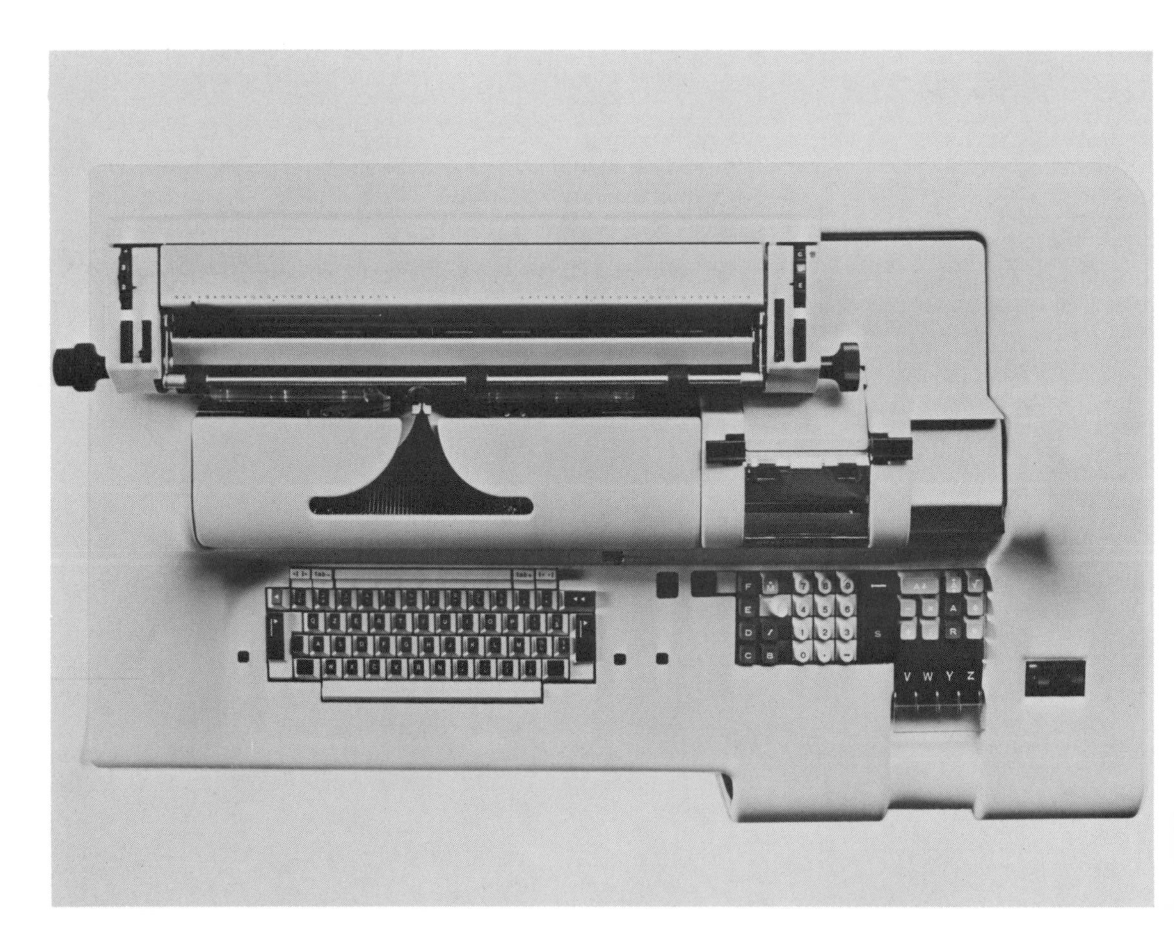

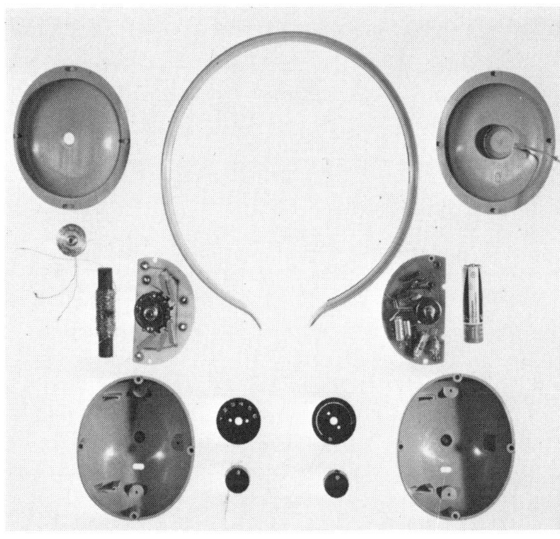

Archille Castiglioni designed this prize-winning transistorized wireless-simultaneous-translator earphone set for Phoebus Alter. Here too, ABS was injection molded and functions as a container. The headband has spring and resiliency so that it will not crack or craze when stretched around the head. (*Courtesy, Archille Castiglioni*)

transfer or plunger molds, because the fibers are macerated through processing pressure and flow unevenly. Another unpredictable example is estimating the amounts of shrinkage of a piece while the part is cooling (mold shrinkage) and then in the long run after twenty-four hours (after shrinkage). Some materials are more stable than others. Close tolerances may be impossible to meet with some plastics. When metal parts are used as inserts with different expansion and contraction ratios, it will prevent some areas of the plastic part from shrinking while other areas will shrink—a fact which causes a distortion or warping. In cooling, sections that are thick will take longer to cure than parts with thinner sections; if this occurs in the same form, warpage, blisters, or sink marks could develop.

SOME PREDICTABLE VARIABLES

Some of the larger manufacturers will help to prejudge the practicability of a design and compare new design solutions with case studies within their experiences. Mobay recently received an award to commend the research they supported in furniture design, with urethanes, even though they do not intend to go into the furniture business. In the plastic furniture field, where design possibilities are being newly ex-

The exterior shells of this suitcase designed by Architect Cini Boeri was injection molded of ABS in white, red, or blue. Two shells of different depth were designed for each size, which made possible three combinations of suitcases for each shell. The overall dimensions have been designed so that it is possible to nest the whole series in a compact package. Nylon wheels partially recessed and a handgrip that is spring-operated so that it will return into the groove in the case minimize breakage through rough baggage handling. The case becomes almost indestructible and is relatively light in weight. (*Courtesy, Cini Boeri*)

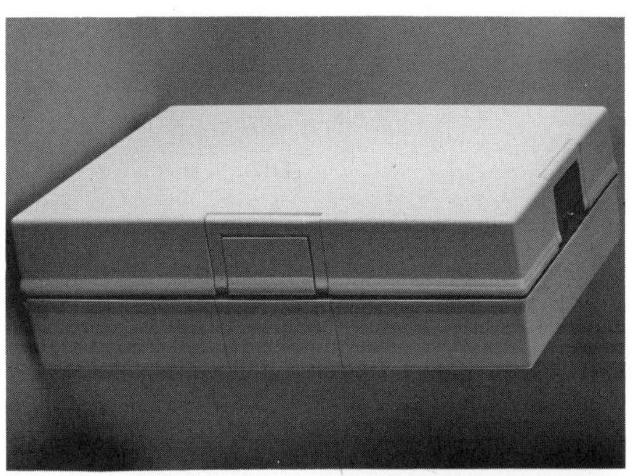

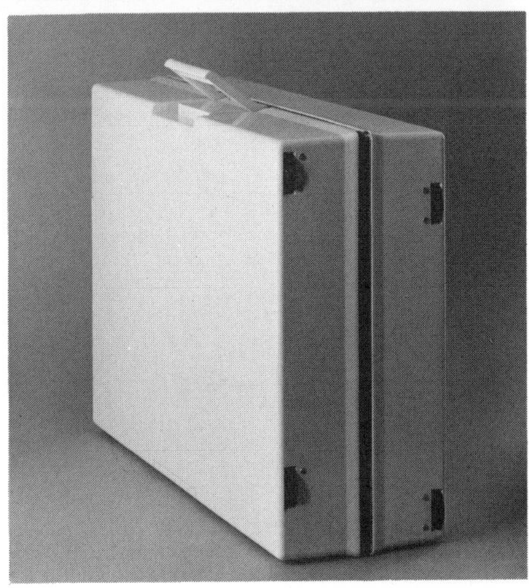

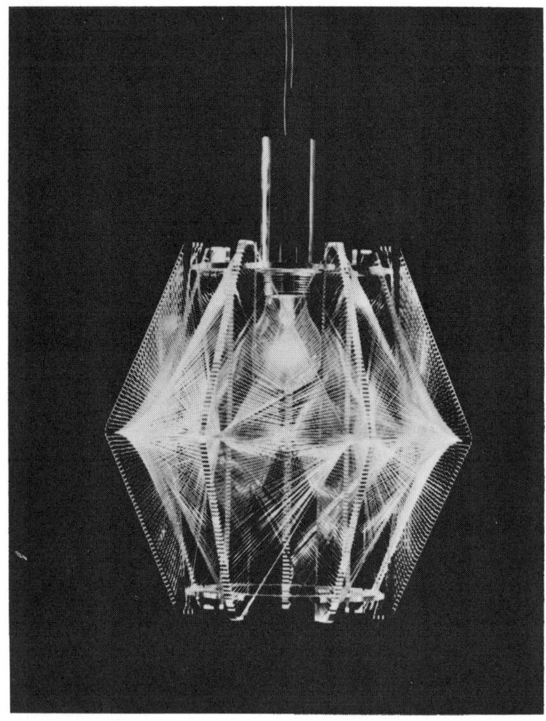

Designing with acrylic requires cutting, thermoforming and/or strip heating of sheet acrylic, attaching and finishing operations. Some small items have been designed for cast acrylic processes. Basically, though, blocks, sheet, rods, and tubes are formed and it is this stock that the designer usually uses as his "raw" material.

"Plexima" by Koch and Lowy, a Scandinavian import, has seen many imitators, possibly because it is easy to copy and very effectively utilizes acrylic's edge lighting potential. Nylon cord is stretched across grooves and also carries light as a linear pattern. The light source is masked, diffused, and transmitted by both the nylon and the acrylic. (*Courtesy, Koch & Lowy*)

Joe C. Colombo and brother Gianni designed a lamp that utilizes Plexiglas and its edge lighting potential by transmitting light throughout its form from its fluorescent light source encased in a metal base. The acrylic shape was made by thermoforming. (*Courtesy, Joe C. Colombo*)

plored, there are many trial-and-error expensive mistakes that help to push up the cost of new designs. Preliminary help could decrease design failure. Failure may not just happen because of design, material, or fabrication but also because of assemblage techniques.

STRUCTURAL DESIGN POSSIBILITIES

In furniture, storage, and lighting design there are certain possibilities. A shape can be spherical, post and lintel, cantilevered, a bent-plate type where the material is folded back on itself for support, a filler type that does not support the thrust of full weight but is used with another material such as metal, a wrapper type where the plastic is used as a covering or container, an integral skin form where structure and surface are one and the same unit and are soft, a molded type where plastics are used as structure and surface but are hard and resisting. Design parameters are demonstrated by what works through deduction and reconstruction of what has been done. What does not work usually becomes proprietary information and remains buried in the morgue of experience. (Some specific do's and don'ts will be covered in discussion of each of the plastics.)

DESIGNS OMITTED

There is no attempt here to illustrate every design idea and exploration with plastics. This would be impossible to do; therefore many excellent forms will not have been included in this volume. There is a deliberate omission, however, and that is within the whole area of design with plastics that attempts to imitate other forms, other materials. The furniture field is full of this plagiarism—a hang-over from the nineteenth century, an era of feeble recapitulation of history's glories. Imitation of wood furniture is one of the design horrors that sells well. More and more companies formerly constructing furniture of wood are converting to plastics—urethanes and polyesters. Their designs, however, have not changed. Technology is stretching its collective brain power to come up with solutions to problems imposed when one material imitates another. What wasted effort!

Perhaps we need a shift in focus, with the designer not as a service element of production, but rather the designer as an inventor who is aware of production skills and as an artist with his form having priority. Plastics should and does open up new possibilities.

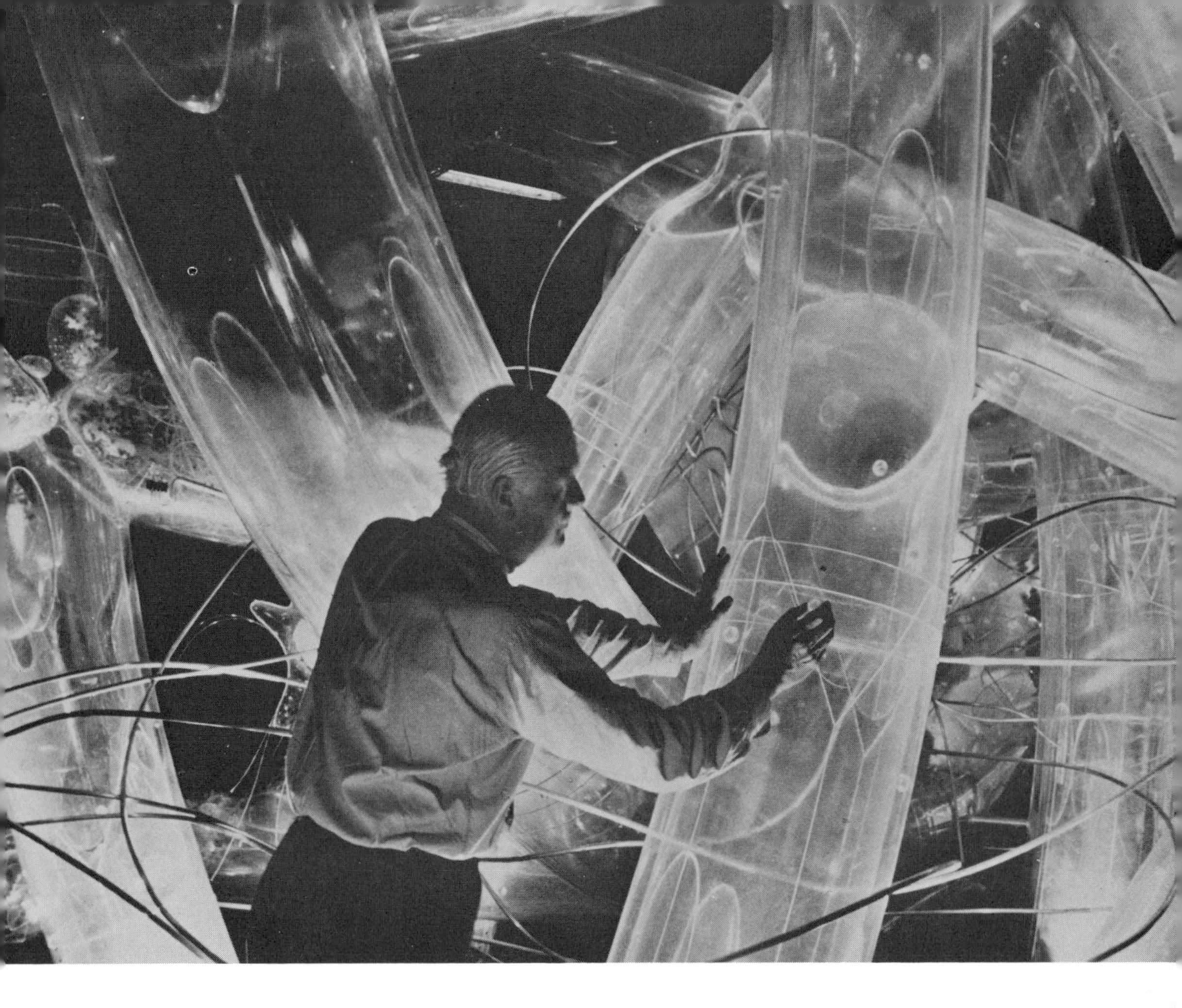

The Upjohn Company designed this display, "Defense of Life," for an American Medical Association convention. It is made of 100,000′ of interlocking tubing of clear Plexiglas to illustrate activity in human blood vessels (enlarged 25,000 times) and to simulate capillaries and venules. In addition, thousands of cast parts of colored acrylic are used to portray red and white blood cells, platelets, and other bodies that exist within blood vessels. The tubing was heated and bent into shape and other parts were attached by cementing. (*Courtesy, Rohm & Haas Company*)

This is a detail of an acrylic light support system utilizing neon as lighting. Acrylic was used both to transmit light and to support the neon tubing in Gallery Visuolita, Milan, Italy, and was designed by Ugo LaPietra. Acrylic transmits the light so evenly that there were no reflections on the paintings. And since neon does not heat up, there is no chance for heat deformation of the acrylic. (*Courtesy, Ugo LaPietra*)

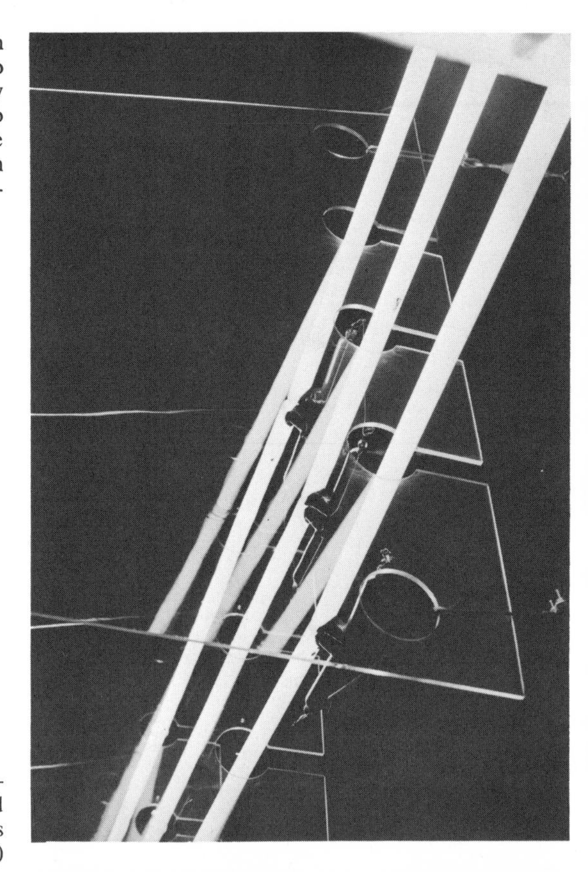

These simple post and lintel tables by Milo Baughman for Thayer Coggin Inc., required only cutting and polishing of the acrylic base pieces. The table top is bronze-colored glass. (*Courtesy, Thayer Coggin, Inc.*)

Used in conjunction with a rosewood pedestal, the ½" acrylic sheet forming the seat area was strip heated and bent. The rosewood base incorporates a memory return swivel; the loose seat cushion and back lumbar bolster are held in place with Velcro because of a slippage problem. Designer, J. R. Strignano. (*Courtesy, J. R. Strignano*)

This series by Milo Baughman, also for Thayer Coggin, employs solid blocks of acrylic attached to a solid lacquer-coated top with especially designed fasteners. (*Courtesy, Thayer Coggin, Inc.*)

The sofa design by J. R. Strignano is made completely of acrylic. The arms are 1¼″ clear acrylic and the seat and back are thermoformed of ¾″ acrylic. Parts were attached by solvent welding; no mechanical fasteners were used. Yet the shear strength is at 2,500 pounds. (*Courtesy, J. R. Strignano*)

Even though this chair by J. R. Strignano is thermoformed of ⅝″ clear acrylic, and the front support is concave to increase strength, I believe there would be and outward movement of the back support as one would sit down. To keep this from happening, a cross-support would be necessary. (*Courtesy, J. R. Strignano*)

Folding back of acrylic here makes for a more stable unit. Neal Small prefers the use of fasteners to solvent welding. These connectors are of mirror-polished chrome. 24″ wide x 32″ deep x 29″ high. (*Courtesy, Neal Small Designs Inc.*)

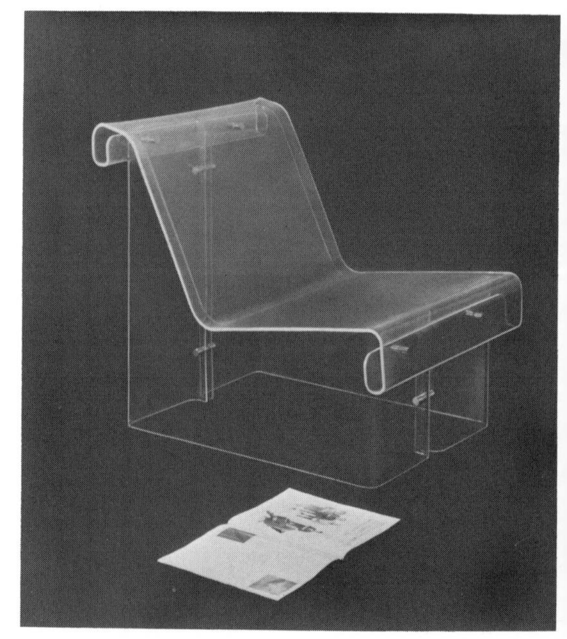

Spiro Zakas also folded back the acrylic sheet to create this wall shelf unit in a bent plate design to add strength. (*Courtesy, Spiro Zakas*)

If the supports for this Spiro Zakas acrylic chair were not folded back, there would be a rocking movement as a person sat down. (*Courtesy, Spiro Zakas*)

The folding back of the sides of this acrylic table by Neal Small increases the structural strength of the table. A thinner sheet can be used, and a larger table span (this one is 30″ x 30″ x 16″) is possible without side-to-side wobbling. The square, a design basis for these tables, is the most economical space with which one can build. Curves lose space. The tables also are made in smaller sizes (16″ x 13″ x 15″) and can be ganged together to form a series as tables or benches. The bends on these forms can be accomplished by using a strip heater. (*Courtesy, Neal Small Designs, Inc.*)

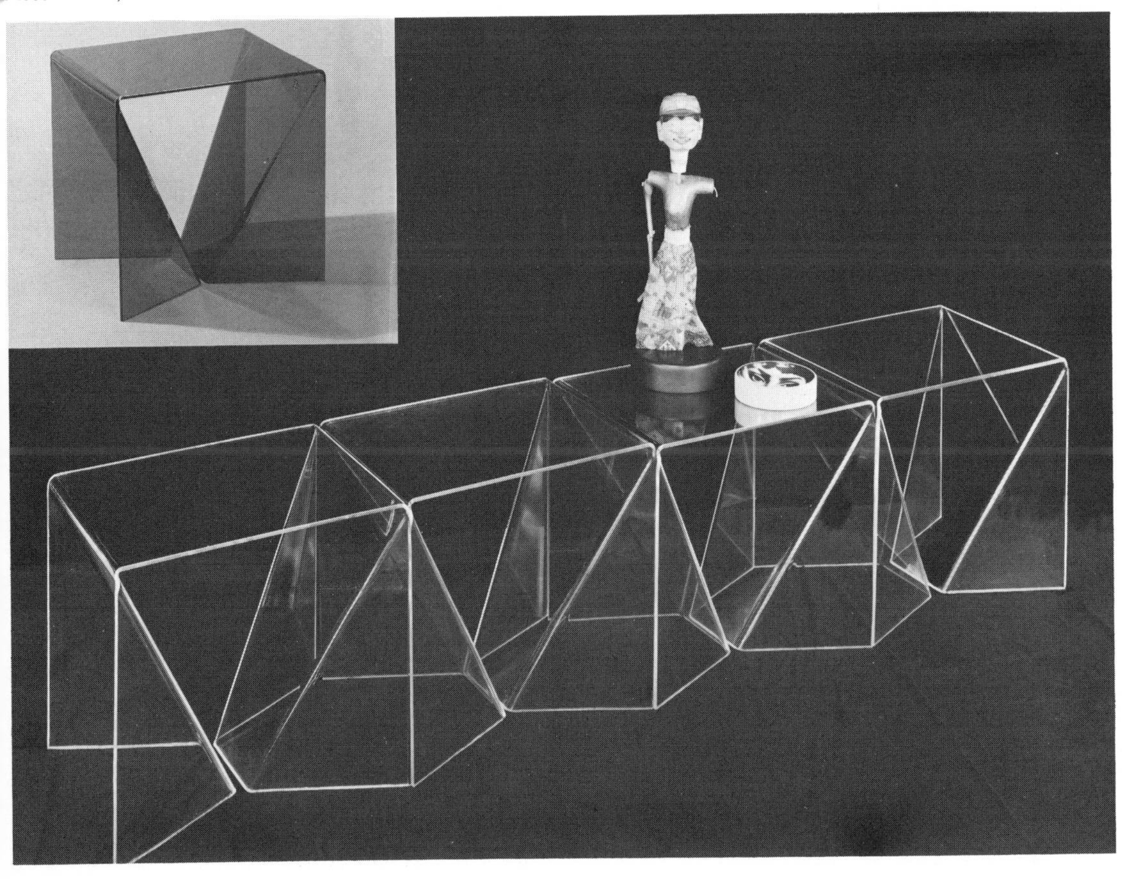

Wendell Castle utilized convoluting both as a strengthening and a decorative method. A scallop shell with its fluting makes for a stronger form of thinner material—a principle that can be applied to acrylic.

A clear acrylic piece can be large and yet appear to take up less space than it actually does because it does not block out floor and background. More seating space could be possible in a small room, therefore, without a feeling of overcrowding. This chair by Blosser for Design Group is a study in the dynamics of balance with minimum use of materials and fastening. A unified form is stronger than one that requires many fastening points. If too much strain is put on the fastener, there will be cracking. (*Courtesy, Design Group*)

Combined with metal, opal Plexiglas only serves to replace fragile glass. Designer, Archille Castiglioni for Kartell. (*Courtesy, Archille Castiglioni*)

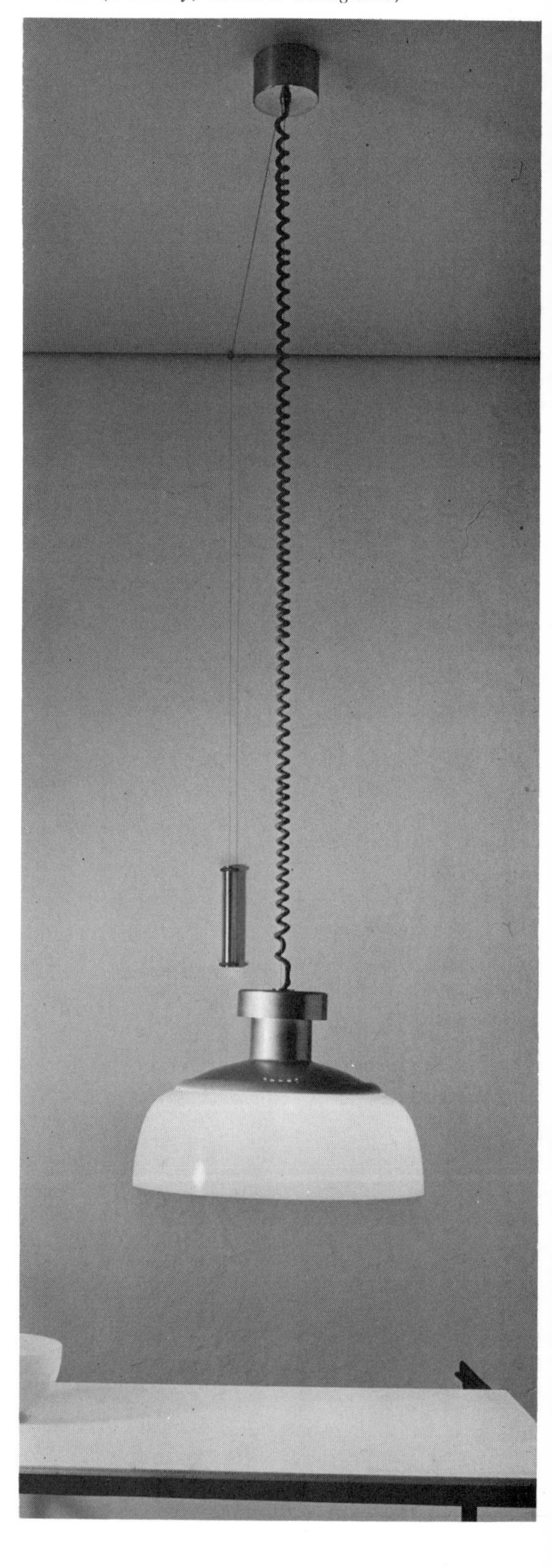

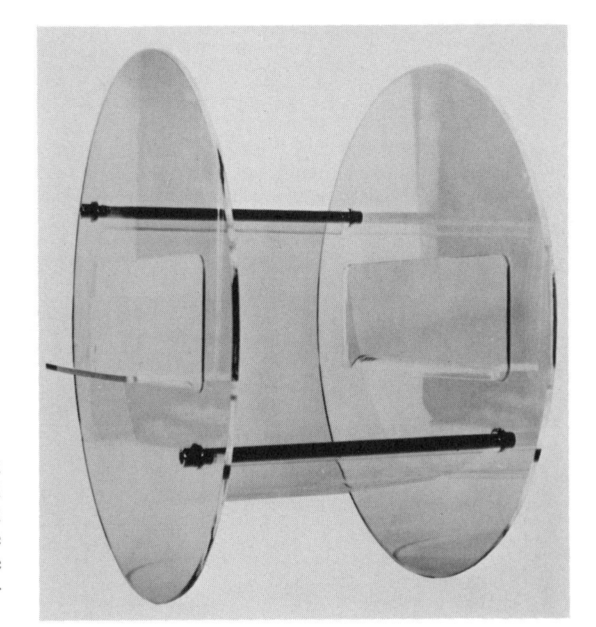

This Blosser Plexiglas rocking chair for Design Group is formed of two clear Plexiglas discs attached with polished chrome fittings. It holds a clear polished vinyl sling (seating area). The width is 32″ from the outer edges of the arm rests, which were cut from the discs and bent. The height is 14″ in the front. (*Courtesy, Design Group*)

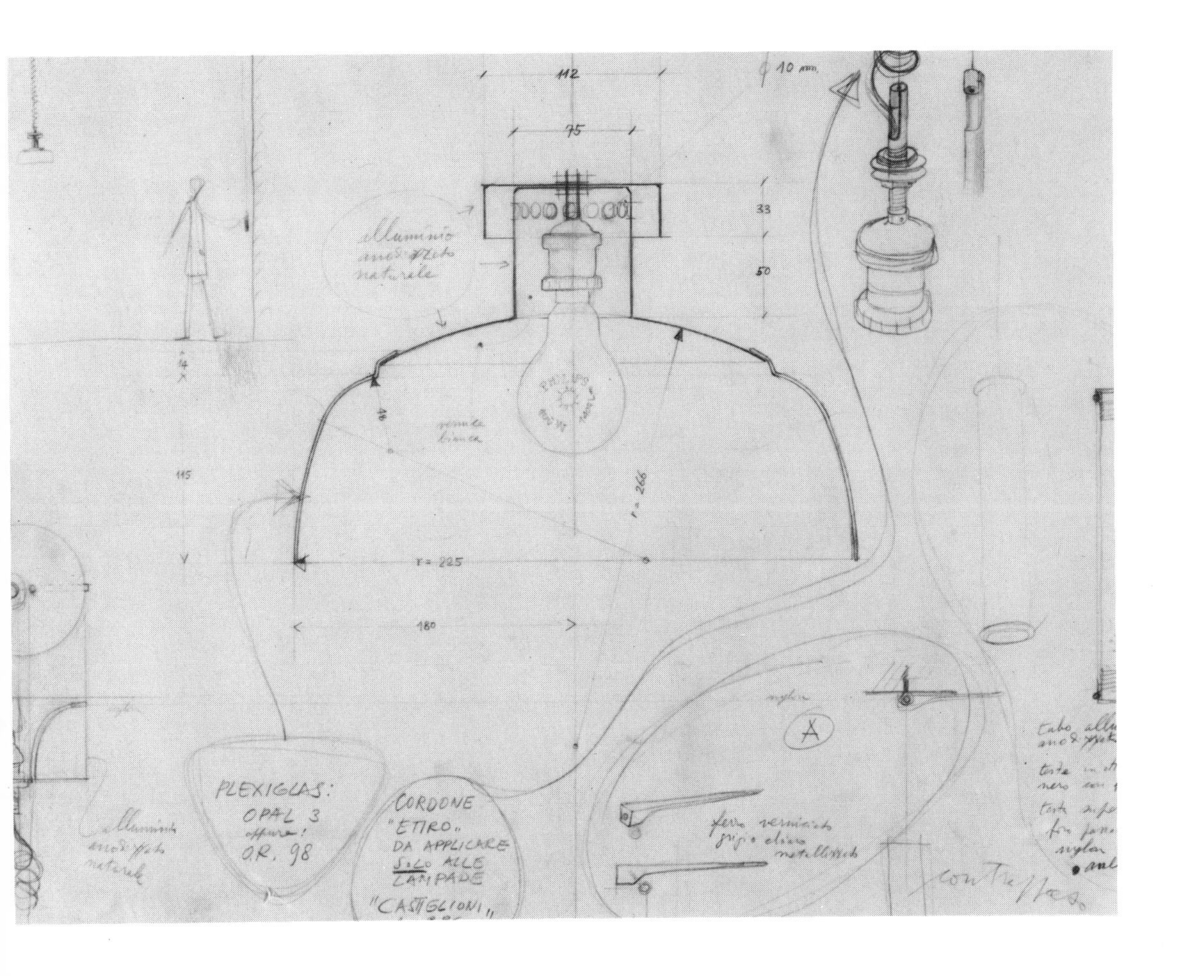

Translucent acrylic diffuses light and otherwise acts as a moldable solid. This area lamp by Neal Small is made of white translucent Plexiglas and is attached with polished chrome hardware. The lamp is 16½" high and 12½" wide and utilizes a 75-watt bulb. (*Courtesy, Neal Small Designs, Inc.*)

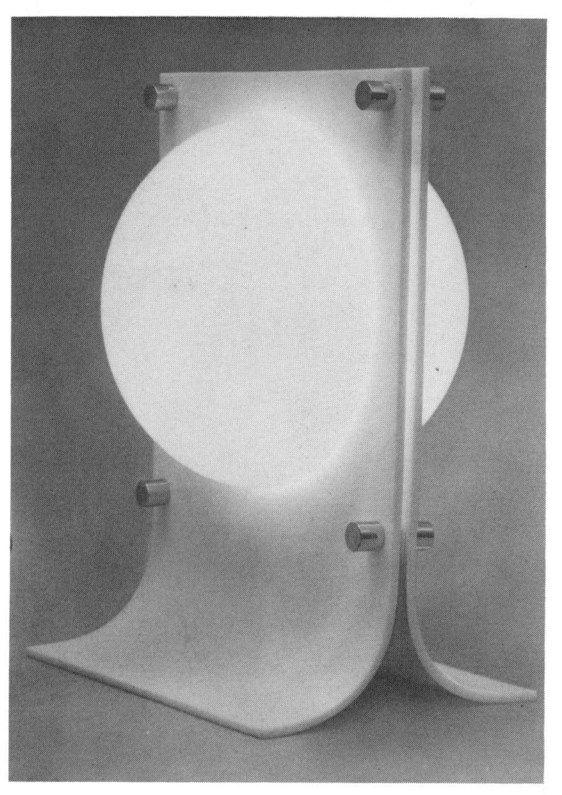

This vacuum-formed solid acrylic writing-desk tra by Architect Sergio Asti cannot be used as an ash tra because heat will discolor the material. That is wh a metal insert is used. Although scratches show up i clear acrylic, they do not stand out in solid colors. A is the case with other opaque materials, scratches ar not defined by light as with acrylic or glass; therefor they are noticeable only after a great deal of wea Produced by Kartell. (*Courtesy, Sergio Asti*)

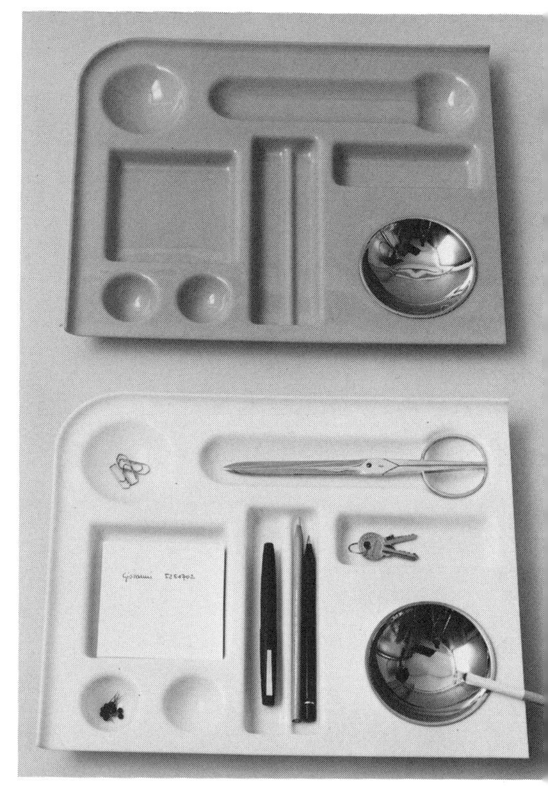

The concave curved form of this acrylic stoc (13½" x 15½") by Neal Small increases the strengt of the thermoformed tube shape as well as conformin to the human form. The cutout acts as a ventilatin area, gives the form flexibility on impact and a handle and releases the design from what could be a common place shape. (*Courtesy, Neal Small Designs, Inc.*)

The support system for this bookcase by Architect Paolo Lomazzi is made of Formica (urea-formaldehyde and melamine) and filled with rigid urethane foam. The Formica is used as a covering, has some of its own structural strength, and is given further reinforcement with rigid urethane foam. Depending on the density, urethane foam can be very strong structurally and will not warp. The Formica shelves rest on pins and conform to the curve of the outside supports. (*Courtesy, Paolo Lomazzi*)

High-pressure white laminated melamine furniture is formed in a process that makes for invisible joints; the furniture appears to be virtually seamless. A special glue and pressure system to achieve these forms is a proprietary process of Thayer Coggin, Inc. Designer, Milo Baughman. (*Courtesy, Thayer Coggin, Inc.*)

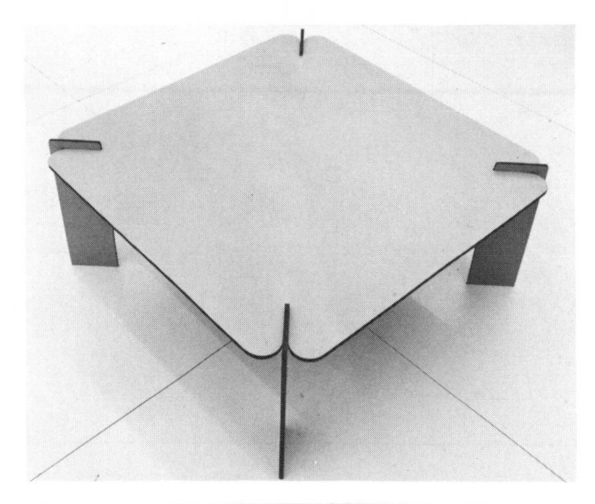

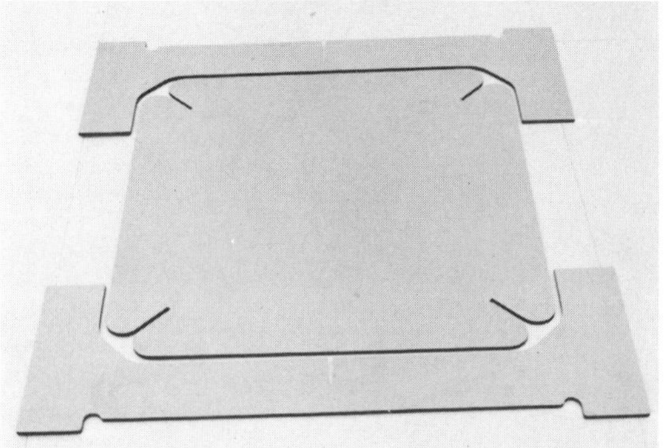

Lomazzi designed this post and lintel Formica table for Zanutta Poltrone. Tolerances remain exact because of the dimensional stability of this laminated material. A wide variety of surface effects can be achieved. (*Courtesy, Paolo Lomazzi*)

Fiberglass reinforced polyester resin can be an even stronger material than ABS. Most fiberglass reinforced pieces have been fabricated through a hand layup or a modified hand layup process; other pieces are stamped from pre-preg sheets one at a time. Therefore labor costs bring up the cost factor. (New injection molding processes have been developed using fiberglass and sometimes polyester.) If a shape is complex, however, FRP may be the best material-process solution.

Architect Sergio Asti's table lamp of FRP tinted pink and clear utilizes the translucent qualities of FRP to diffuse light. Note that the texture of the fiberglass mat is accentuated by the light. (*Courtesy, Sergio Asti*)

Dishwasher-washable plastic dishes are usually injection molded of melamine because this material can withstand high heat without distorting and has a hard surface. For that reason melamine was chosen by Giorgio Soavi for Olivetti's ash tray. (*Courtesy, Olivetti*)

This is a hand layup chaise made of fiberglass reinforced polyester resin as designed by Architect Alberto Rosselli for Saporiti. The form is a shell contoured to fit the body. The "cantilever" seat back can withstand backward pressure because of the larger, heavier shape at the forward section and the great strength of FRP. (*Courtesy, Alberto Rosselli*)

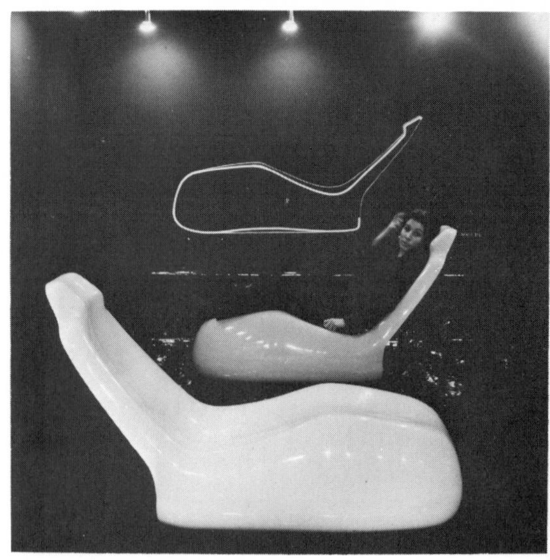

Polyester can impart rich gemlike colors, patterns, and textures.

Screen (detail) by Ted Hallman for the Coo
Union Museum. Acrylic shapes are woven into a pa
(*Courtesy, Ted Hallman*)

Delicate acrylic rods form a necklace based on m
ular concepts. Designer, Nancy Thompson.

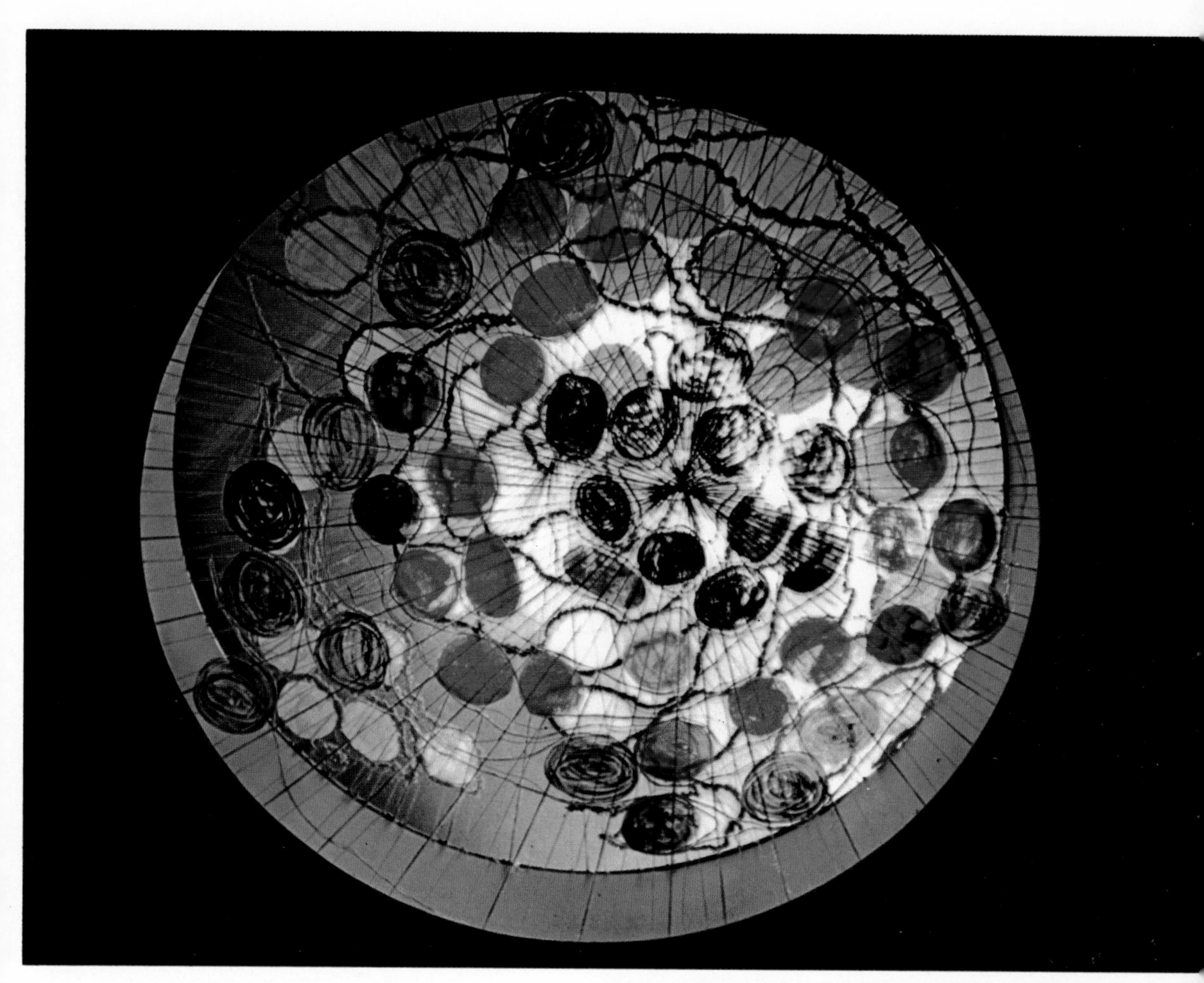

Spider weblike dome screen of fibers and dyed acrylic circles by Ted Hallman. (*Courtesy, Ted Hallman.*)

Hundreds of acrylic rods pierce acrylic globe and pipe light through their edges from a light source in the stainless steel base. Sculpture-light by Thelma R. Newman.

Detail of Ted Hallman's dome screen. (*Courtesy, Ted Hallman*)

Many designs are executed at first in polyester-fiberglass; if the form proves successful and warrants increased production, it can be injection molded later on. This was the intention of Kartell (designers, Anna Castelli and Ignazio Gardella). Meanwhile, a pre-preg consisting of a sandwich—layers of polyester reinforced with fiberglass—is stamped out with heat over a male mold in a large press. (*Courtesy, Kartell-Binasco*)

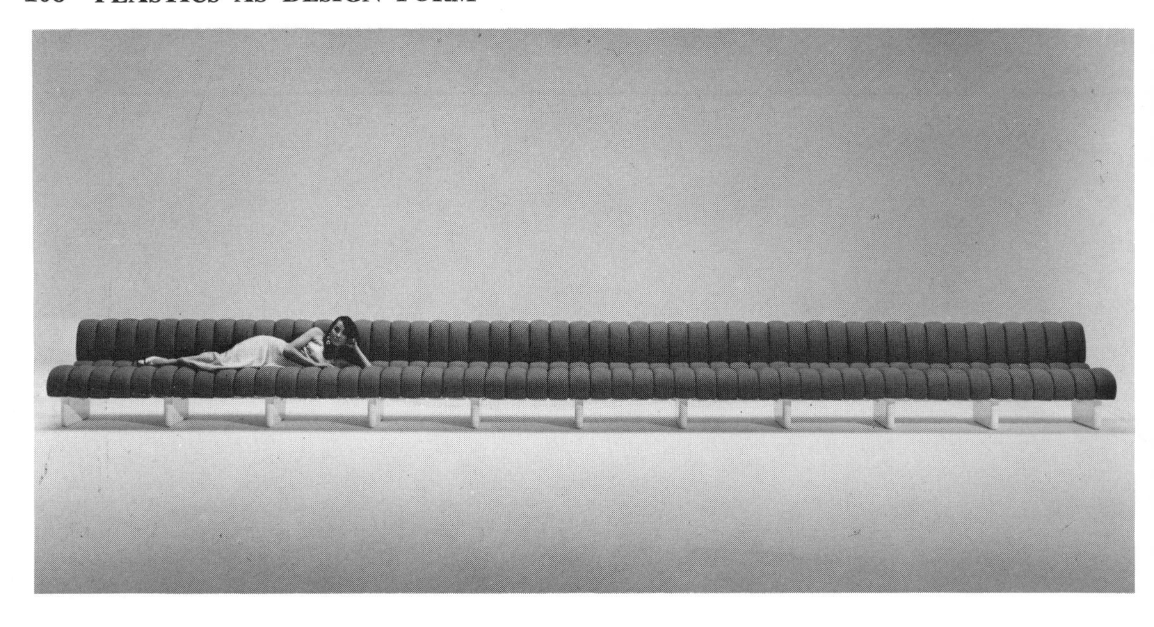

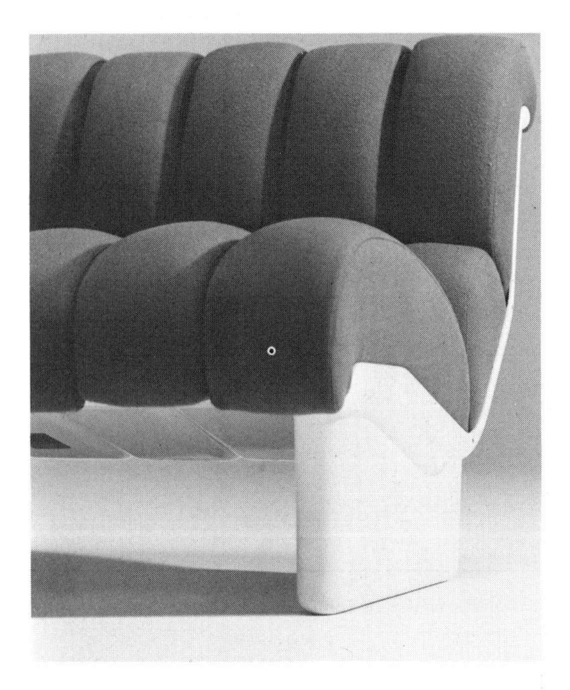

This couch system, designed by Architect Marco Zanuso, was made the same way by C & B Italia (with prepreg). At this point it is very expensive and not very many have been made. (*Courtesy, Marco Zanuso*)

TALIA
C&B
ITALIA

FRP can be used indoors and outdoors. Colors can be very bright and fast (this chair is made in white and orange). The rounded archlike area in the front of the chair is necessary so that downward seating pressure does not crack the legs at the seat juncture. A single, solid shell like this can take a great deal of abuse and, if damaged, can be repaired without perceptible change. Sergio Mazza designed this lounge chair to stack. (*Courtesy, Moreddi*)

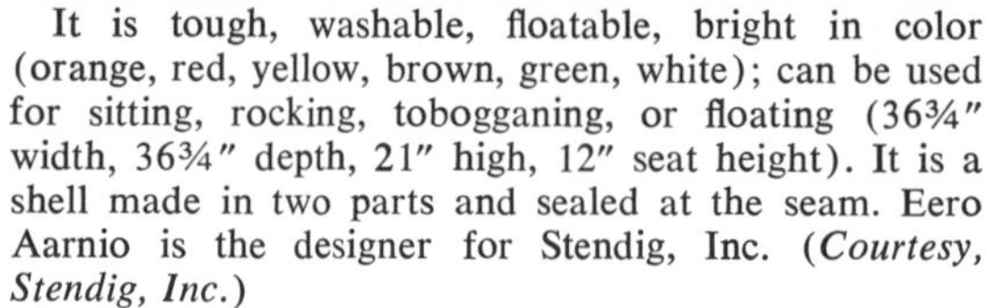

It is tough, washable, floatable, bright in color (orange, red, yellow, brown, green, white); can be used for sitting, rocking, tobogganing, or floating (36¾″ width, 36¾″ depth, 21″ high, 12″ seat height). It is a shell made in two parts and sealed at the seam. Eero Aarnio is the designer for Stendig, Inc. (*Courtesy, Stendig, Inc.*)

Eero Aarnio of Finland also designed this spherical large fiberglass reinforced shell that swivels on an enameled steel pedestal base. The inside of the shell is completely upholstered and provides a quiet escape. When sitting in the chair you cannot hear noise from the back, only from the front. There is a psychological quality of escape or protection, a throwback to the womb position, that envelopes one when he is relaxing in the "Ball Chair" (49″ high, 38½″ deep, 15″ seat height, 43¼″ diameter of shell). (*Courtesy, Stendig, Inc.*)

These stacking benches or stools were pre-preg stamped of FRP for C & B Italia and designed by Architect Mario Bellini. Here again the corners are rounded. Indentations at the corner edges hold the stacking chairs in place. (*Courtesy, Mario Bellini*)

Injection molded stackable chairs of polyethylene make for a very economical combination that is nearly indestructible, resilient to banging and falling. The open ribs in the back reduce weight and allow little fingers to grip it; its convolutions increase its strength. Marco Zanuso designed this for Kartell. (*Courtesy, Kartell-Binasco*)

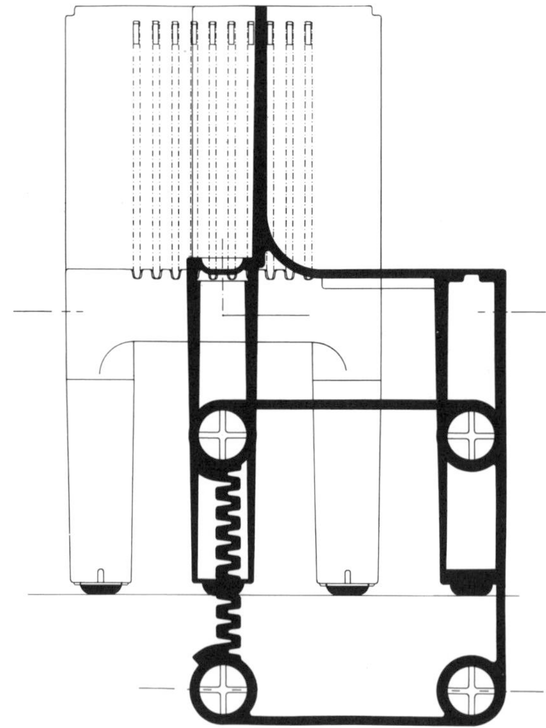

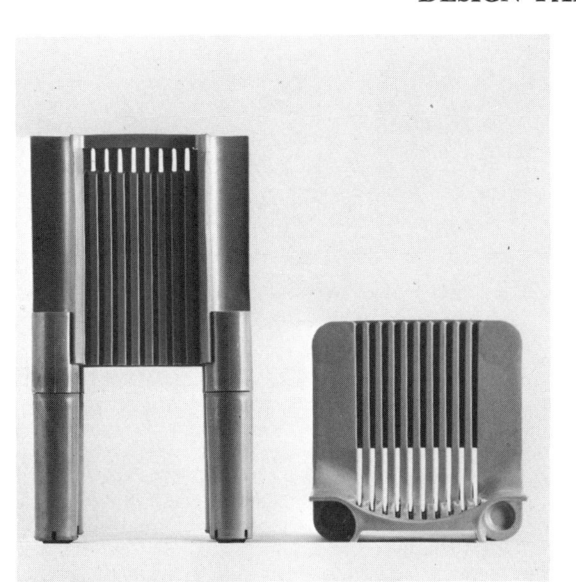

Wendell Castle made this FRP bench via the hand layup process. He uses open mold units interchangeably, creating different variations as in his "light" form, which is covered with bright medium-blue flocking and topped with a blue bulb. (*Courtesy, Wendell Castle*)

The extreme ductility of the PVC joint makes it possible to form and make combinations of all the spatial dihedrals. The fixed panel joint is inserted into the panel itself in such a way as to fix the section, adjust the panel, and complete the job. Spring clips hold the joints together.

Architect Angelo Mangiarotti has explored and developed prefabricated systems that would permit unlimited variations and combinations. His modular support unit called "Cube-Eight," made of PVC, permits interior space division with storage capacity. (*Courtesy, Angelo Mangiarotti*)

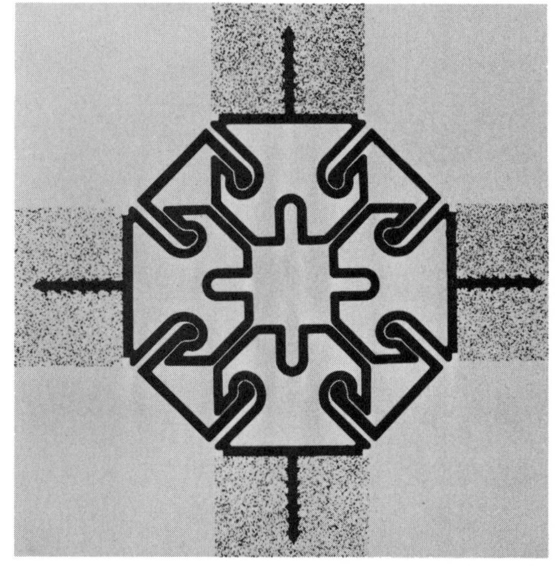

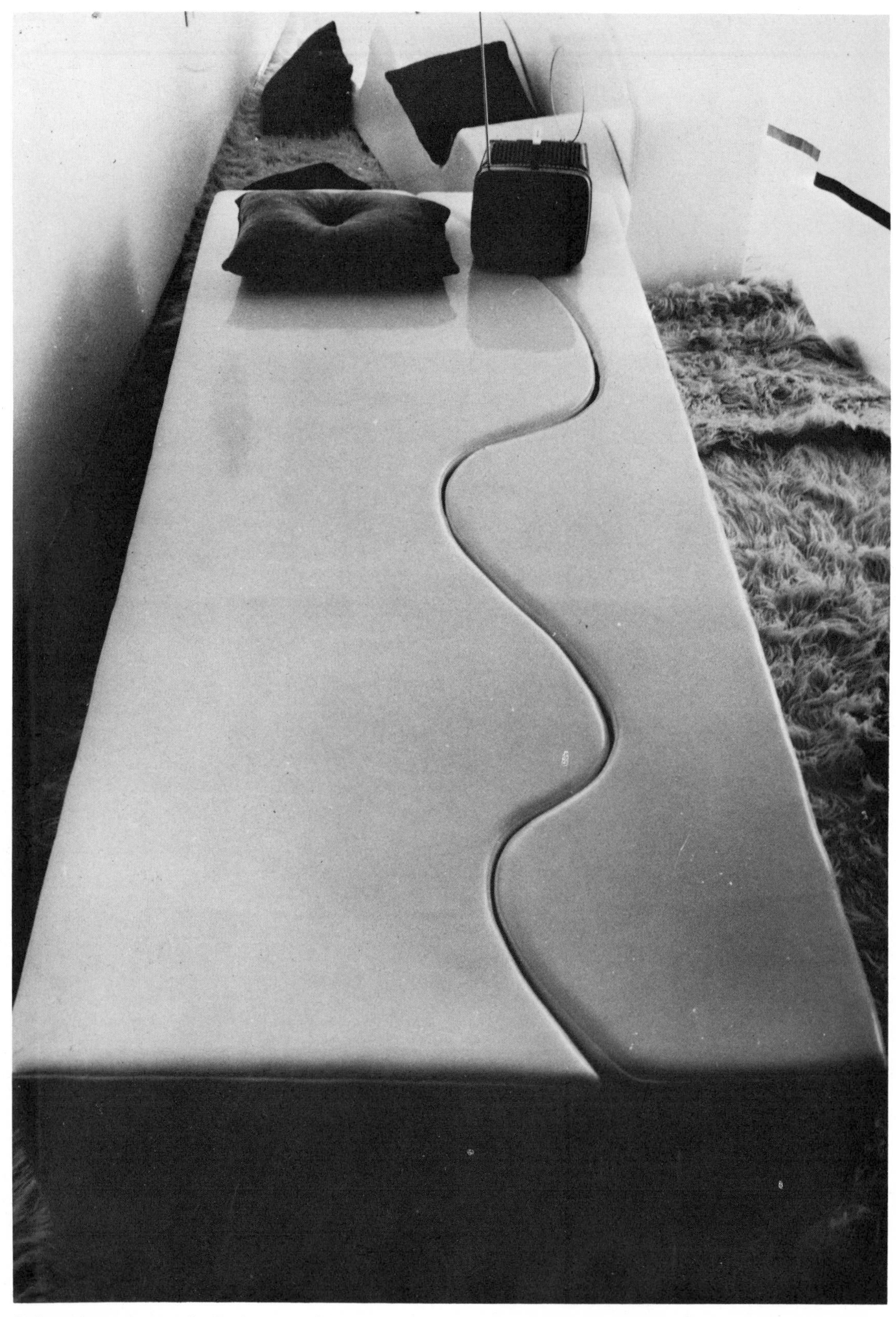

Polyurethane is practically indestructible, does not easily degrade. Its foam can be rigid or flexible and can be foamed to any density from exceedingly hard to very soft and pliable. A single unit can be constructed of dense and soft flexible foams in one piece—for instance, density in the base, softness on the surface. It is easy to cut shapes from blocks or mold the foam in a mold form.

Rotationally molded, white translucent polyethylene spheres stack to produce a lighting sytem. Each 16″ sphere uses a 60-watt bulb, but the light yield is much lower because of the density of the polyethylene. The form, according to Designer Neal Small, is 63″ high by 32″ wide. (*Courtesy, Neal Small Designs, Inc.*)

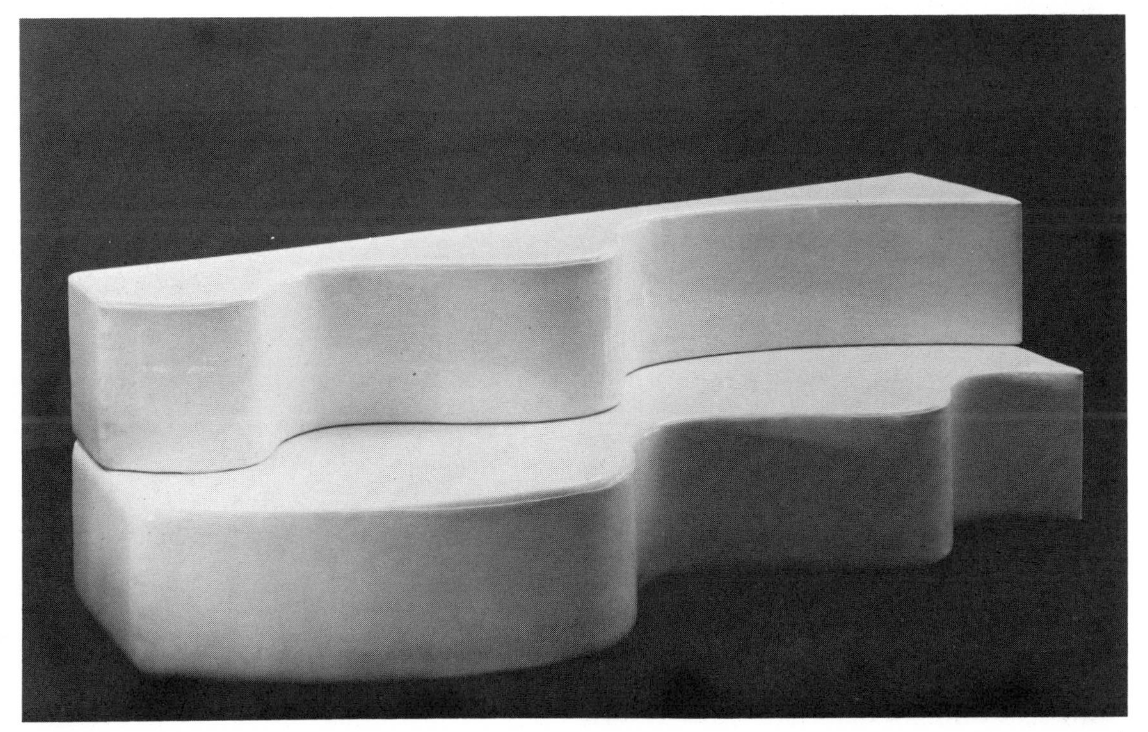

The Archizoom Group for Stendig, Inc., used two blocks of expanded polyurethane foam for the jigsaw couch, bed, chaise lounge—forms that can assume many different combinations and roles simply by changing the relationship of the two blocks. The covering is black or white glossy patent vinyl. (*Courtesy, Stendig, Inc.*)

A self-skinning urethane that needs no upholstery but has an integral skin over varying densities of foam promises to cut costs particularly for institutional furniture. Architect Angelo Mangiarotti designed "In" for Zanotta. The cantilevered armrests and pedestal base are very firm and dense, while the seat and back surfaces are a bit softer. No inner supports or framework are necessary. There are no joints or fasteners to break away. (*Courtesy, Angelo Mangiarotti*)

Architects Decursu, DePas, D'Urbino, and Lomazzi designed "Ciuingam" (pronounced "chewingum") after the flexible shape of its namesake. The polyurethane chair encloses the body softly as one cuddles into it. (*Courtesy, Decursu, DePas, D'Urbino, and Lomazzi*)

This is a design idea translated in three different ways by Architect Cini Boeri. Each polyurethane form has a different function—for normal sitting, for relaxing, and for resting. (*Courtesy, Cini Boeri*)

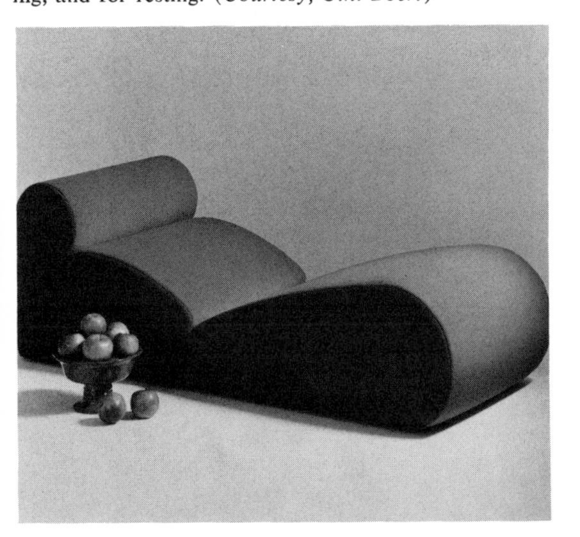

This lamp is one of a group designed for Atelier International, Ltd. The translucence of the white vinyl surface sprayed over a cocoon webbing of fiberglass acts as a translucent diffuser emitting adequate, yet soft light. (*Courtesy, Atelier International, Ltd.*)

Vinyl cocoon lamps, originally designed by Henry Miller at least eighteen years ago, still have a market. Shapes are determined by an inner network, usually of wires, to hold the cocoon spray as it meshes a web over the skeleton. Lighting fixtures and mothballing of naval vessels have been its chief applications, although the vinyl cocoon has been used as a protective covering on buildings as well. (John Zerning recommended use of the vinyl cocoon as a mold for FRP.)

Vinyl has been used for clothing, wall and floor covering, upholstery, but in each of these instances the application is as surface covering. The first departure from this is the use of air as a filler cushion and support for inflatable buildings and furniture. Costs are so low that popularization has prostituted the original chair concept claimed by Lomazzi of Italy (*Courtesy, Paolo Lomazzi*)

4

light and transparency

Light is an essence. Without it we cannot see. As it refracts and reflects off surfaces, it reaches our eyes in various wave lengths. Because of this we make distinctions of color, shape, and movement as each relates to what we know. Until recently our concern with light in art form was capturing illusions, depicting luminescence, atmosphere, form, and color as light in painting, in sculpture, and in stained-glass windows. But little significance was paid to transmission of light as it passed through transparent objects, interruptions in its path, bending, rearranging, diffusing, and modulating. Transparent plastics project new possibilities and light can become as important as color and, without color, the defining element. Light can reach beyond the perimeters of a form by filling an entire room with visual echoes in endless variations of the form itself. Light may be blocked and then allowed to pass through etched or textured surfaces, piped through holes and projections. It may be interrupted by opaque solids or refracted by lens shapes or defracted by gratings. Light plus time adds other possibilities. If time consists of intervals and an ordering of events, then light interrupted, synchronized, alternating, constant, repeating, accelerating, fast, slow becomes another dimension. Light for its own sake, then, embodies elements in itself such as rhythm, repetition, temporal form; when it is combined with transparent plastics as a defining "container," with plastics' own volumes and edges, we have entirely different aesthetic elements with which to express ourselves. Black-and-white photography depends on light and traps light into two-dimensional form consisting of white, grays, and black. But, with transparent plastics as three-dimensional form, light lives in space and time —it can be created and it can die.

Moving light combined with reflective surfaces, plastics with internal stress and strains, distorting lenses, metallized surfaces, mirrors and shadows that define, contrast, and move with its high lights and middle lights create optical illusions of movement. Movement instead of being a purely physiological experience becomes both physiological and psychological. The slightest changing of optical direction on the part of the observer causes a transformation—gradual, sequential change, a metamorphosis from one image or pattern to another, sometimes with dizzying effects. Film can play with these elements, but with only the illusion of three dimensions, not in tangible, physical three-dimensional form. Environmental effects, artificial spotting of light, natural light changes caused by clouds, shadow forms, nearness and farness, a light reflecting background, or a light absorbing background will alter the mood of a piece.

In order further to define these aspects of light with transparent plastic, let us explore, through a primer of some problems, how light works operationally. In following these possibilities, let your mind spring off from where we are to other potentialities and try to project what would happen if you changed some aspect or component of these questions.

At first, to explore we need to gather together available shapes of transparent objects of either plastics or glass, some filled partly with water, and arrange them on a white surface in a darkened room. Using a flashlight or lamp we can play with the light by pointing the light source at different angles and observe. What happens to edges of forms? What variations would make a texture, or a pattern? How do forms relate to one another in terms of light and shadow? How is volume defined? What differences are found between thickness and thinness, among bent or straight edges and planes? What happens to curves and their reflections? How far can a shadow or a reflection stretch without losing a relationship to the original form? With a slow movement of light and a quick flash or by turning the light on and off, with a fixed gaze at one point there is one effect, and with head movement there is another visual image. What happens when a continuous and then a discontinuous light source is used? What occurs when light passes through a transparent hollow form as compared with a transparent solid form? Light can sharply define curves, angles, hardness, density, or light blurs and then obliterates texture and form. With light we can achieve form without materiality. Insideness and outsideness in a kind of X-ray vision is a consideration that sculptors using solid opaque materials never had to regard. With transparency as an entity there can be an ambiguity to the random, ever-changing internal-external relationships that "visibilize." Sometimes fussy little patterns of dark and light obliterate the external-internal structural symbiosis. With controlled use of planes reflecting and refracting at properly designed angles to pattern light and dark or with internal solids injected through lines or larger shapes, the eye can grab onto a defined shape that will integrate and connect the internal with the external. This is what we need to consider if we wish to chain our attention to the form itself. But light can also reach out beyond the perimeters of a transparent object by filling an entire room with all kinds of visual echoes. Fragments, variations, and transformations emanating from the object potentially can dance about, stretching to the outer reaches of the room and enveloping everything. Light can dissolve all solids into a unitary visual continuum, starting with *the* object.

The large problem here is how we control light and form to express the effects that we want. And a corollary problem is how we can design a form to take advantage of the material's maximum capabilities.

All plastics are not transparent. The rules, principles, elements, and their variables apply, however, to all materials, whether transparent or opaque, but there is a variable set of criteria that relate to transparent forms. We should understand how these visual design forces effect a transparent-translucent form. Solid materials have been amply covered in our literature. Later on in this chapter we will consider some particular design parameters peculiar to various plastics. Since the design form covered in *Plastics as Design Form* is mainly three-dimensional, we will concern ourselves with those visual forces that pertain to three dimensions. Line, plane, the part and the whole, space, balance, volume, size, shape, density, rhythm, position, tension, proportion, movement, and space relationships number among the principles and elements slanted toward understanding transparent plastics. We will also concern ourselves later with tools, materials, and their processes, in order to know how to carry out our concepts.

LIGHT

When do we notice light? When the light bulb burns out, when it flickers, when color changes, when storm clouds darken the daytime sky, when we move from a darkened theater into the sunlight.

Light from the sun illuminates the earth; yet the space through which it travels from sun to earth is lightless—dark because in space there is nothing to reflect light. There are no dust particles or bits of water vapor. Light is visible

1. Regular transmission through clear plastic.
2. Diffuse transmission through milky white acrylic.
3. Spread transmission through acrylic that is frosted on one side.
4. Mixed transmission through ribbed acrylic.
5. Scattered transmission through pebbled acrylic on underside. From some angles the light sources can be seen at nearly full brightness.

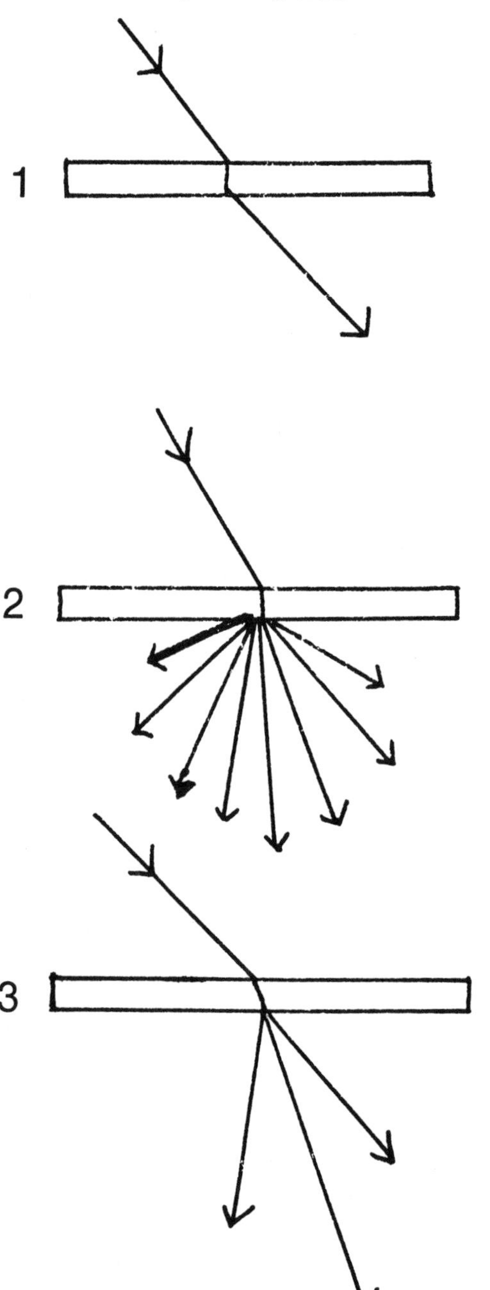

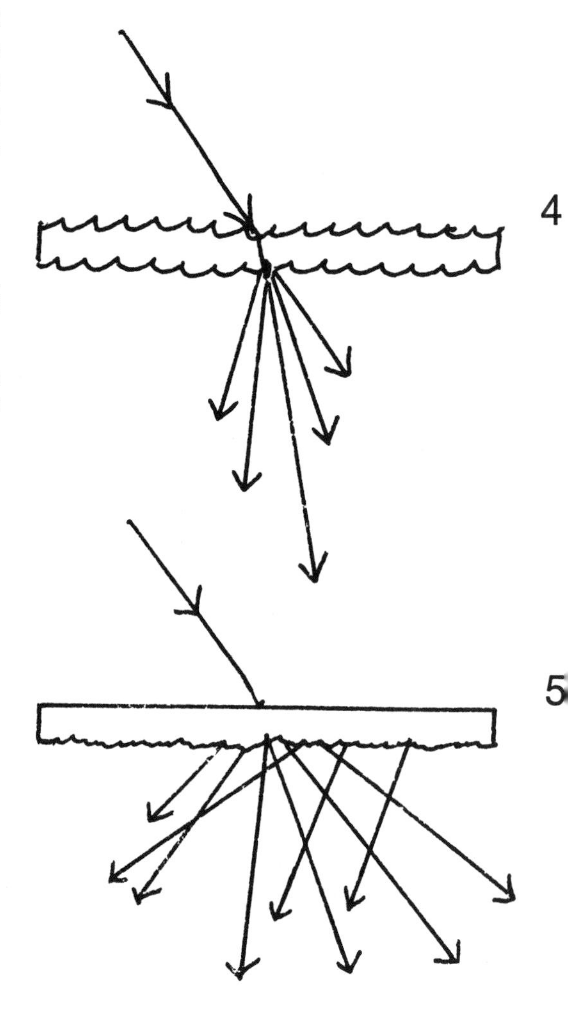

only when reflected from a surface, as when the reflection of a street lamp cloaked with raindrops shines into a puddle at night and sparkles shoot out. Each drop sets up ripples which catch the light of the lamp. Or, grass with dew on it appears gray from a distance, but when we look closely we find that droplets of dew are clear and transparent. There is a layer of air between the drops and the blade itself. The gray appearance is caused by reflection of light both inside and outside the droplet. Sometimes we see a silvery glint because light is reflected back to us from the flat, back surfaces of large droplets. The metallic sheen in bird feathers and their brilliant colors are not due entirely to coloring matter but are caused by light passing through thin layers of outside cells.

Light travels at different speeds through different objects—quickly through air, more slowly through water, and even more slowly through glass and plastic. When light strikes a piece of

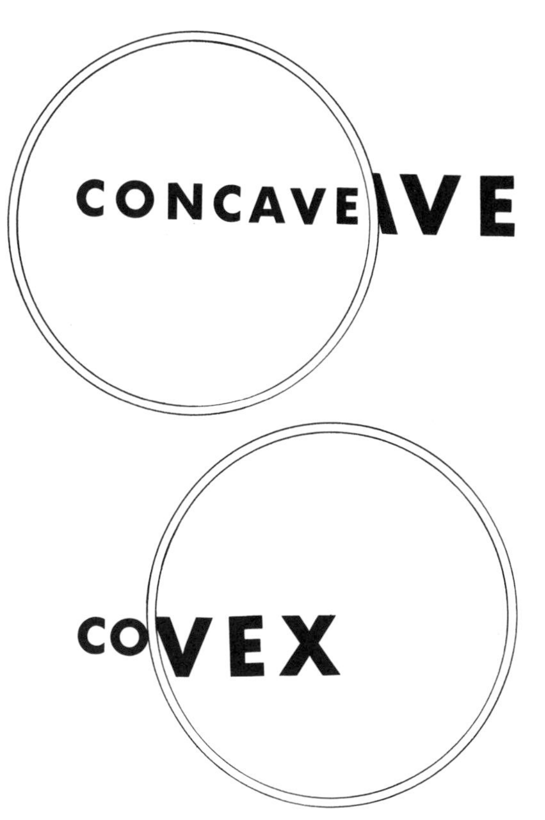

Concave lenses reduce print because all rays that do not pass through the center of the lense are bent upward toward the thicker part of the lens and all rays that pass through the center are straight and unbent. At the point where the two rays meet, a virtual image is formed which is right side up and greatly reduced in size. Conversely, the convex lens magnifies print.

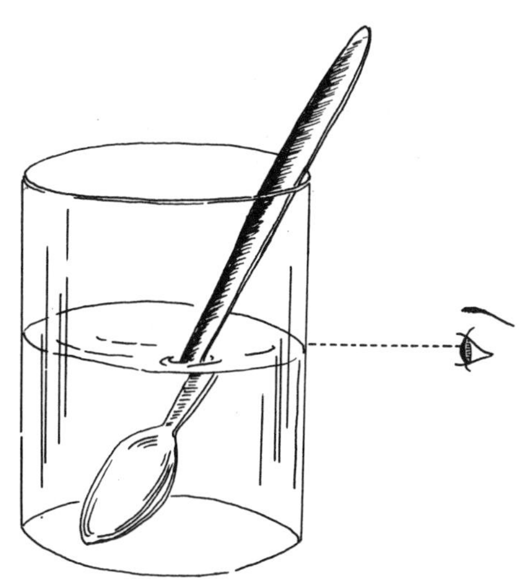

The handle of the spoon appears to be disjointed when viewed at the water line because light refracts and appears to "bend" the handle several degrees.

glass or a container of water at an angle, the light is bent because its speed changes. Bending of light is called "refraction." The prism is a classic form of refractor. It can redirect light at any angle. Boundaries between air and plastic can be arranged so that light can be directed to any desired angle. Accurate light control is possible by means of refraction. Prisms can be incorporated into shapes of transparent plastic as well as into relatively flat surfaces. Actually, a lens is simply a built-up system of prisms.

When images are seen in parallel mirrors, the image is reflected repeatedly. The second mirror reflects to the first mirror; the first mirror returns the image back to the second mirror, over and over again. Again, when two surfaces of window glass are not completely parallel, the objects seen through the windows appear to stretch as if they were made of rubber. When we move our head from side to side we see the distortions from the curves in the glass. Some curves enlarge images or parts of it and some minimize the image.

Curved shapes filled with liquid act as lenses and magnify; so do drops of water act as magnifiers. Light is bent as it travels through water. Mineral oil [1] used as a filler in transparent plastics such as acrylic slows down light and, depending on the shape of its container, can bend the light to create optical effects, as do lenses.

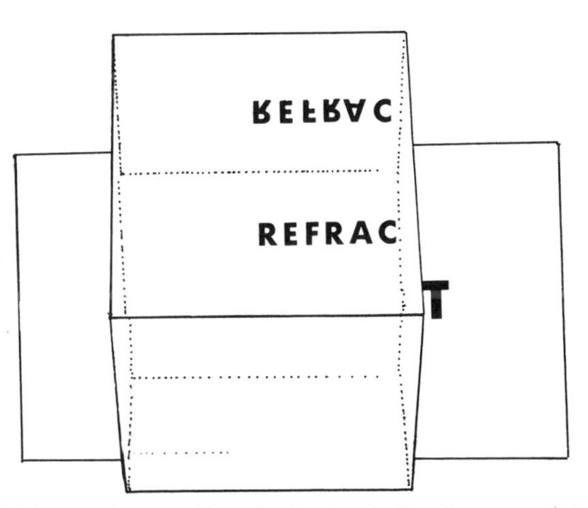

When a clear acrylic cube is placed directly over print, and you look down on it, the print will appear to be raised up from the rest of the paper. Looking at the cube over the top, at an angle, the print will be reflected on the side of the cube even higher than before because of the angle of refraction of light.

Light is refracted or bent in various degrees depending on the angle of view as it passes through clear plastic or any dense transparent medium.

[1] Yellows within five years. There are some industrial silicone "oils" that might not discolor.

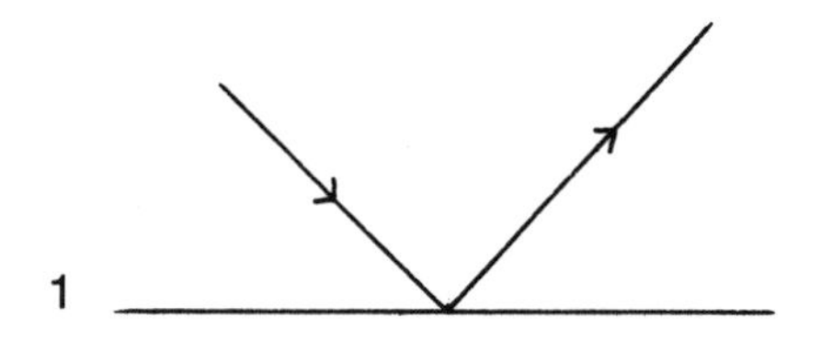

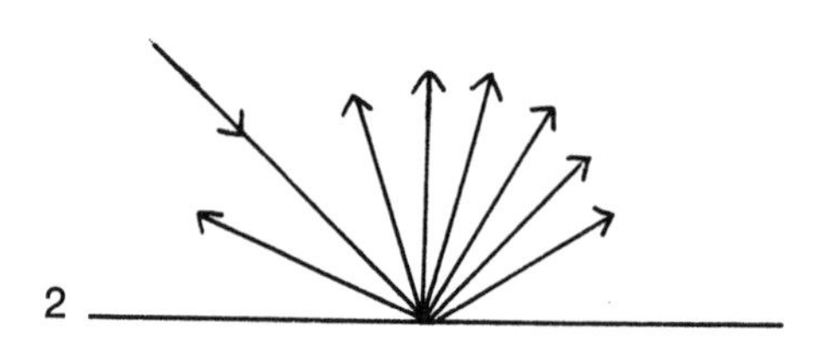

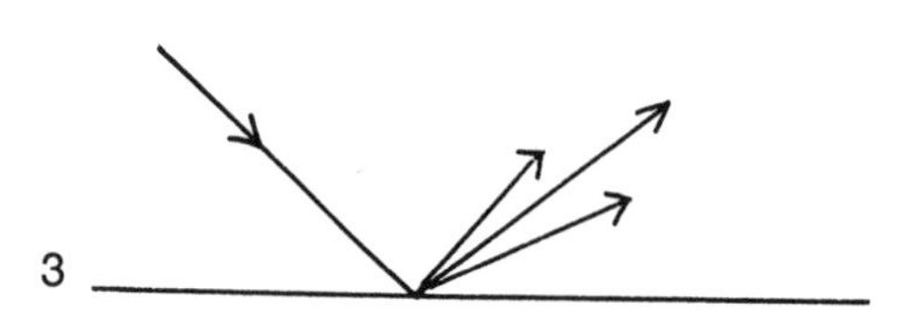

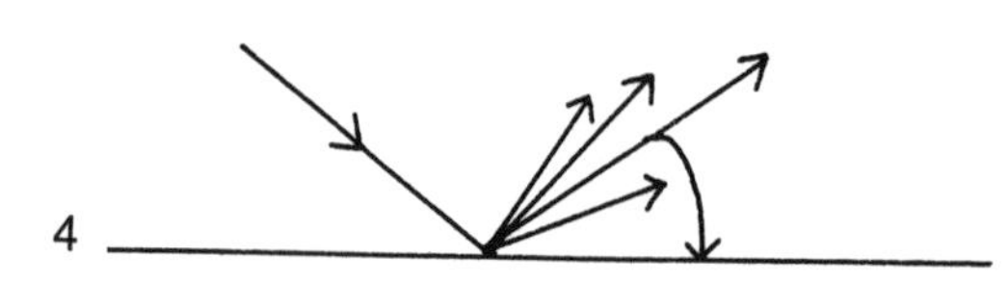

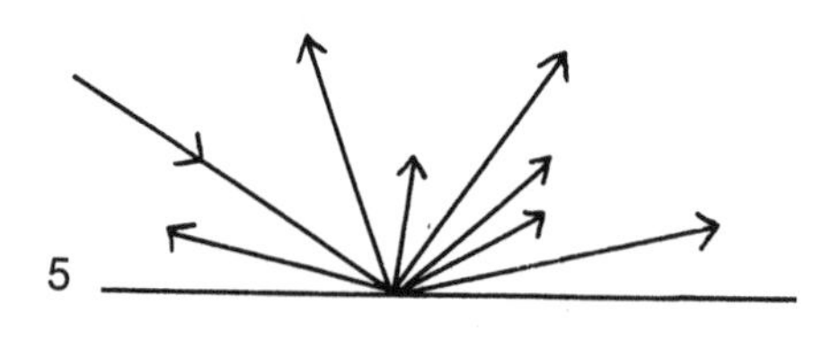

1. Regular or specular reflection is in a mirror reflection. A ray of light hits the material and is reflected off.
2. Diffuse reflection is when reflected light is scattered in all directions.
3. Spread reflection is usually of a matte surface.
4. Mixed reflection occurs when part of the light is diffused and some is regularly reflected, as off a shiny painted surface.
5. Scattered reflection happens when tiny portions of the surface act like mirrors and cause a multiplicity of separate reflected rays, as in rippled plastic or crinkled metal foil.

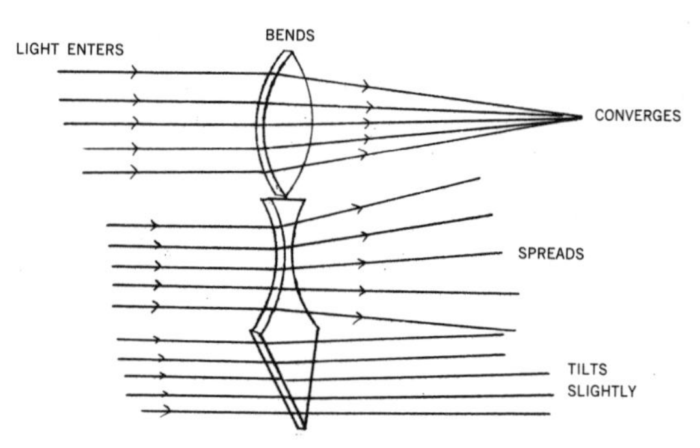

Not only do carved lines glow with piped light, but curved and straight acrylic edges transmit light in angular directions.

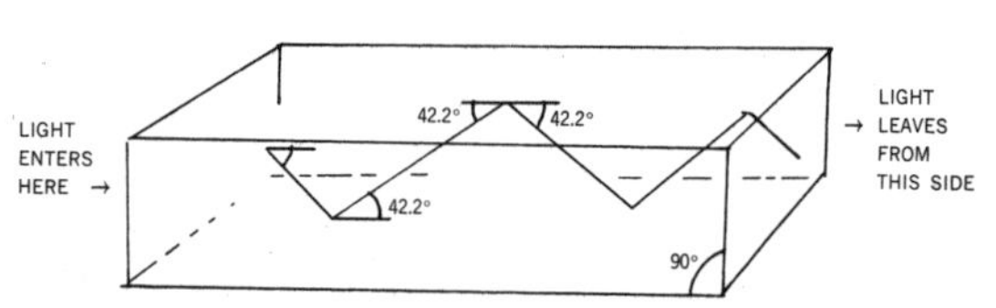

92% of light entering an unscratched sheet of acrylic will be transmitted by internal reflection to the opposite edge. If light inside the piece of acrylic strikes the surface at an angle greater than 42.2 degrees, it is bent so that it cannot escape into the air but must be reflected back to the other side of the plastic piece. The light will bounce back and forth until it reaches the far end, and will glow.

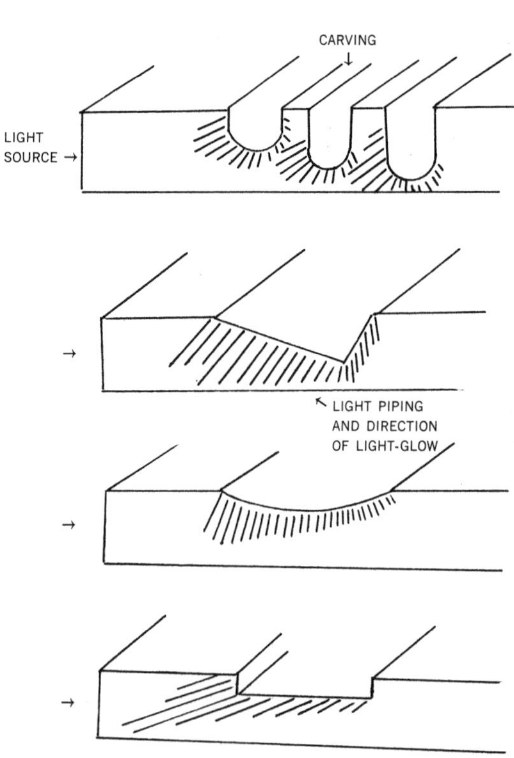

When the surface of a plastic sheet is carved, some of the light reaches the air and causes a glow. Incising lines at different angles and into varying shapes, as in this enlargement of carving shapes, will cause different qualities of glowing light.

Concave lenses reduce size because all rays that do not pass through the center of the lens are bent upward toward the thicker part of the lens and all rays that pass through the center are straight and unbent. At the point where the two rays meet, a virtual image is formed which is right side up and greatly reduced in size. Conversely, the convex lens magnifies size. Solid transparent objects, i.e., acrylics, can effect distortions through reduction in size of environmental images surrounding them, or can magnify parts of forms. Variations can be achieved not only in size and shape of the acrylic piece but also in the proximateness and size of objects.

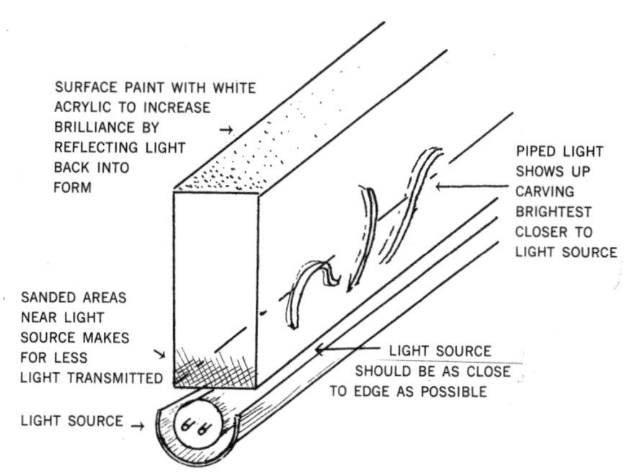

Light source and a carved acrylic block.

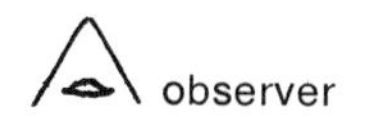

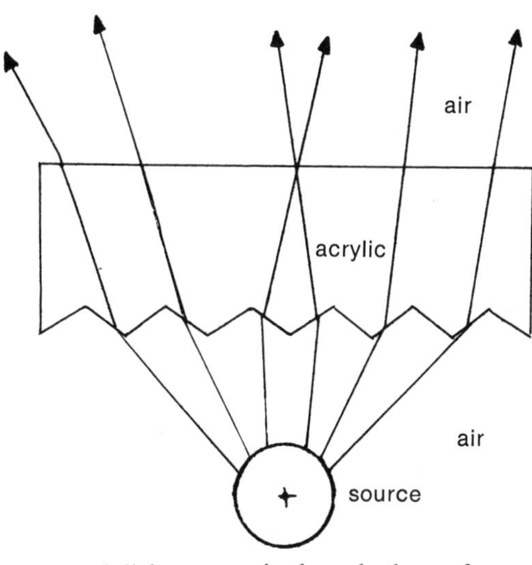

Concentrated light rays exit through the surface opposite that from which they entered. (*Courtesy, I. K. DeBileu, Plastics Specialist for E. I. du Pont de Nemours & Co., Inc.*)

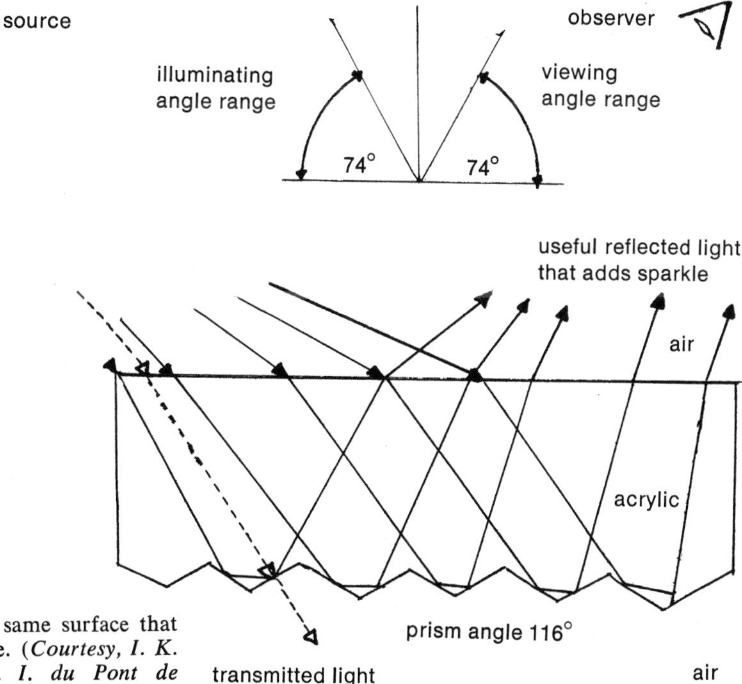

Light rays changed to exit from the same surface that they entered, so as to cause a sparkle. (*Courtesy, I. K. DeBileu, Plastics Specialist for E. I. du Pont de Nemours & Co., Inc.*)

When we see ourselves in a polished table, these reflections happen because light waves strike the surfaces and bounce off again in the same manner as a ball that is bounced off a wall. There are several different kinds of reflections when a material is opaque, translucent, or backed by an opaque surface. The reflecting of an image from a polished surface is called specular or regular reflection because most of the rays hit the material and bounce off again in the opposite direction. But if there are variations in the surface, light waves will bounce in different ways and therefore create various effects. An evenly pebbled texture will cause diffuse reflection, when light rays scatter in all directions. Or, when light bounces off a matte surface, it travels in a single direction and then spreads out. Mixed reflection occurs when part of the light is diffused and some is regularly reflected as off an unevenly painted surface. Rippled plastic or crinkled metalfoil will cause scattered reflection as tiny portions of the surface act like mirrors, and cause a multiplicity of separately reflected rays.

Clear plastic such as acrylic will permit most of the light to be transmitted. But when the acrylic is milky, or if the polyester resin casting is very thick, the transmission diffuses much like diffuse reflection, except as it passes through the form it diffuses upon exit. If the acrylic is frosted only on one side, however, light will spread after it passes through the frosted layer. A ribbed piece of acrylic will create a mixed transmission with light spreading, diffusing, and some reflecting as with a mirror. And if the underside of acrylic or polyester is a pebbled surface, the light scatters at different angles as it transmits, sometimes sparkling in nearly full brightness.

Multilens sheeting (thermoplastic), sometimes with 10,000 parabolic lenses per square inch, creates moiré effects as light reflects and diffuses through its translucent and clear patterns. When the smooth surface of one sheet is placed against that of another, floating or sinking moirés result. A pack of multilens sheeting creates a strange illusion of a three-dimensional moiré.

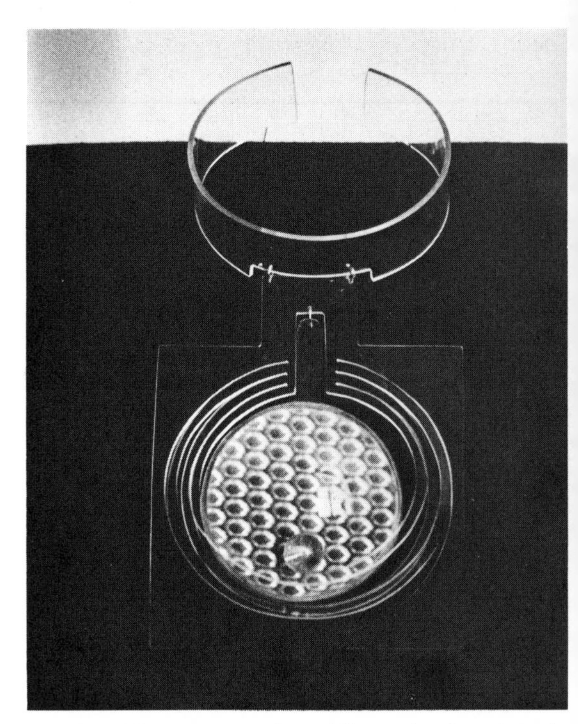

Carolyn Kriegman used multilens sheeting for the round disk in her body ornament. By slicing into the acrylic sheeting she creates a linear pattern through edge lighting.

Then there is literal light which is the actual use of an artificial light source in a form. Neon, reflected light patterns on a screen, flickering light bulbs, vapors filling the atmosphere with light beams passing through are some forms used by artists and designers. Liquid light is the projection of a color-water design with the light of an overhead projector passing through to create ever-changing, pulsating background. This system has been used to fill the environment of rock concerts and light shows. Ugo LaPietra, an Italian designer-artist-architect, uses light, transparent plastic and sound to create an interaction of what he calls a synaesthetic process. Light-sound, visual-tactile, tactile-perception, visual-acoustical all interact in what appears to be a luminous shifting space sometimes filled with

Cross-section of multilens sheet.

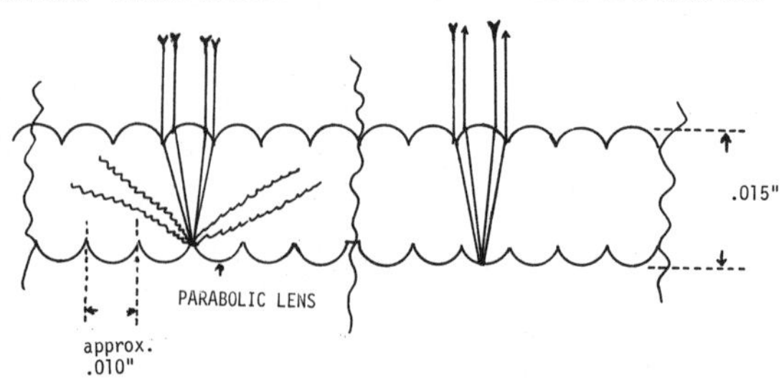

One end has a light in a large environment; the rest of the area is dark. Sound vibrations interact with light transmission to create more vibrations that take the shape of light patterns as light-sound strikes concave reflectors in Ugo LaPietra's audio-visual environment. (*Courtesy, Ugo LaPietra*)

transient color. Paths traced by rays of light strike concave reflectors, and in some of his environments interruption of sound waves causes modulation of light and sound. The entire sensory system interacts with the visual experience.

Normally, on a lesser level of excitation man is shifting and changing continuously. Man walking changes pace, shifts weight and direction, redistributes his balance with the speed of his forward thrust and pauses. Eyes are always moving. Even so-called static objects are continuously being transformed through speed and light. Stimuli are never static but can be considered serial transformations while static, immobile forms are continuous nontransformations. Rapid transformations seem to dematerialize an object, such as the blur of a speeding train or the spokes of a wheel. Light, in a sense, is movement (if we consider light waves as movement), but light *play* depends on movement to create transformations of shadows and reflections from projected beams thrown onto complex moving surfaces.

Transparent plastics play a significant role here because they can interrupt light and change its course, add other qualities such as patterns and textures without blocking the light source. L. Moholy-Nagy was one of the early users of acrylic in sculpture and in light machines. Before 1947, in *Vision in Motion* [2] he wrote, "In my pictures I have tried to follow this line of space-time articulation by painting on water clear, transparent plastics, introducing light effects, mobile reflections and shadows, indicating a trend away from static pigmentation of surfaces toward a kinetic 'lightpainting.' " Indeed, Moholy-Nagy predicted and practiced more than manipulations of light, but presaged the future importance of plastics as art and design form as well.

The following expands our vocabulary in the application of light with transparent forms. Design principles were used in these exercises by the author to demonstrate how design aspects translate to transparent forms, using light as a definer, modulator, expander, modifier, et cetera. Where designers-artists successfully employed these ideas, their forms were used as examples in this description.

[2] Page 252.

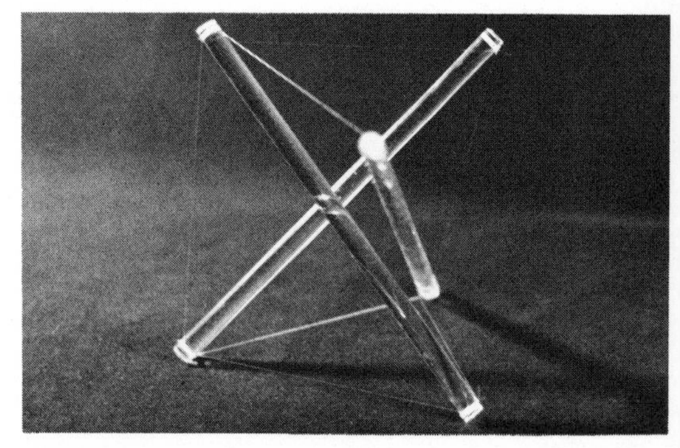

Tension figure of three acrylic rods that are free standing exists in space by taut nylon strings pulling at their top and bottom edges.

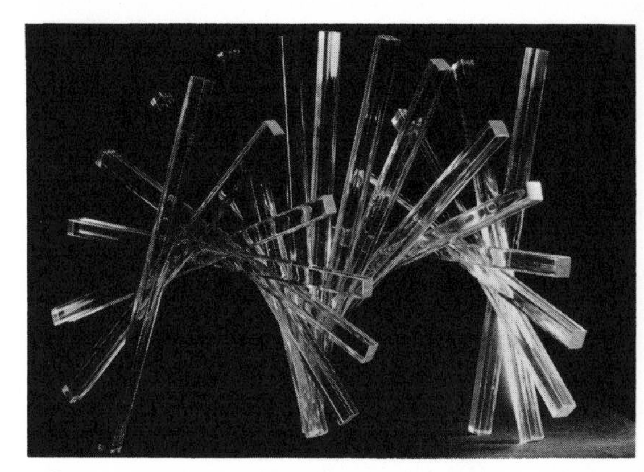

Interpenetrating sticks form a rhythmic recurrence with a connectedness of dynamic continuity. The form can go on and on. (These are extruded acrylic square-sided rods that have clearly defined corners.) Even as they rotate they appear to spin like a baton, dancing in circles because the viewer is able to see movement in a continuum like Duchamps' "Nude Descending a Staircase."

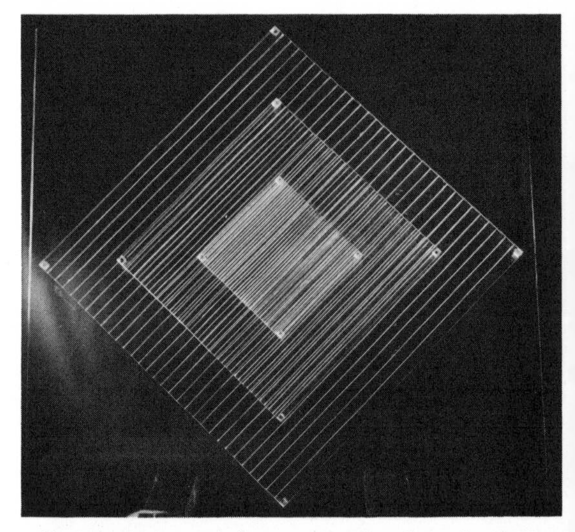

Transparency and interpenetration here is shallow spatial depth achieved through differences of density. Change in gradients inscribed by lines etched into the acrylic states planes that at once separate and interpenetrate; as we move, lines vibrate and produce an optical effect through variable spacing. Transparent

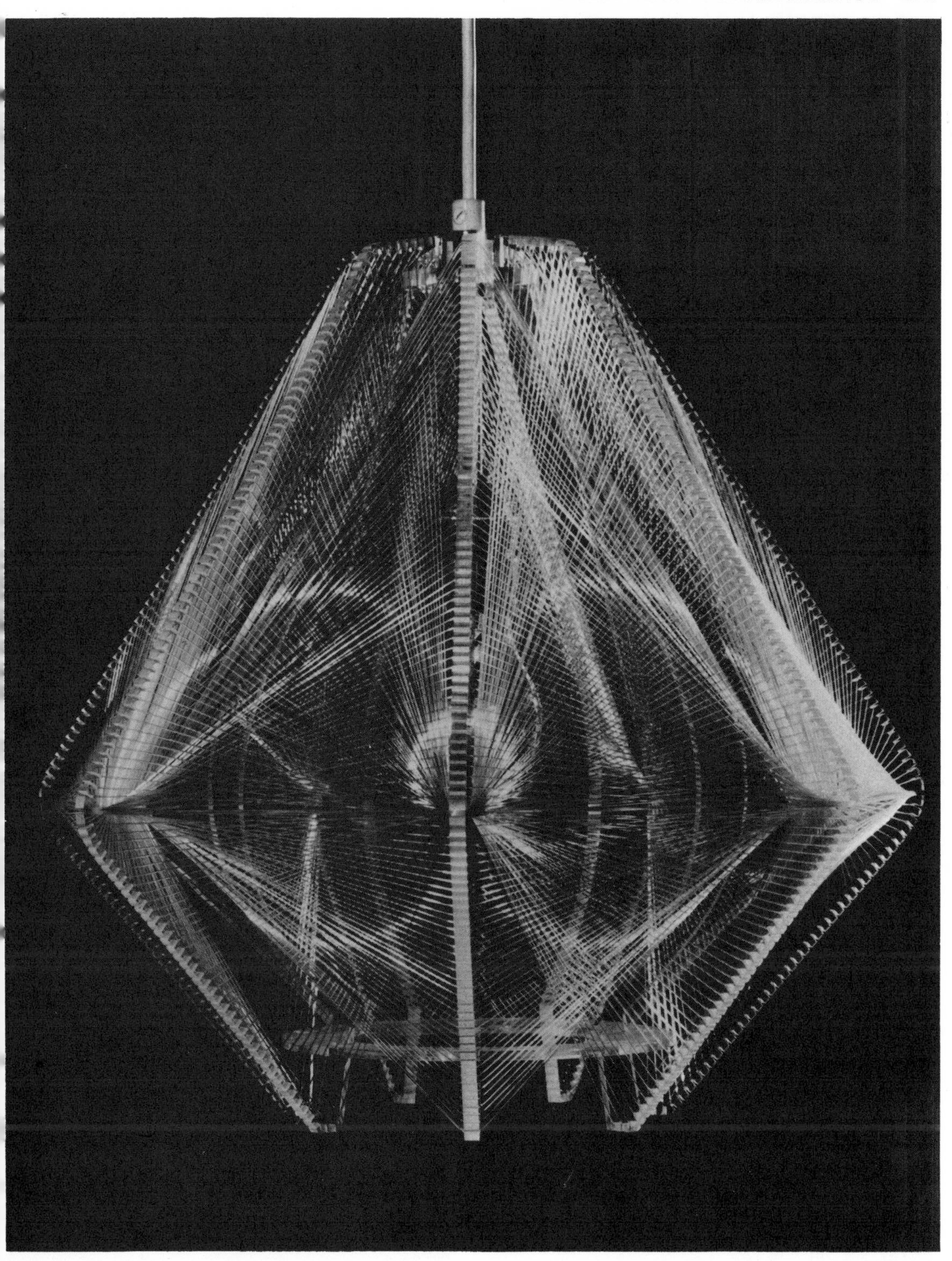

Design and Light in Transparent Form—Focus on Acrylic—Tension and Compression. There is a sensation of somewhat tentative equilibrium between two attracting forces that are being pulled away or together by a common relationship of shape, space (between and around), mesh, or web. Position and direction are held together and created by taut strings. Only edges seem to pull at edges. Space becomes caged and then escapes because of the clarity of the positive and the negative. (*Courtesy, Koch & Lowy*)

overlapping planes exhibit no fixed depth because all levels can be seen simultaneously. The etched lines serve as cues to the spatial positioning of the planes.

Planes on an axis look static in a photo, but movement and time interact with edge lighting to produce a variable rotation around an axis, because the rings can move sideways as well as in a circular orbit. The process of transparent movement affects our emotions. In this case, as our human energy spins the rings, there is an internal sense of unwinding. As the edge-lighted rings spin around their axis, there is an illusion of transparent volume caused by transparent movement.

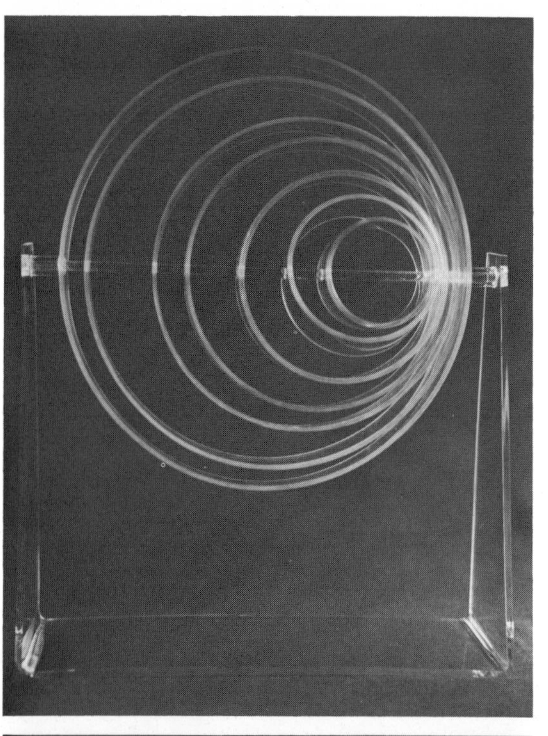

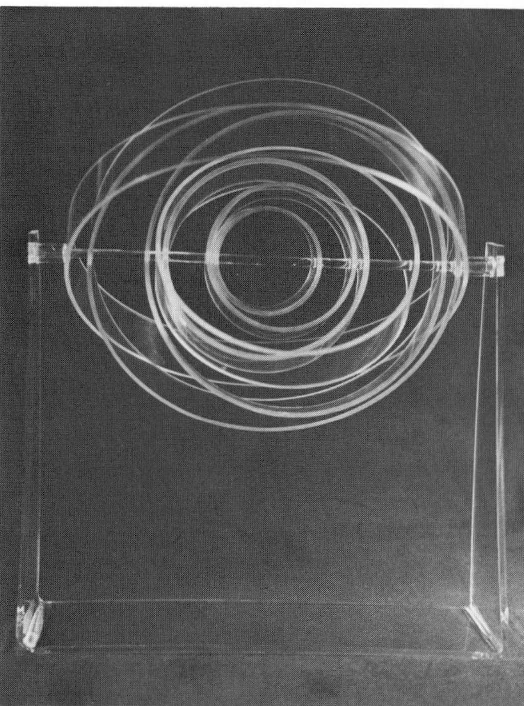

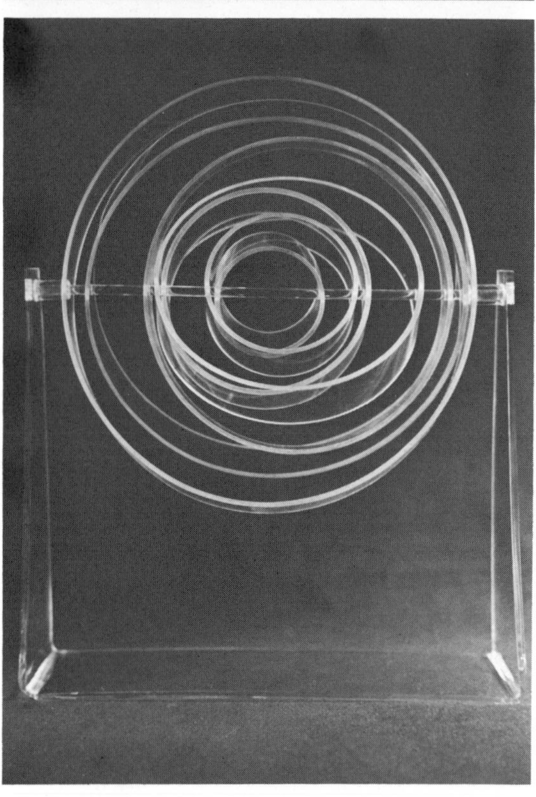

Regularity and irregularity is created at one and the same time by the irregular grouping of regular shapes. Intervals surrounding the acrylic cube become invisible space except for background reflections that act as crystal shadows. Each cube influences the next form by refracting and reflecting light off of and into planes. The light modulates and transforms through varying intensities of light, even though each form is the same size and shape.

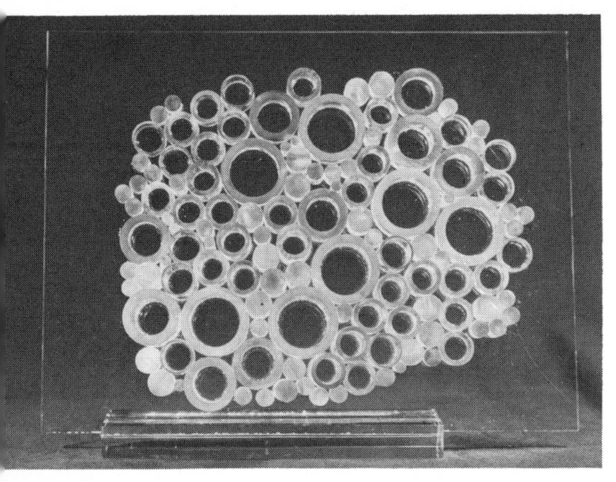

Density relief with cross-sections of acrylic tubes of various sizes becomes a pattern of order. Differences of density are arranged and counterbalanced in three directions, which produces a sense of equipoise. Negative spaces and positive (edges) are distributed at varying levels. Edge lighting is more intensely described when the walls of the tubes or the rods are thicker in a strong pattern.

Progession here uses regularity in a balance structure. Regular shapes as intervals build with repetition into a network of an equipoised structure. Direction of movement is stated by the linear edge lighting of the forms. The shape is strong, but the transparency of the material imparts a delicate, almost fragile illusion that seems almost to break away. (*Courtesy, Thelma R. Newman "Moduples" for Poly-Dec Co., Inc.*)

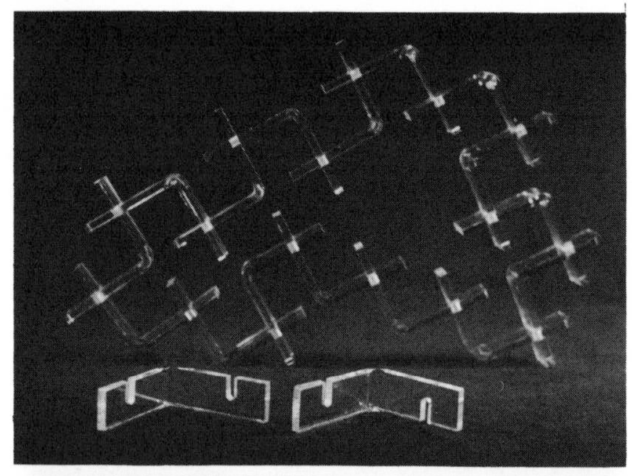

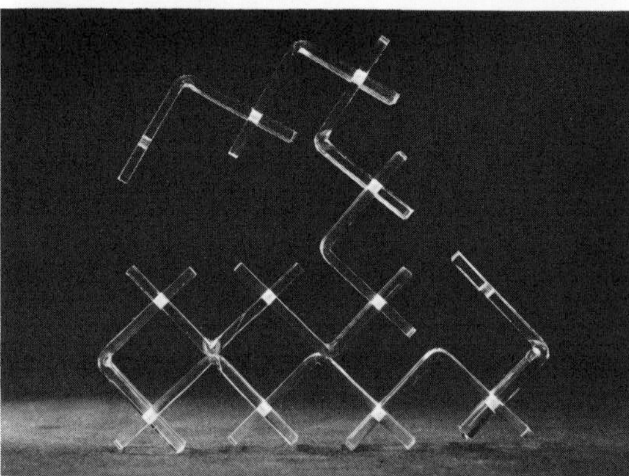

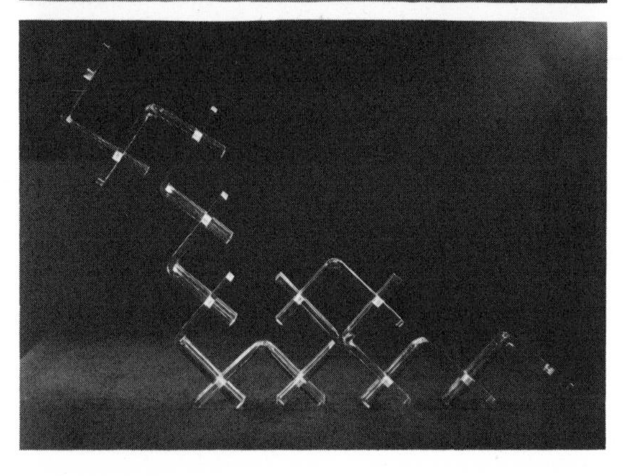

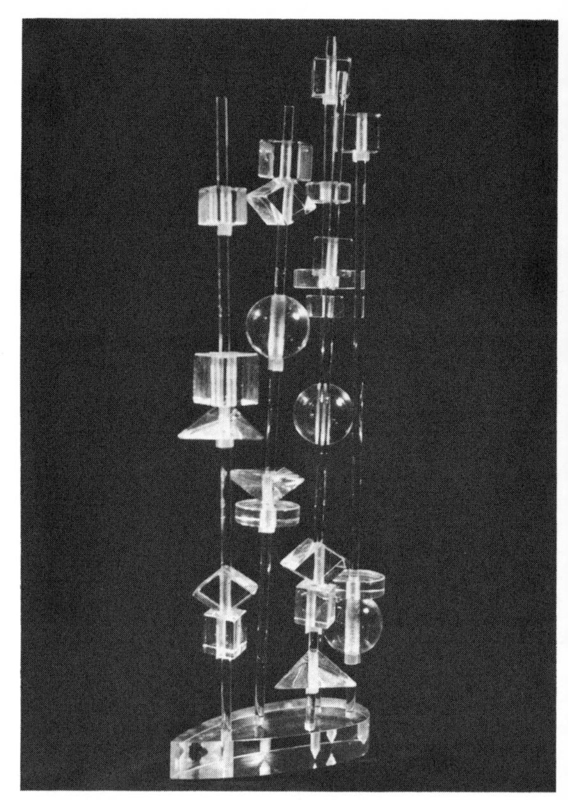

Changing relationships of geometric shapes build on a vertical axis a pretty form because of the exciting gemlike play of light on their planes. (The forms can be restrung and moved up and down on the rods.) The transparent quality of the rod-axis almost disappears and leaves a floating relationship of transparent volumes. The planes of the geometric shapes are modulated by the transmission, refraction, and reflection of light.

Form links to form in a rhythmic progression by repetition of similar elements repeated at regular or related intervals. Position, direction, and shape and their logical orientation as a rhythmic recurrence creates a time-unifying whole. Through a progression of edge-lighted planes volume is built up to indicate depth, while front, back, and interior are seen all at the same time. (*Courtesy, Neal Small Designs, Inc.*)

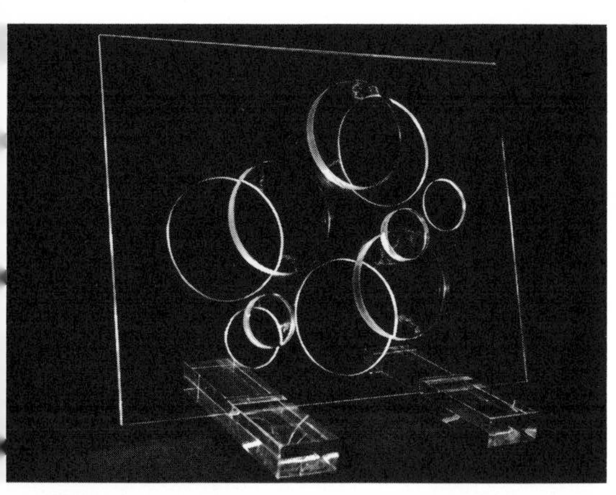

Negative and positive, figure and ground have a new relationship because negative space becomes as important as transparent positive; they are both the same shape and color. It is only the glowing linear edge that defines these shapes and unifies because of repetition of a similar contour—the circle. If these were translucent shapes, they would subtract from, fragment, and be divided from one another.

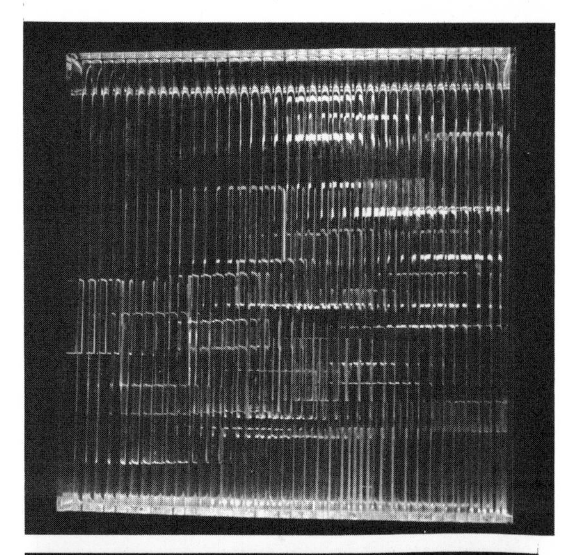

Asymmetrical Volume and Optical Movement. Solid blocks of acrylic complete with edge-lighted planes become a ghostlike illusory structure because of the moving optical effects caused by refracting light in the vertical acrylic rods. The rods are fitted closely together both in front and back of a clear acrylic frame. Movement on the part of the viewer changes the apparent direction of the linear transmission glow of light and also subtracts from the caged, blocklike structure.

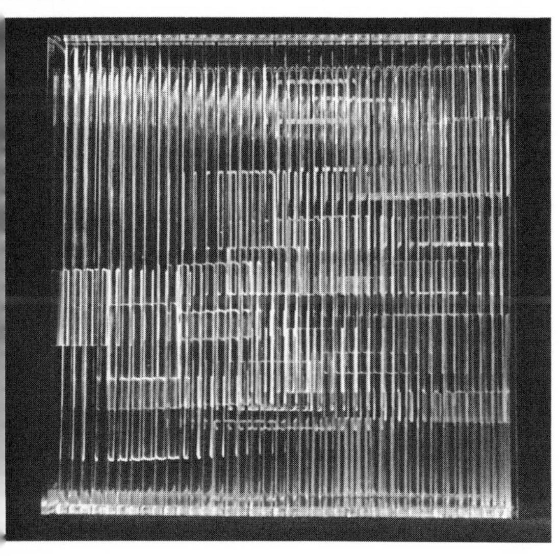

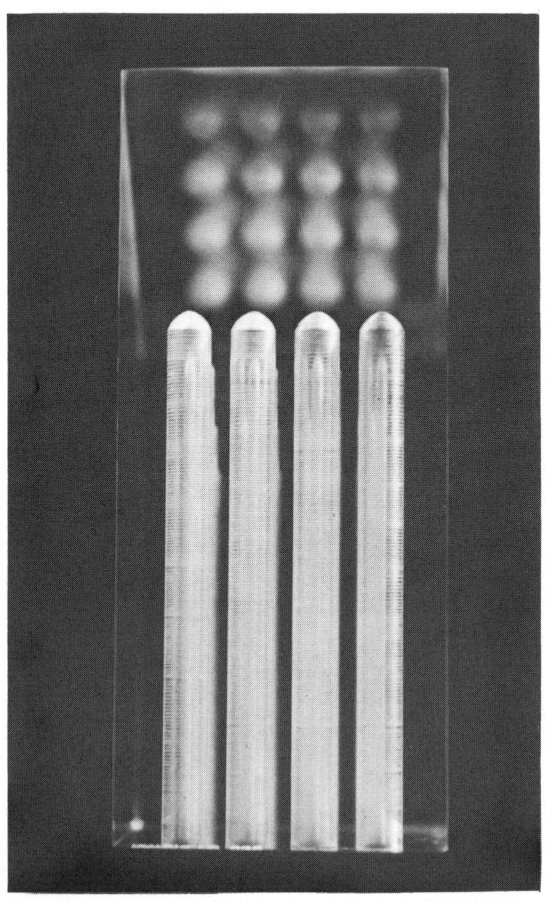

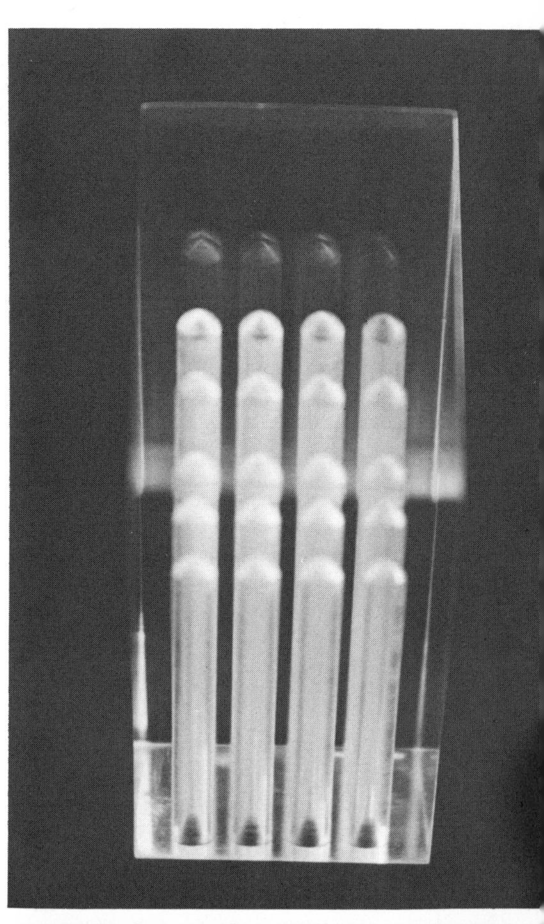

Voids are negatives or blank areas in solid three-dimensional forms; but when one is working with transparency, voids can become "solid" references because of edge lighting. This block of acrylic, part of a sculptural ordering of the same form by Ugo LaPietra, is a series of drilled holes that look like solids mirrored by the angles of the block and refraction of light. A second piece by LaPietra, greatly enlarged, dramatizes light refraction. In a third series of sculptures, Ugo LaPietra builds up geometric solids with drilled holes and again designs for the refraction potential. Some of these forms are carved with a single pyramid of drilled holes; others contain two triangular areas.

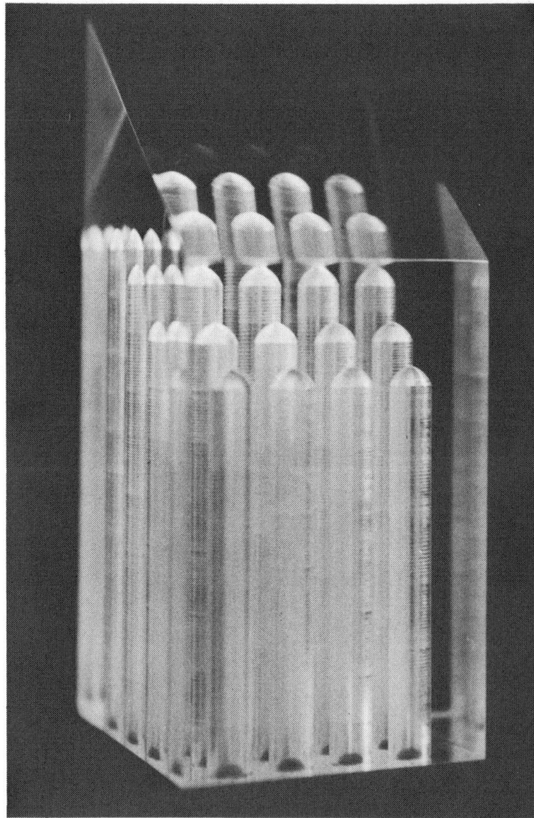

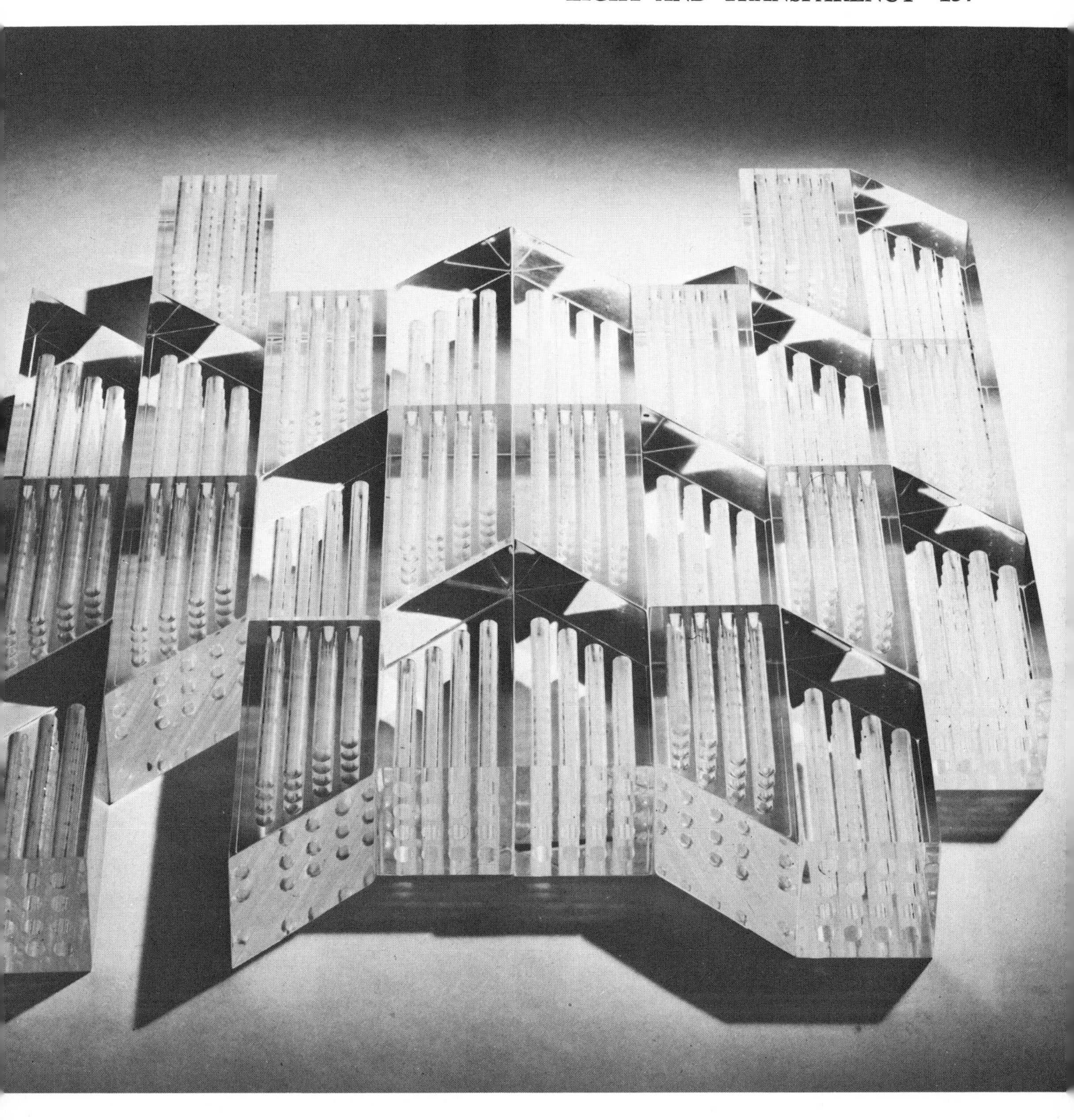

(*Courtesy, Ugo Lapietra*)

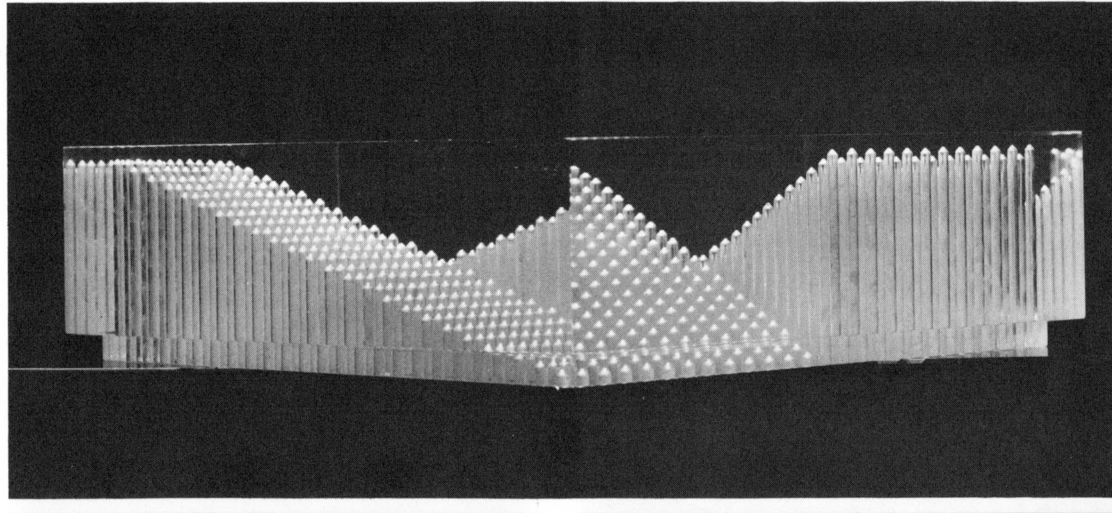

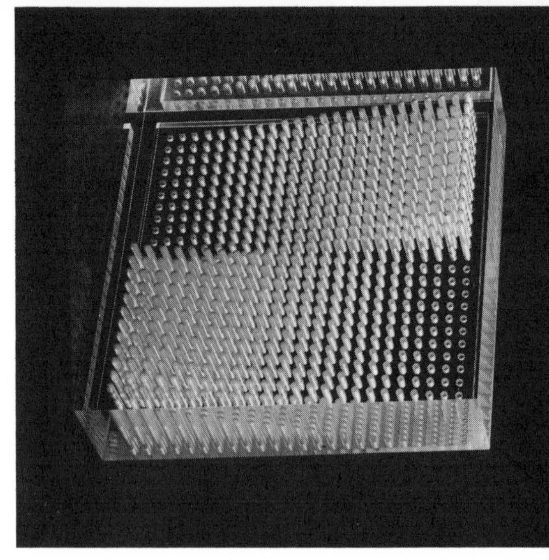

(*Courtesy, Ugo Lapietra*

"Yantra" is Peter Nicholson's contemplation environ- ment made of reflective Mylar surfaces in the shape of rhombiculoetahedron with warm and cool controlled illumination. 5′ x 5′ x 5′. (*Courtesy, Peter Nicholson*)

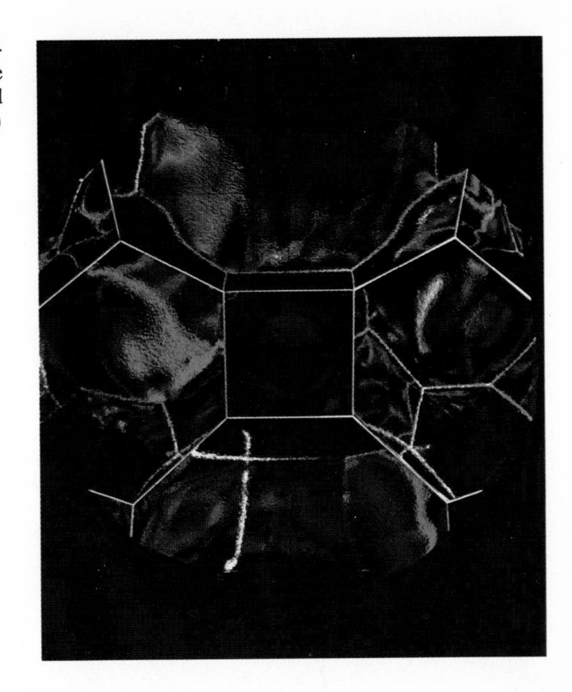

Prismatic light at it refracts through transparent m. (*Courtesy, Sandoz Color and Chemicals Divi- n*)

Machine-hooked wall hanging called "Sunrise" in nylon and Zefran acrylic by Nell Znamierowski. 24″ x 50″. (*Courtesy, Nell Znamierowski*)

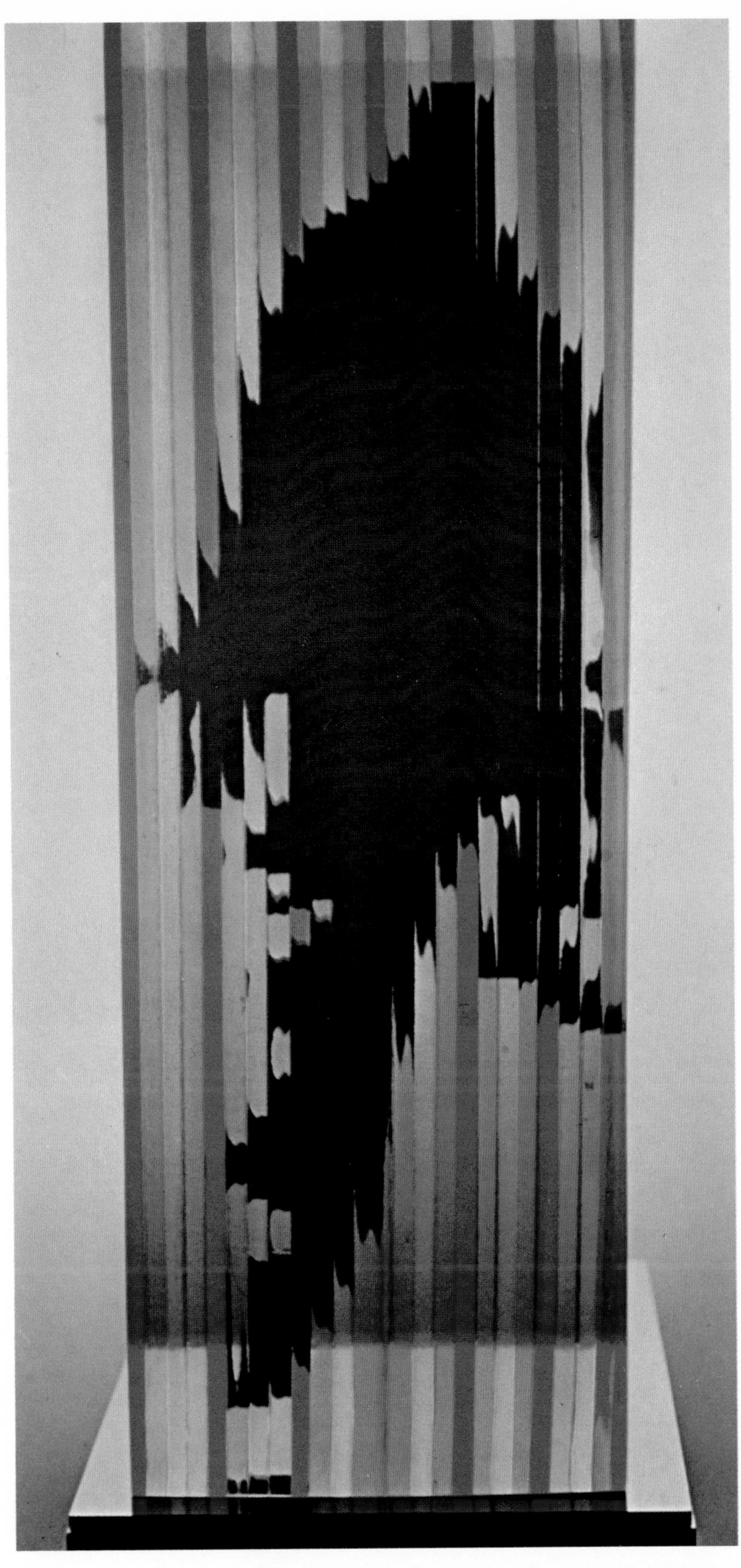

Optical effects of movement are created in Eugene Massin's polyester "Man in Purple, Red, and Green." Photo by Ray Fisher. (*Courtesy, Eugene Massin*)

"Rocking Men" by Eugene Massin in polyester is a motion study in optical changes. Photo by Ray Fisher. (*Courtesy, Eugene Massin*)

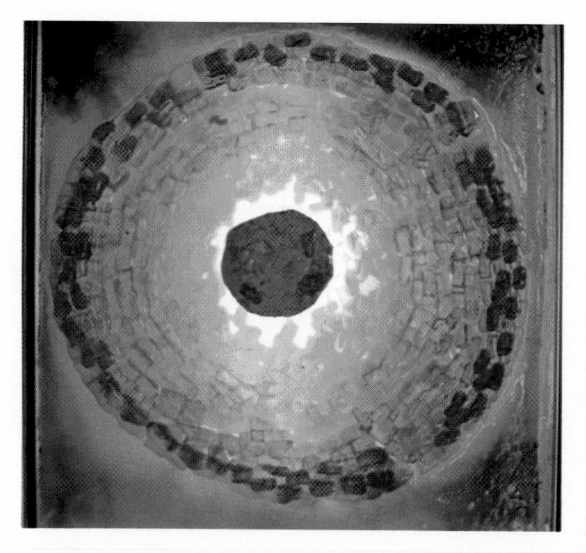

"Sun Spot" of fused Polymosaics and polyester resin is a free-standing light box by Thelma R. Newman.

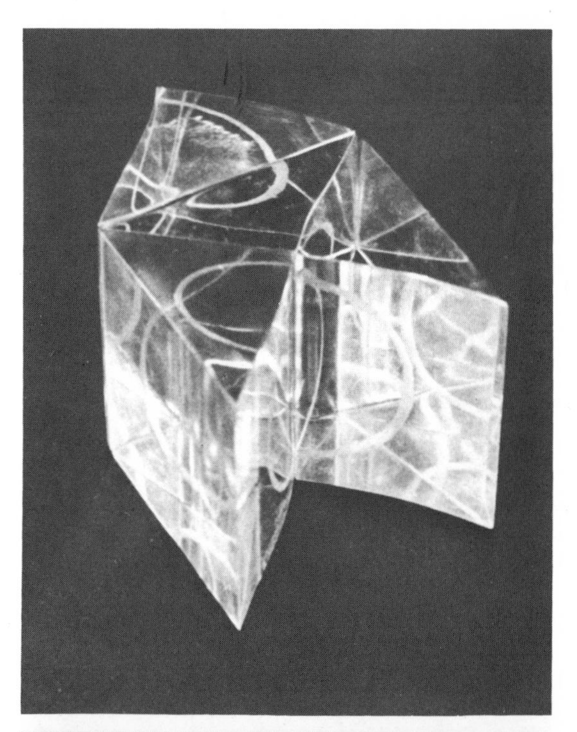

With solids, shapes may be inner or outer surfaces, may be articulated from inside boundaries or outside. But when one uses clear acrylic, both the inner and outer are articulated at the same time. There are obvious edges and surfaces to a cube; for instance, if you place a finger on an opposite plane, you can clearly see it. But if you slice the cube into equal triangles and inscribe a form on two of the planes, the sides act as mirrors and repeat the design as well as continuing to retain their original clarity.

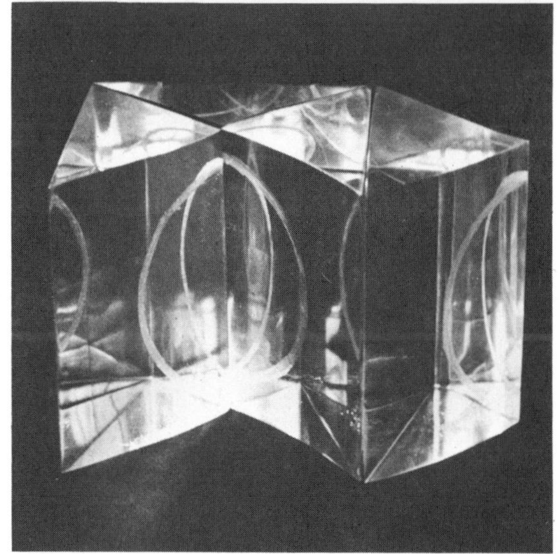

By drilling holes and cutting shapes into a progression, many illusions of volume can be constructed through repetition with variations. In this series Neal Small frames his concepts within an acrylic cube box. While he takes advantage of edge lighting to build texture, pattern, and volume, he does not, in these particular pieces, play with light refraction and the illusory repetitions of forms in the same way as Ugo Lapietra, although he does achieve some echoing of form as one moves around the cube—each corner and each side produces another image. All these works are a tour de force. (*Courtesy, Neal Small Designs, Inc.*)

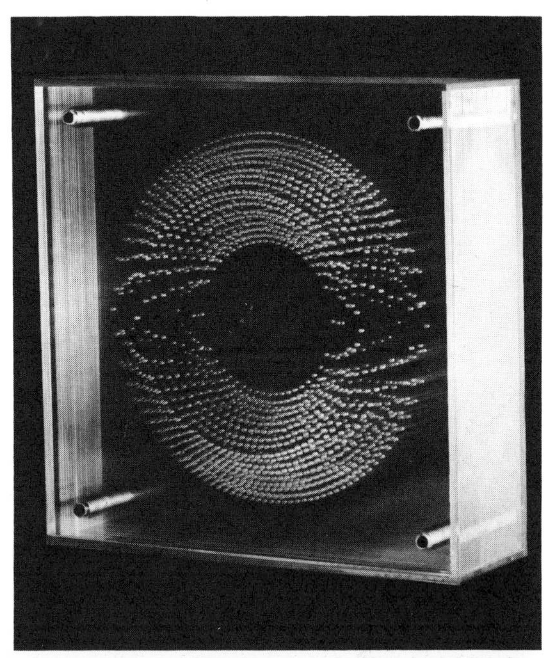

Still other effects have been created by Neal Small in incising, carving, and then highly polishing internal forms. Light crazily transmits, refracts, and reflects all at the same time, bringing in more distorted aspects of the environment than with use of edge-lighted "solids." (*Courtesy, Neal Small Designs, Inc.*)

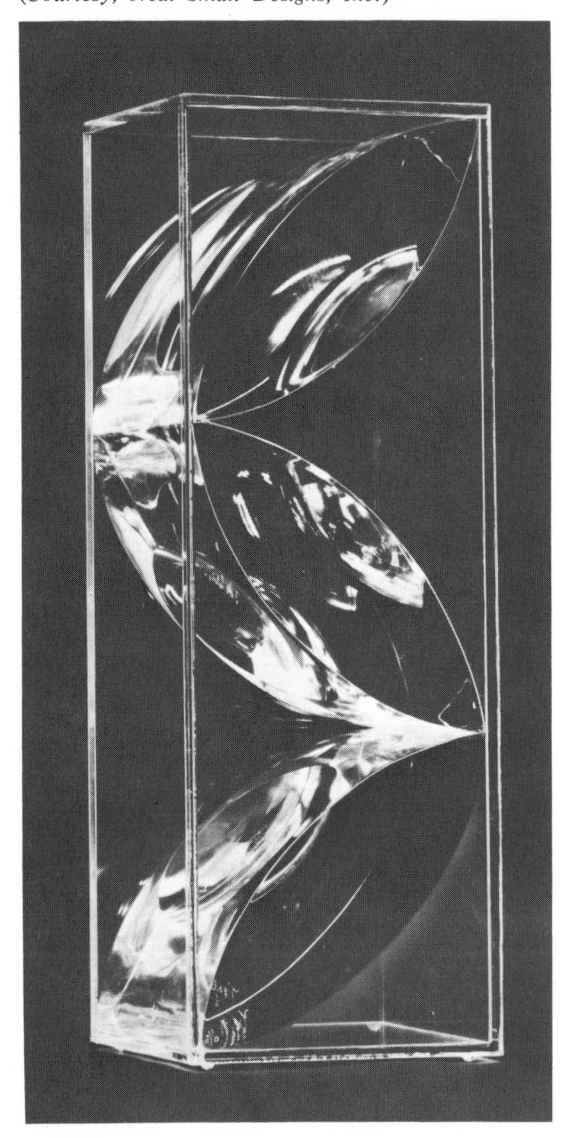

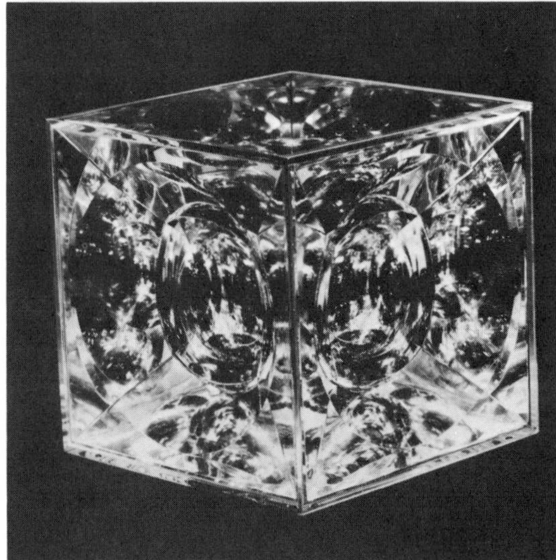

LIGHT IN TRANSPARENT FORM (A FOCUS ON POLYESTER)

Acrylic is not the only plastic that can create optical effects as an integral quality of its potential. Although polyester is not quite so optically clear as acrylic, it still performs in very much the same manner if the polyester used is water-white. There are some polyesters on the market approaching the refractive index of acrylic. But in most polyesters light waves move more slowly and edges are not quite so sharply defined.

Eugene Massin has been particularly concerned with optics, kinetics, principles of light reflections, refractions, mirror images, and multiple-surface views and distortions which become transformations from one configuration to another. His colleague Julia Busch wrote in *Art Journal* [3]:

> At first glance, one reacts to the mirrored surface of mylar . . . , the rainbow striations of color, and the hard glittering surfaces like huge jewels. But a closer look within reveals the silhouette of a man. He may resemble melting ice, a machine part, a row of paper dolls. He is seen . . . [as] now-you-see-him-now-you-don't, now you see three and a second later maybe twenty-three, appearing and disappearing all within the same piece. The sculptures which at first glance seem glassy and cool turns out to be infinitely complex.

In Eugene Massin's "Man in Purple, Green, and Red Progression" color, mirror images, multiviewpoints caused by refraction of light and the break-up of the image inclusion are caused by what appears to be a "bleeding" of the forms into separately cast and differently colored layers of polyester resin. As one looks around the sculpture, mirror images and multiviewpoints appear and reappear. Color progresses from solid amethyst to varied tones of amethyst to blurred shades of green and purple to distinct bands of purple, red, and green that fracture the "running man."

Massin's slowly rotating "Chinese Puzzle" utilizes the same principles as "Man in Purple, Green, and Red Progression" but is more complex. Eighteen figures, according to Julia Busch, are contained in layers of polyester resin colored red, violet, avocado, and clear. As the piece moves, the color bands change from distinct layers to misty greens and intense reds, while the figures multiply due to mirroring. At times fifty or more figures can be counted. The figures, both opaque and clear and left flat as cut, function as mirrors reflecting light back from the piece. The clear inclusions create the illusion of etchings on the interior walls—etchings that are infused with

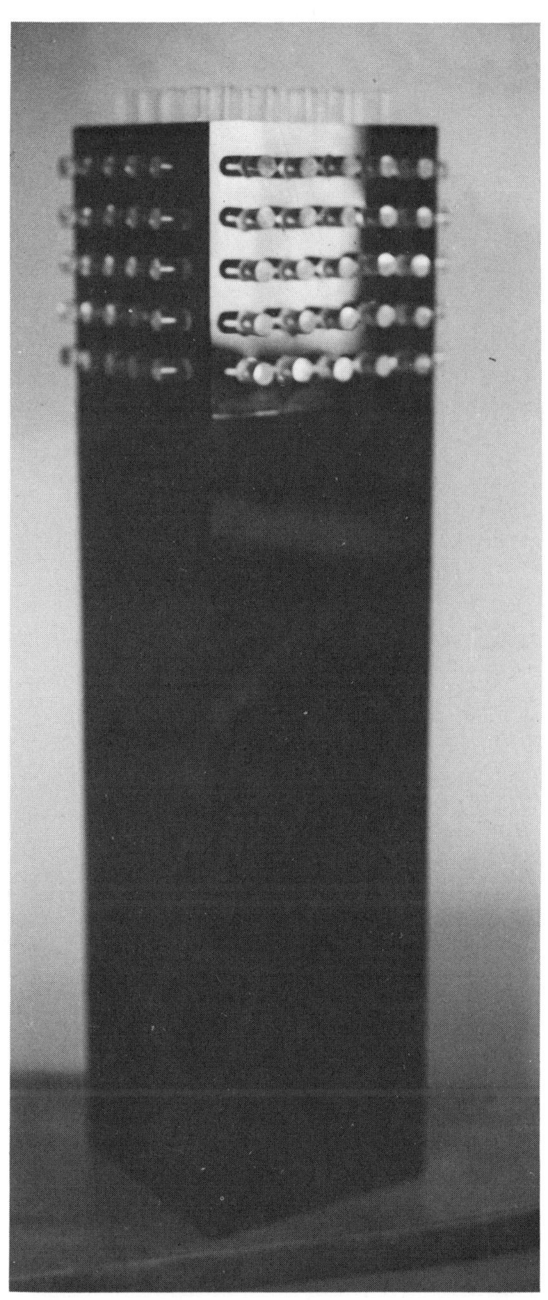

There have been a rash of boxes on the market, programmed alternating lighting patterns contained within an opaque box and transmitted through projections of various sizes and patterns of acrylic rod. In all interpretations of this effect light fluidly moves and transforms through color, direction, and pattern via an internal rotary mechanism. The effect is pacifying and entertaining because of the great variety of patterns that emerge.

[3] Summer 1970, p. 438.

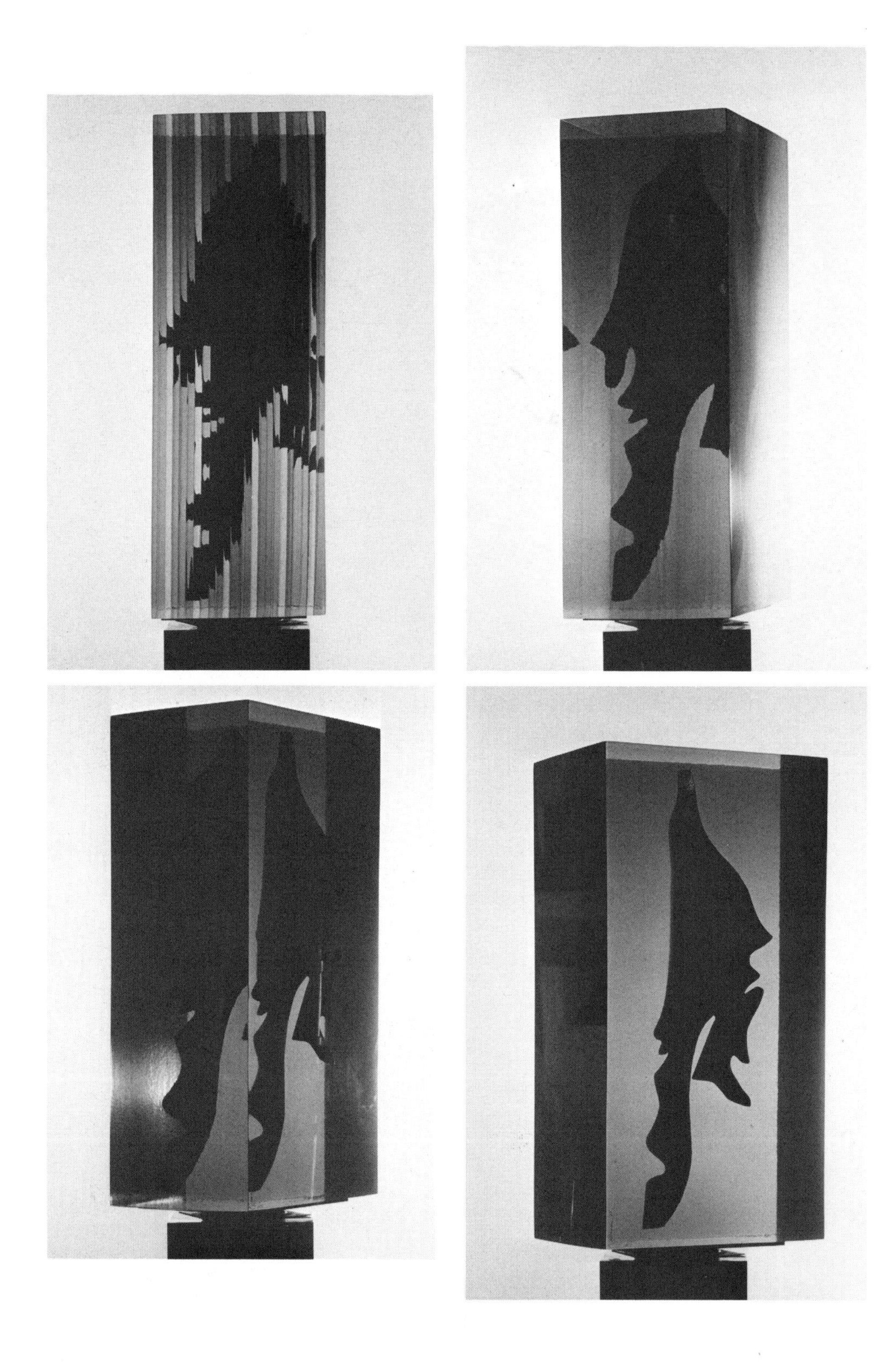

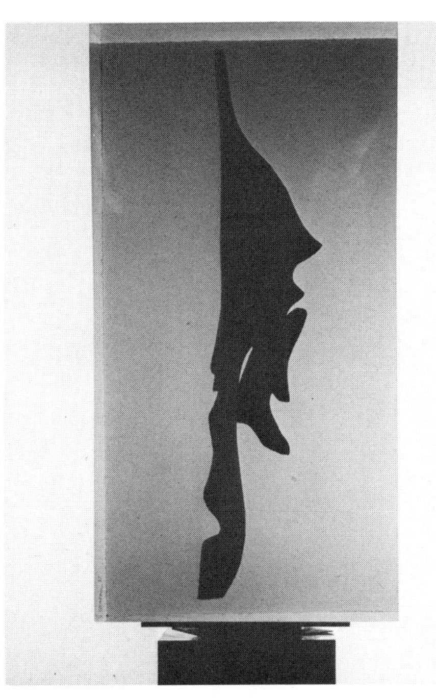

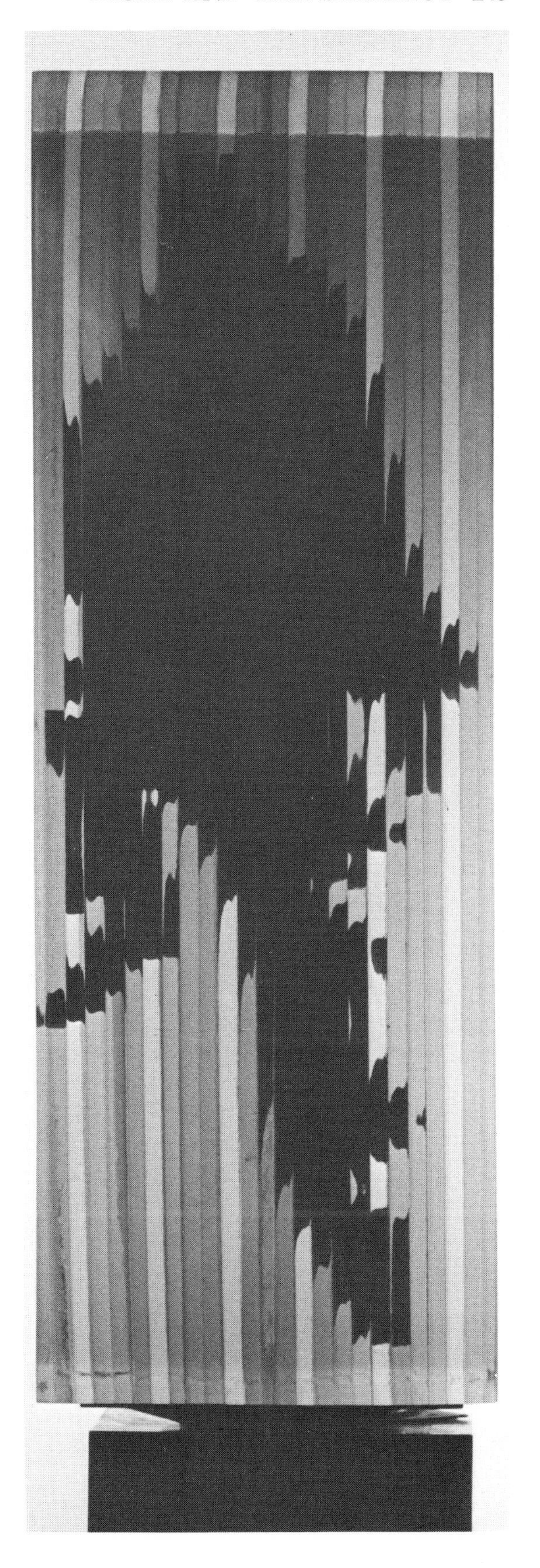

In Eugene Massin's "Man in Purple, Green, and Red Progression," color, mirror images, and multiviewpoints caused by refraction of light break up the image inclusion caused changing illusions. (*Courtesy, Eugene Massin*)

"Chinese Puzzle" by Eugene Massin, while utilizing the same principle as "Man in Purple, Green, and Red Progression," is more complicated because 18 figures are encapsulated in alternating layers of polyester resin, each a different color. At times as many as 50 figures can be counted as one moves around the sculpture.

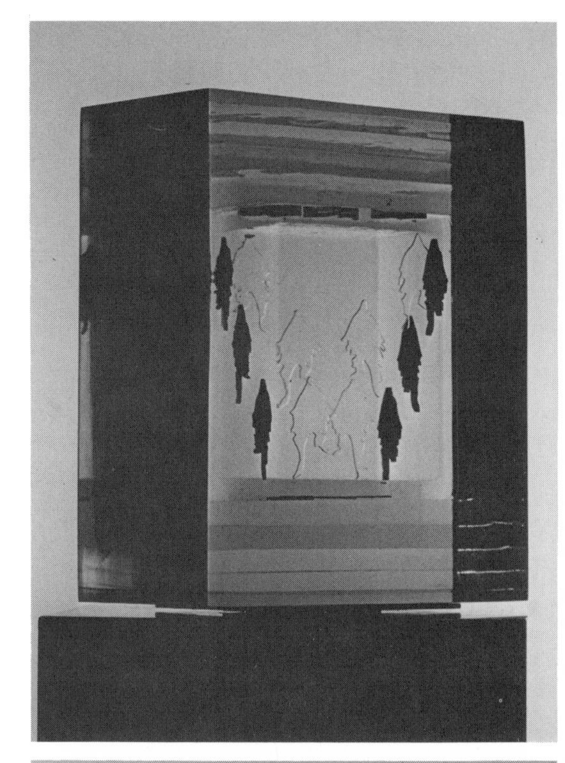

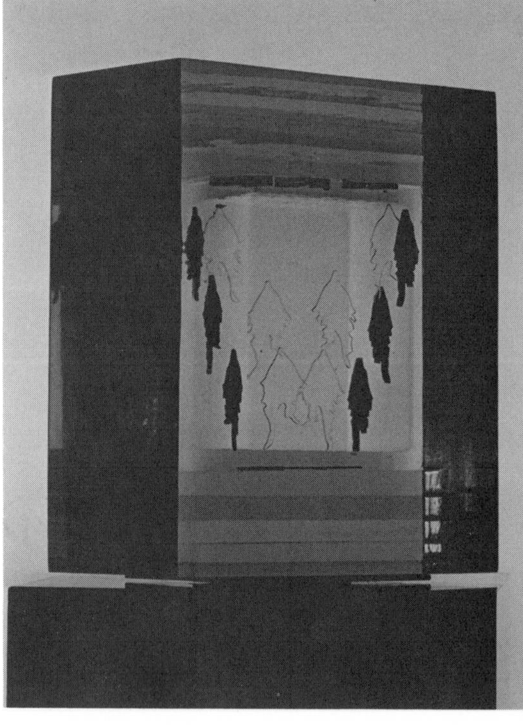

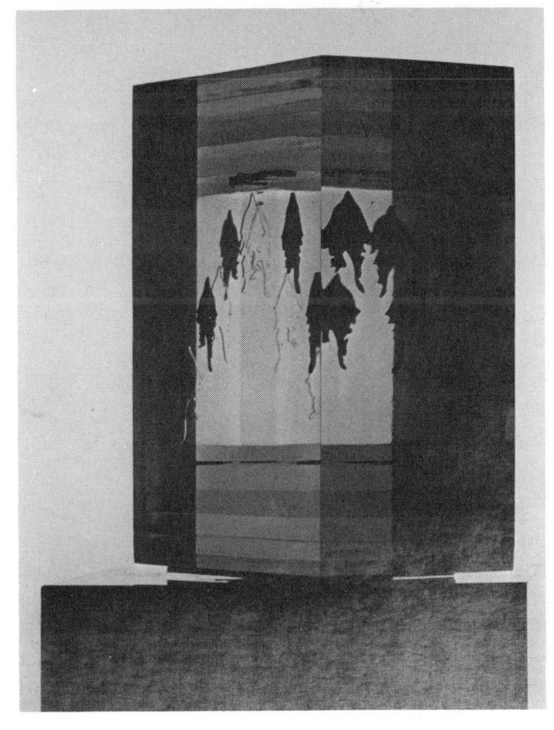

Optical illusions play with "Rocking Men" by Eugene Massin, diminishing and enlarging their apparent size. (*Courtesy, Eugene Massin. Photo, Ray Fisher*)

In "Four Yellow Balls" by Eugene Massin black silhouettes of men are placed in the center of four balls which turn imperceptibly and thus create an illusion of men walking. (*Courtesy, Eugene Massin. Photo, Ray Fisher*)

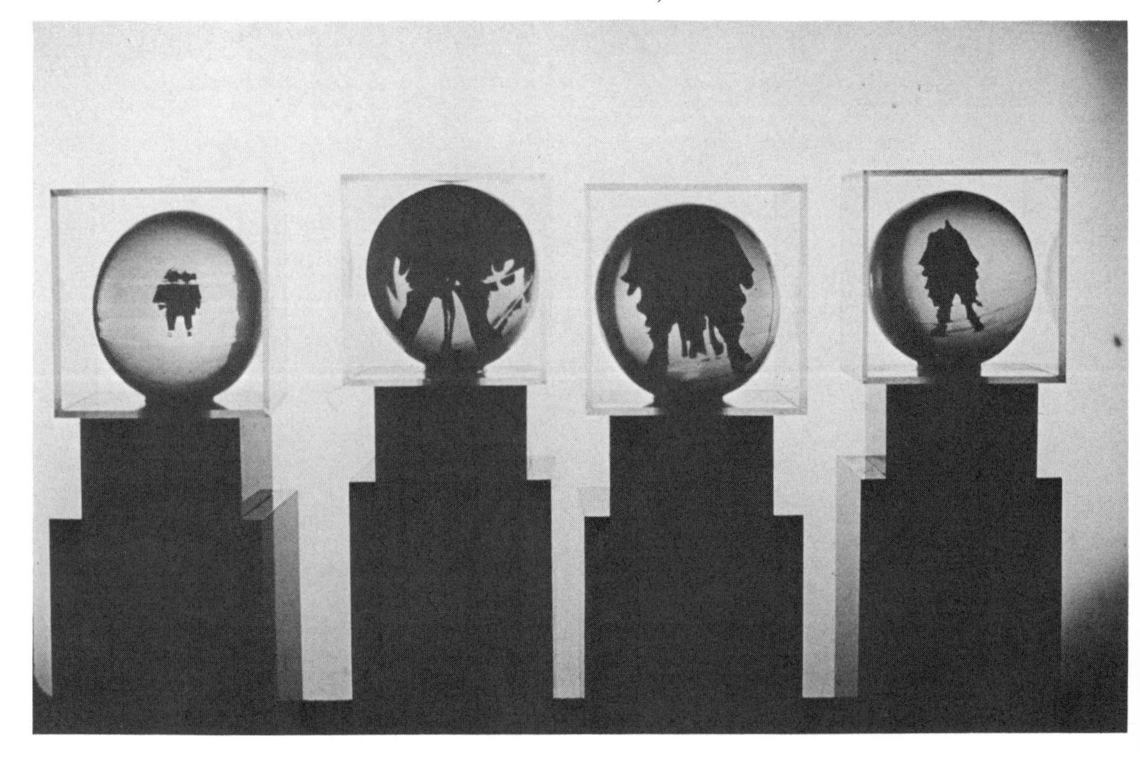

light or that completely disappear according to the point in the rotation. The opaque black silhouettes mirror light and appear to be electrically charged.

"Rocking Men" by Massin employs optical illusions caused by curved surfaces—concave shapes that diminish and convex that enlarge. The outer form is generally cylindrical with machined areas to create concave, convex, and flat facets. The flat facets add multiple views and mirror images, as in the above series. In addition, these cylindrical forms are small enough to rock with the mere touch of a finger, and as the form rocks the figures in an illusory way are set into motion, stretching, distorting, relaxing, dancing, running, and jumping. Also, color changes occur within the piece as the layers merge and blend because of the movement. Incidentally, the actual size of the "men" is approximately one half the diameter of the cylinder, and they are identical.

Curved surface optics plus movement and light play in Massin's "Four Yellow Balls." Black silhouettes of men are placed in the center of four balls which turn imperceptibly. An illusion is created when the men appear to walk while the balls seem to be immobile, while, in fact, the reverse is true. Subtle surface distortions heighten the illusion of interior movement, as do flashing lights hidden in the bases and that are directed to the legs of the "walking men."

"Three Brown and Clear Blocks," another Eugene Massin feat with optical effects, consists of single silhouettes, black, gray, and clear, slightly bent and tightly encased in cubicles striped brown and clear. As the viewer moves around them, the interior walls reflect back the image of the inclusions which are multiplied simultaneously by corner facets. At the same time, color changes occur as brown and clear bands fracture the inclusions into jagged or spiny images and blend to tones varying from topaz to amber. In the bottom cubicle, the clear inclusion appears to disappear at certain vantage points and reappears as an etched effect at other angles.[4]

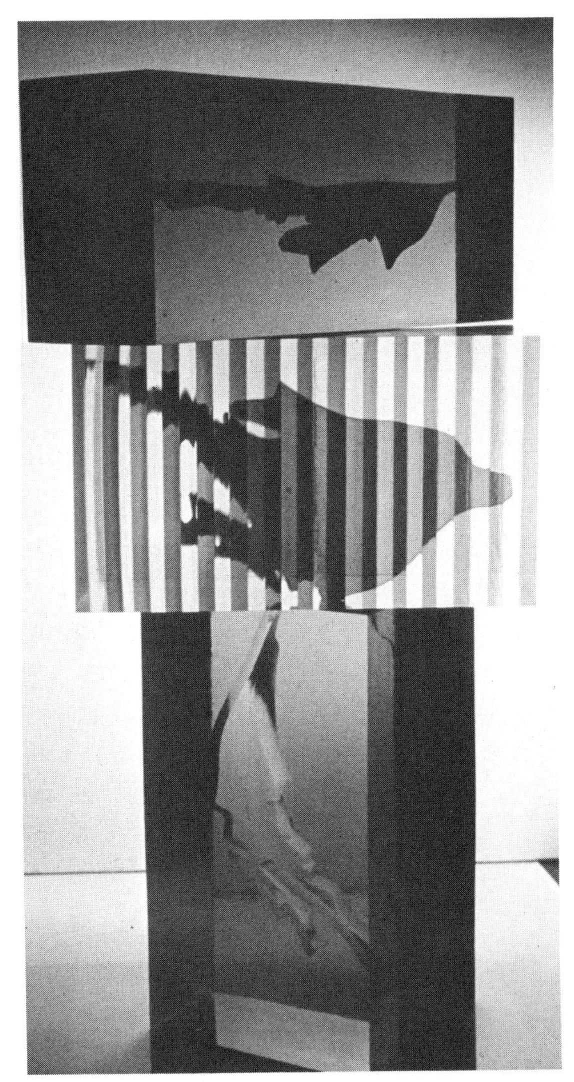

Interior walls of the three blocks reflect back images of inclusions in Eugene Massin's "Three Brown and Clear Blocks." (*Courtesy, Eugene Massin. Photo, Ray Fisher*)

LIGHT AND MYLAR

Metallized Mylar (polyester film), cousin to the mirror, can be formed to create its own shimmering mirages or, when married to acrylic, can produce the illusion that an acrylic shell is a solid block.

Mirror-finished silver Mylar .075 mils thick was used to translate paper models into decorative form by Jude Johnson. Mylar can be scored, cut, glued, and, in fact, handled like paper. When rotating on an axis, the mirror surface distorts fragments of the environment into continuous transformation.

LITERAL LIGHT AND METALLIZED ACRYLIC

When Mylar is adhered to clear acrylic, in the creation of this prototype, the surface is not mirror-smooth, but has small unevennesses evenly

[4] Special thanks to Julia Busch for her descriptions of colleague Eugene Massin's work.

distributed. If the surface were mirror-smooth, the effect of producing an illusion of a solid acrylic form from a shell would be more perfect. If metallized plastic [5] is placed on adjacent planes on only part of the shell, it acts as a refractor, and mirror images repeat in the same way as an acrylic block does. Strips of Mylar multiply the red neon pattern of the author's sculpture—light intensifies and also causes the other clear acrylic surfaces of the case to reflect the neon form as well.

[5] Rohm & Haas Company manufactures a beautiful mirrored Plexiglas that can be cut, with edges glued very much like Plexiglas sheet.

Metallized polyester films can be formed like paper into reflecting volumes. These constructions are by Jude Johnson. (*Courtesy, Jude Johnson*)

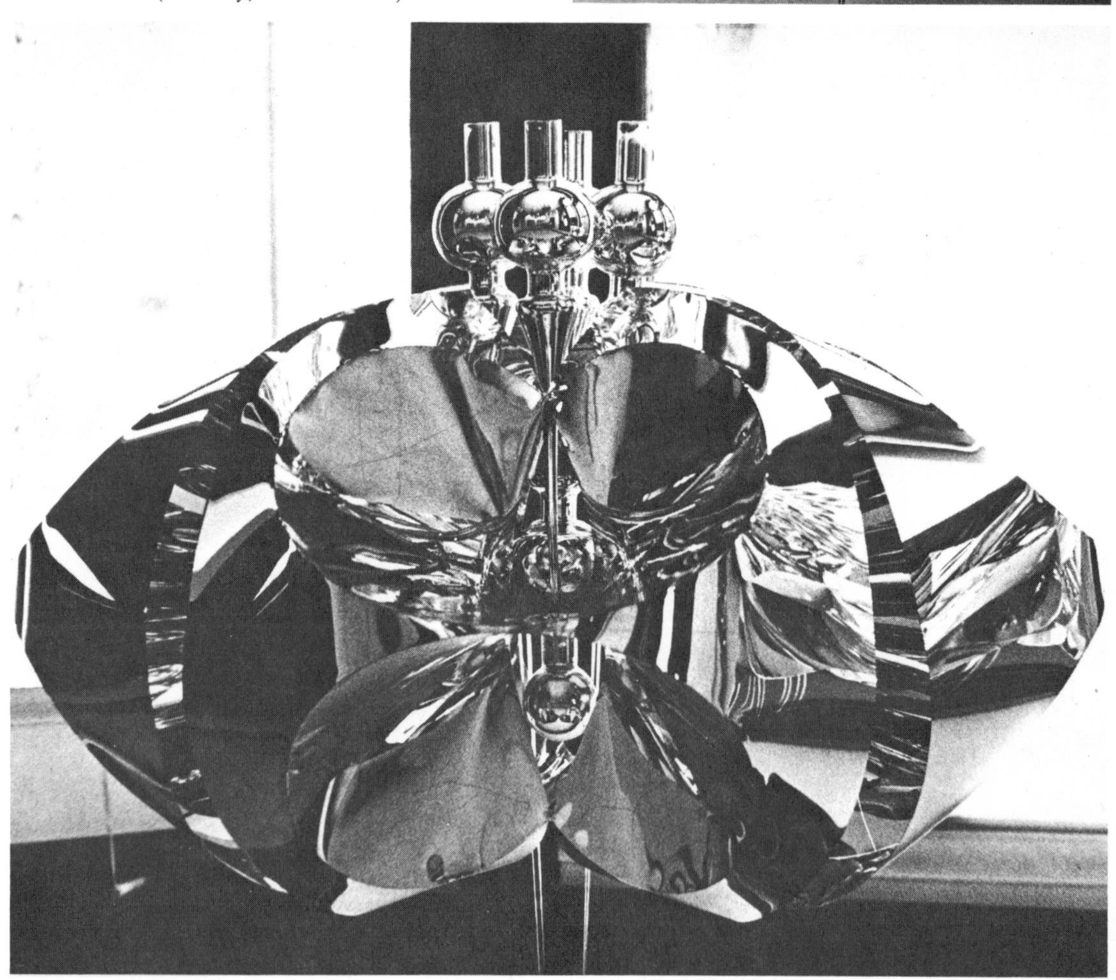

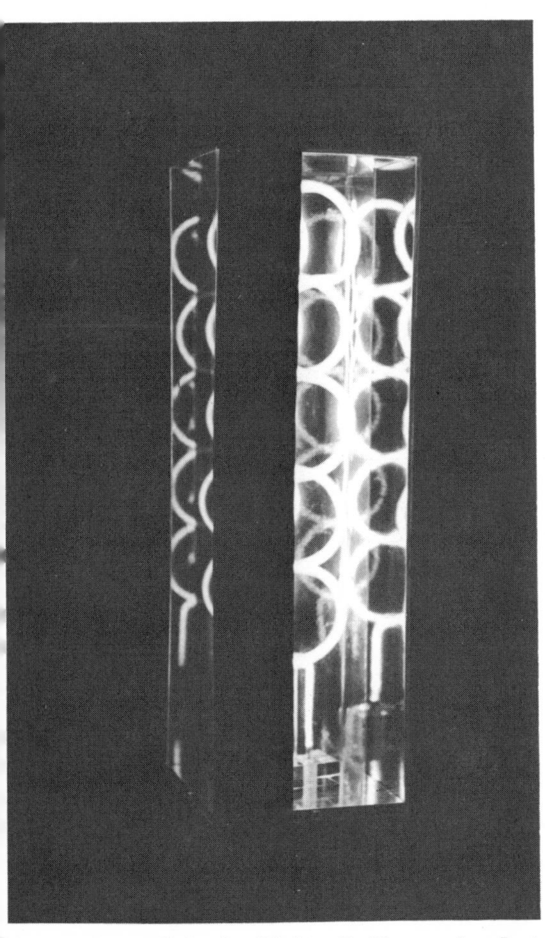

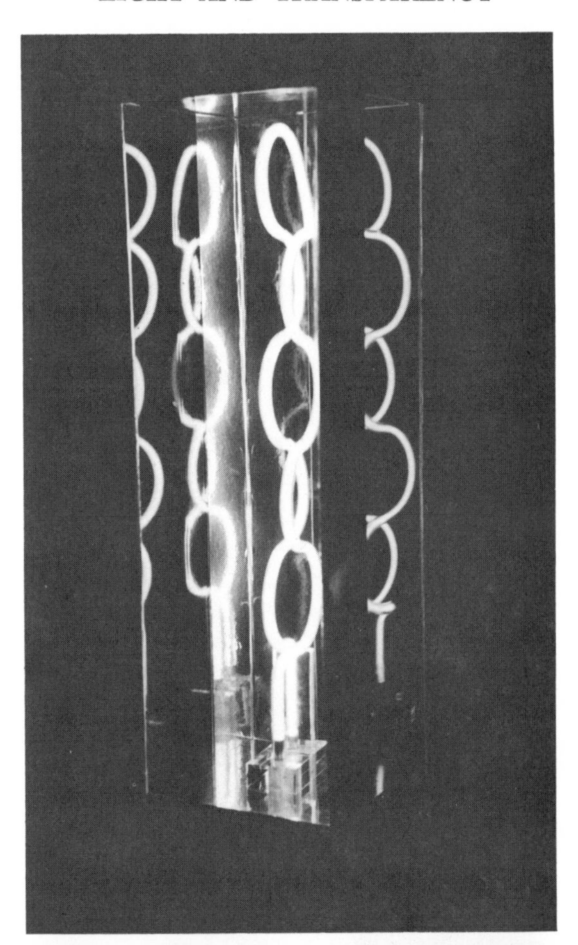

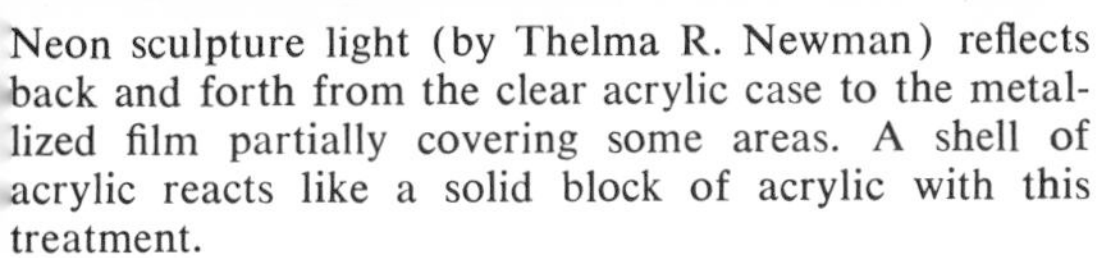

Neon sculpture light (by Thelma R. Newman) reflects back and forth from the clear acrylic case to the metallized film partially covering some areas. A shell of acrylic reacts like a solid block of acrylic with this treatment.

"Yantra" is an environment created by Peter Nicholson of rhombiculoetohedron-shaped reflective Mylar surfaces. Warm and cool lights, manually controlled, illuminate the environment. In close-up the surface reflects like soap bubbles on a mirror.

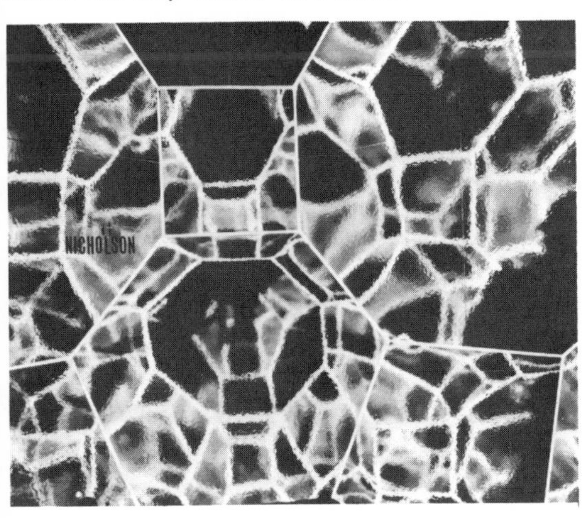

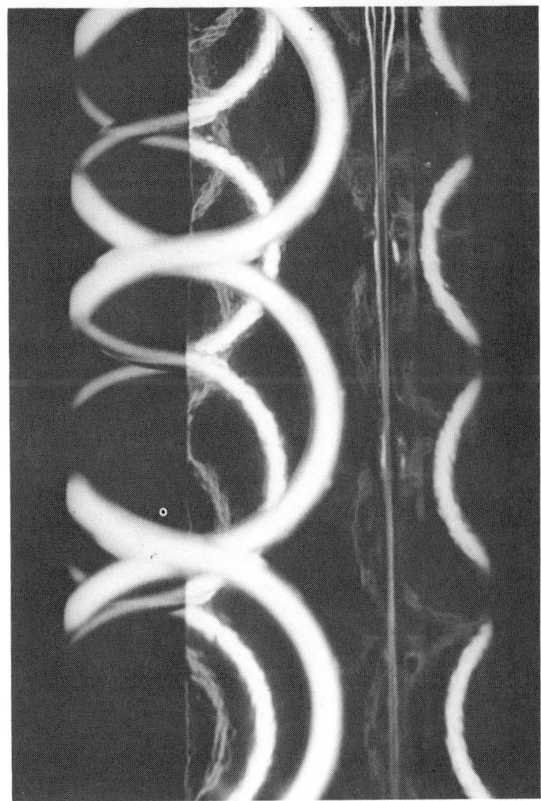

Acrylic sheet comes in a variety of textures and patterns as evidenced by this Plexiglas assortment. (*Courtesy, Rohm & Haas Company*)

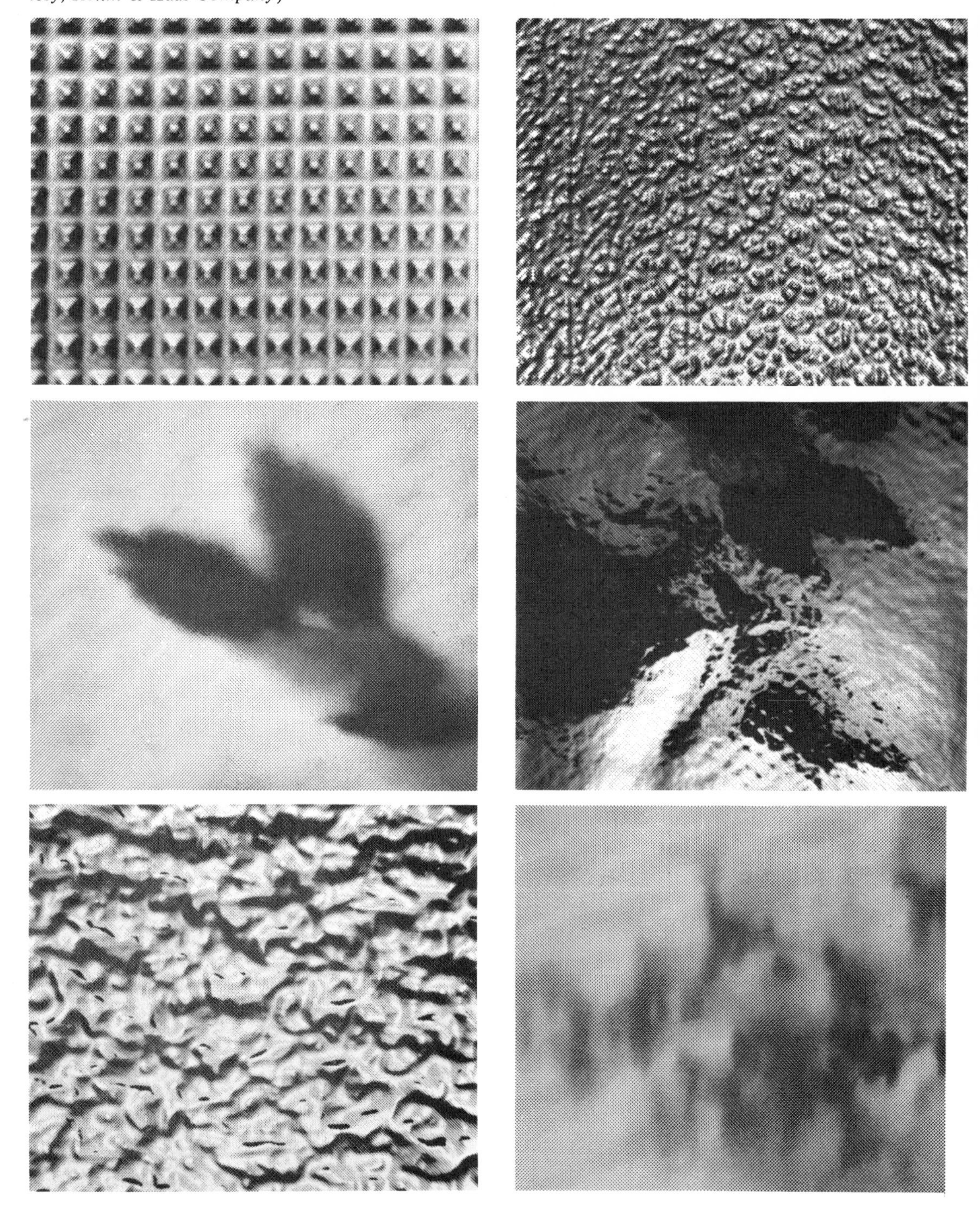

5

designing with rigid sheet, block, rod and tube

Sheet, block, rod, and tube are generally thermoplastic, characterized by moldability and remoldability through heating, softening, and shaping. These forms are manufactured by injection molding, extrusion, or casting processes. From the manufactured shape, designers then fabricate their own forms.

Although sheet, block, rod, and tube are available in a variety of polymers, such as rigid polyvinyl chloride, polystyrene, polyethylene, cellulosics, polypropylene, acrylonitrile-butadiene-styrene (ABS), nylon, solid polyurethane, fluorocarbons such as TFE, polycarbonate, polymethyl methacrylate (acrylic), only one solid will be explored in detail—acrylic. Generally, techniques satisfactory for one kind of thermoplastic plastic are standard for other kinds. Some notable differences will be indicated.

Design possibilities, as can be seen by the photographs, are vast and range from jewelry to furniture, lamps, eyeglasses, toys, building components, architectural models, boxes, bowls, et cetera. The plentiful use of these plastics can be attributed to simplicity of fabrication, economy, design flexibility and potential range of possibility, light weight, breakage resistance, attractiveness of color and texture, and the possible use of transparency and optical properties.

GENERAL FABRICATION CONSIDERATIONS

Although fabrication techniques for plastics are similar to those employed with metals and wood, there are important differences. All thermoplastics can be machined. Standard woodworking and metalworking equipment and tools can be utilized, occasionally with some slight modifications and with different techniques to accommodate for different characteristics. The most outstanding of these characteristics is the generation of frictional heat. Heat generated by surface friction builds up during machining and causes a variety of problems. If heat is allowed to increase, the surface of the plastic will expand, increase friction, and result in poor tolerance, finish, discoloration, sticking of particles to the surface. Most of the heat is transmitted to the tool. For these reasons temperature must be controlled by use of air, coolants, and by varying the machine speed and length of time a tool is used.

DATA CHART FOR ACRYLIC

Appearance: Excellent clarity, transparency, pipes light; will transmit over 90% of daylight; transparent in slabs over three feet thick (glass is almost opaque at thicknesses of six inches); does not cut off ultraviolet rays; texture is warm and pleasant to touch

Physical Properties: Odorless (in final form), tasteless, nontoxic, strong and rigid, lightweight (one half as heavy as glass); good resistance to sharp blows; surface scratches easily seen because scratches catch the light; easily machined, molded, and polished

Moisture: Negligible absorption; dimensional stability and electrical properties remain good under humid conditions

Heat: Deforms at 150°F to 340°F; cannot be steam-sterilized, decomposes on heating and burns slowly but does not flash-ignite (burning rate for ¼″ thickness 1.1 inches per minute); does not become brittle at low temperatures

Solvents and Chemicals: Resistant to weak acids and alkalis, e.g., ammonia, but can be attacked by strong acids, alkalis, oxidating agents; resistant also to salt water; dissolves by aromatic solvents, e.g., turpentine, benzine, toluene; chlorinated hydrocarbons, e.g., methylene chloride, carbon tetrachloride; ketones, esters; some organic acids, such as acetic acid. Perfumes, gasoline, cleaning fluids, acetone, chloroform attack acrylics

Light and Environments: Actinic radiation—sunlight, ultraviolet light, bleaching, odor change, chalking, cracking affect it very little; weathering does not appreciably affect transparency, dimensional stability, or electrical properties; inert to fungus, bacteria, animal, and rodent attack; decomposes in ionizing radiation and high vacuum

Shrinkage: Generally upon heating to forming temperature and cooled unrestrained to room temperature, acrylic shrinks slightly over 2% in length and width, and over 4% in thickness

Available Forms: Solutions and emulsions for surface coatings, paints, adhesives

Sheet: Various thicknesses from clear (and in colors), translucent, opaque, patterned, and corrugated surfaces

Rods and Tubes: A range of thicknesses and diameters

Molding and Casting Compounds: Range of clear to colored grades for compression and injection molding, extrusion, and casting

Plastics also are relatively resilient when compared to metals and wood; therefore machining procedures should incorporate proper support of the form to minimize distortion. Elastic recovery occurs both during and after machining, so provisions must be made in the design of both the tool and the shape to provide for clearance. It is important to keep tools sharp. High-speed or carbon-tipped tools are efficient and economical in the long run.

Basic Equipment and Supplies

Hand-powered tools can be used, but power tools make the job much easier. A basic workshop should contain most common woodworking and metalworking hand tools. Hammer, handsaw, screw drivers, level, files, pliers, square, ruler, router, sabre saw, band saw, radial arm or circular saw, drill or drill press, buffing wheels, sanding wheels, vacuum cleaner—these are basic. Special forming techniques can include strip heater, heat gun, circulated hot air oven, hydrogen torch for flame polishing, vacuum forming machine for molding. Bonding requires syringes, C-clamps. Supplies range from sandpapers and polishing compounds to adhesives, tapes, flannel cloth, polishes, cleaners, and coolants.

CUTTING

Saw blades designed for brass, copper, or aluminum should be used. They must be kept sharp and free from nicks and burrs. The piece to be cut also has to be held firmly to prevent chattering and chipping. Chipped edges are sensitive and cracks often start at a chipped edge. *Use a fast tool speed and a slow feed.* A slow speed will reduce the necessity for a coolant. Compressed air or an atomized spray of water-soluble coolant will prevent overheating and improve the finish of the machined surface if heat cannot be minimized.

Circular Sawing

For average use, in a table circular saw, hollow-ground high-speed blades are recommended.

All teeth should have uniform height, be the same shape and uniform "rake." Circular saw blades should be just slightly higher (¼″)

ACRYLIC THICKNESS TO BE CUT	BLADE THICKNESS	TEETH PER INCH	TYPE OF BLADE
1⁄16″, 1⁄8″, 3⁄16″	1⁄16″ or 3⁄32″	6–14	Hollow Ground
1⁄4″, 3⁄8″, 1⁄2″	3⁄32″	4–6	Hollow Ground
5⁄8″ and over	1⁄8″	3–4	Hollow Ground or Spring Set
1″–4″	1⁄8″–5⁄32″	3–3½	Spring Set or Swaged

than the thickness of the acrylic and held firmly to prevent chipping. Carbide-tipped blades do not have a tendency to bind when thick sheets are cut; therefore feeding of the material can be faster. Circular saws should be operated at approximately 3,400–4,000 r.m.p. (the normal speed of the woodworking table saw). Minimum motor size recommended is one H.P. for cutting sheets under ¼″ thick; 1½ to 2 H.P. for thicker sheets. Binding and twisting of the blade can be caused by an uneven feed, an improperly set saw fence, masking paper adhesive.

When cutting unmasked sheets, place felt on working surface and keep it clean of chips and dirt. Keep blades clean by wiping with mineral spirits, naphtha, or alcohol. For cooling the blade, if you use Johnson's TL-131, dilute it 40 to 1.

Traveling saws, such as radial arm saws, are excellent because the sheet remains stationary, does not slide, and does not get scratched.

Use of circular table saw for cutting acrylic. (*Courtesy, Rohm & Haas Company.*)

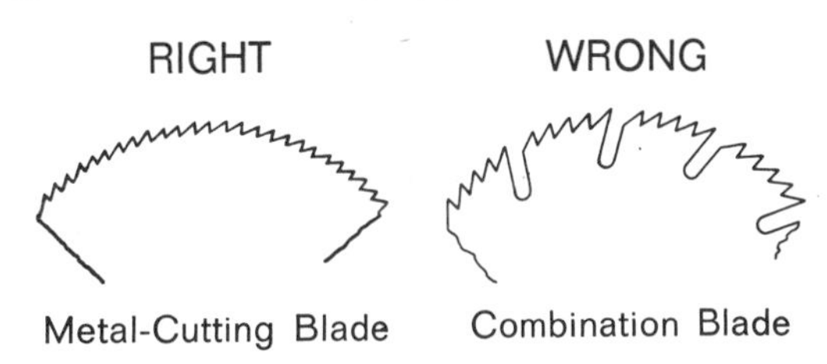

circular saw (table saw)

(*Courtesy, American Cyanamid Co.*)

Band Saws

Band saws are used when flat forms are to be cut into irregular shapes such as curves. With the aid of a guide bar, straight cutting operations can be accomplished. Only metal cutting blades should be used. The upper guide should be lowered so that it is no more than one inch away from the material. The recommended speed is 2,500–4,000 surface feet per minute, depending on the thickness of the material (a 14″ band saw is usually designed to operate at approximately 3,000 feet per minute if a standard 1,725 r.p.m. motor is used).

ACRYLIC THICKNESS TO BE CUT	MINIMUM BLADE WIDTH	TEETH PER INCH	SAW PITCH
⅛″	3⁄16″	18	Between 24 and 10
¼″	3⁄16″	14	10
½″	¼″	10	6B[1]
1″	⅜″	8	5B[1]

[1] See "Sources of Supply."

Use of a band saw to cut acrylic. (*Courtesy, Rohm & Haas Company*)

SABRE SAW

Distance "A" between guide board and line of cut is equal to distance between blade and edge of saw base.

HAND SAW

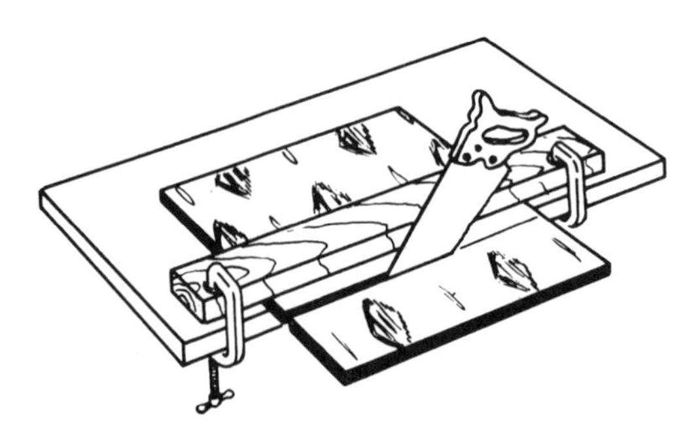

Cut close to edge of bench

(*Courtesy, American Cyanamid Co.*)

Jig Saws and Sabre Saws

Jig saws are used for cutting closed holes, inside cuts. Because the stroke is short and the blade does not get a chance to clear the chips, the blade heats up quickly and tends to soften and fuse the plastic. Fine-toothed blades with about fourteen teeth to the inch should be used. Work should be supported to minimize vibration with both the jig saw and the sabre saw.

Handsaw and Scriber

A crosscut saw can be used if work is clamped to the workbench to minimize flexing. A specially designed scriber can be used to score acrylic. A scratch repeated many times will partially cut into a piece and allow it to snap into two when pressure is exerted. To break, place a wooden dowel under the length of the intended break, hold the sheet firmly with the palm of one hand, and apply downward pressure on the short side of the break with the other hand.

Routing and Shaping

Stationary or hand routers can be used for slotting, beveling, rabbeting, rounding edges, and for trimming. Two- or three-fluted endmills with straight or spiral cutting edges will work better than wood router bits. The speed is usually 10,000 r.p.m. on stationary cutters (feed the piece into the tool). Shapers using veneer blades can be used to trim formed pieces.

Drilling

Standard twist drills can be modified to prevent grabbing by creating a blunt cutting angle (see illustration) and grinding small flats onto both cutting edges. Specially modified drills can now be purchased. Smaller drills are run at faster speeds. Holes larger than one inch in diameter can be cut with a hole-saw. Feed should be slow but steady. Coolants can be used for deep holes, as this gives a better finish.

DIAMETER OF HOLE	SPEED IN R.P.M.
⅛″	4,000
3⁄16″	3,500
¼″	2,000
⅜″	1,500
½″	1,000
⅝″	900

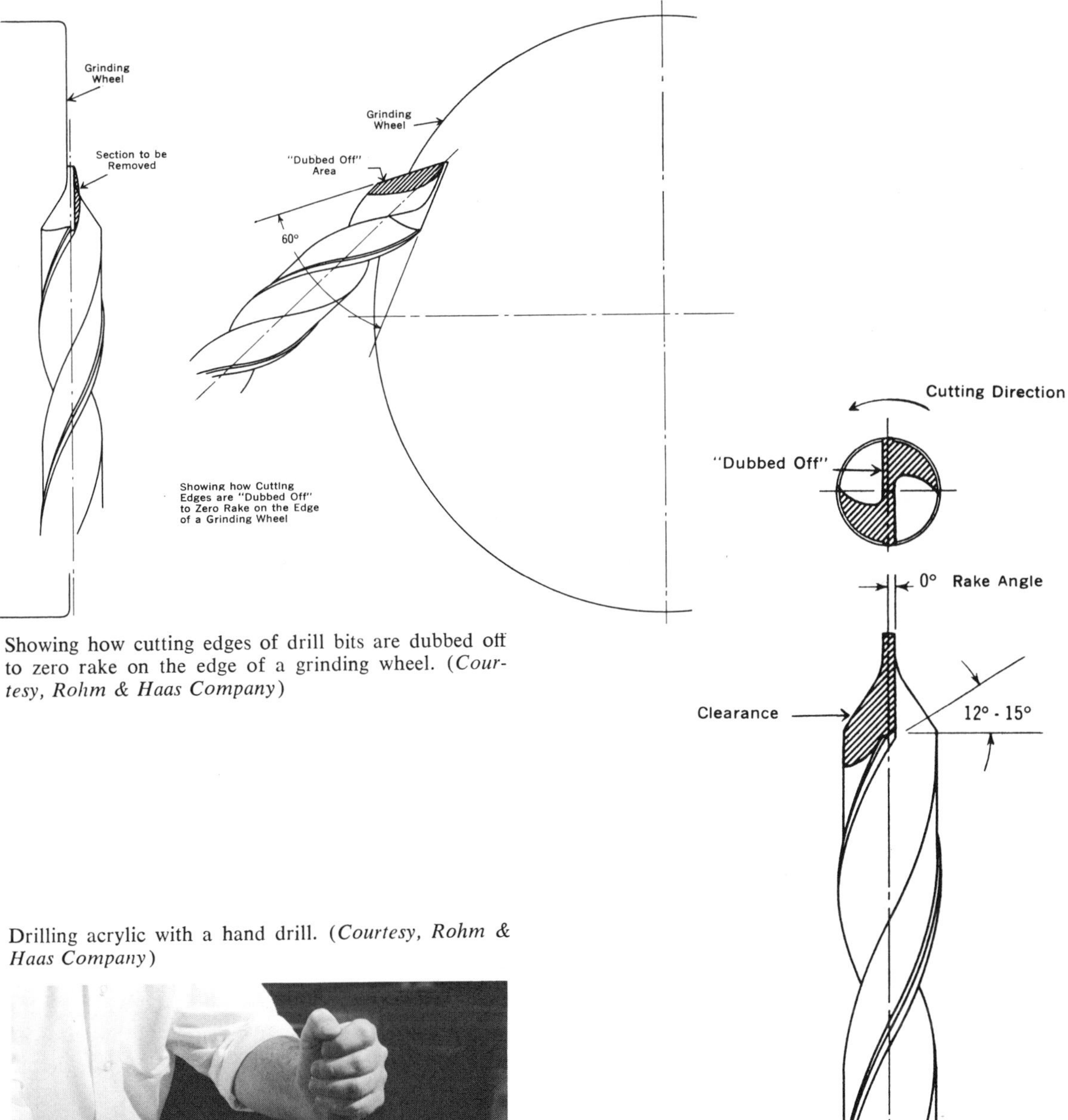

Showing how cutting edges of drill bits are dubbed off to zero rake on the edge of a grinding wheel. (*Courtesy, Rohm & Haas Company*)

Drilling acrylic with a hand drill. (*Courtesy, Rohm & Haas Company*)

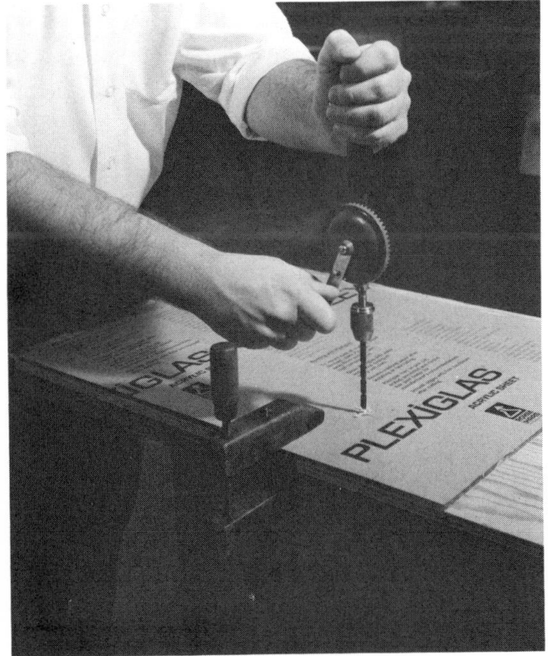

HAND ROUTER

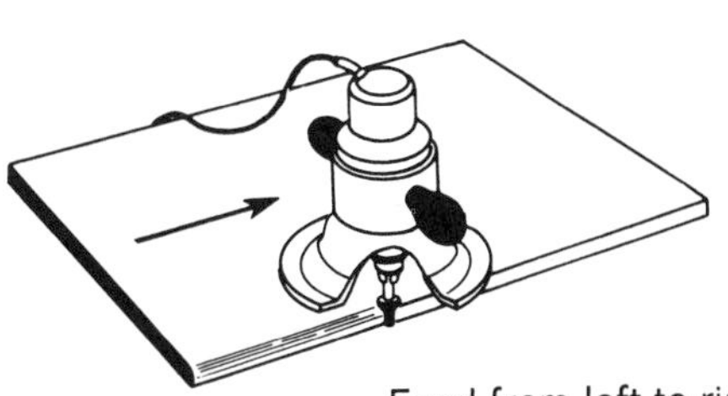

(*Courtesy, American Cyanamid Co.*)

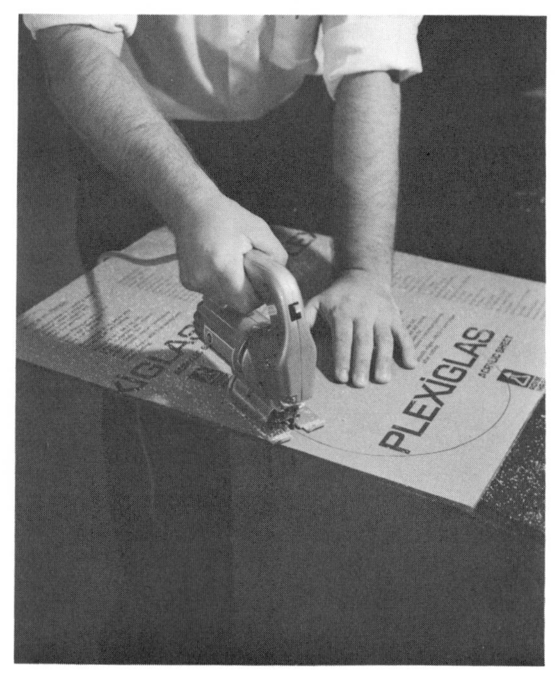

Cutting acrylic sheet with a sabre saw. (*Courtesy, Rohm & Haas Co.*)

TURNING

Acrylic can be turned on a lathe, to give a semimatte surface, at speeds of 500 feet per minute with very slow feeds. Disks can be made from square stock on a lathe by turning.

REPAIRING CRACKS

To stop a crack, drill a small hole at the end of the crack as soon as possible. Cementing a small piece over the hole will reinforce the area, or feed solvent cement into the crack and apply a little pressure for a few minutes until the solvent evaporates.

BONDING AND FASTENING

SCREWS AND TAPPING

Wood screws or self-tapping sheet-metal screws should never be forced into acrylic because internal fractures will occur. Machine screws with coarse threads are safer. A soap solution or wax dressed on the screw will keep the threads from burning. Parker-Kalon Type "F" Tapping Screws [2] and Shakeproof type 23 and type 25 thread-cutting screws [2] can be used. Handmade acrylic screws are excellent because they expand and contract at the same ratio as acrylic sheet.

[2] See "Sources of Supply."

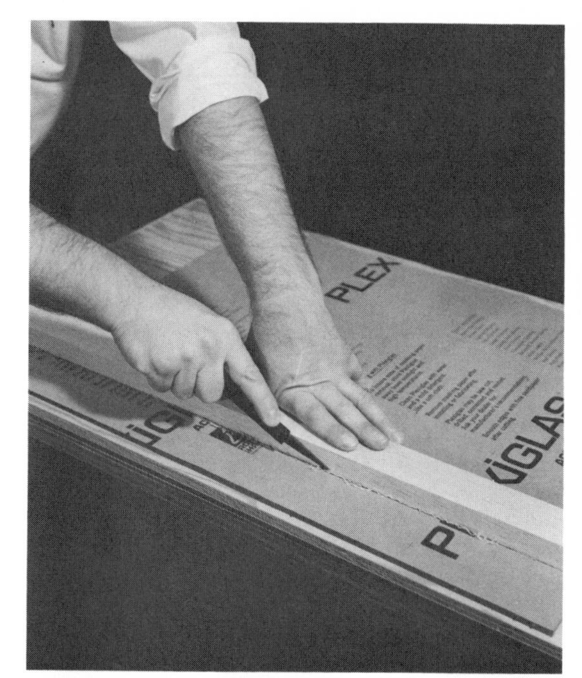

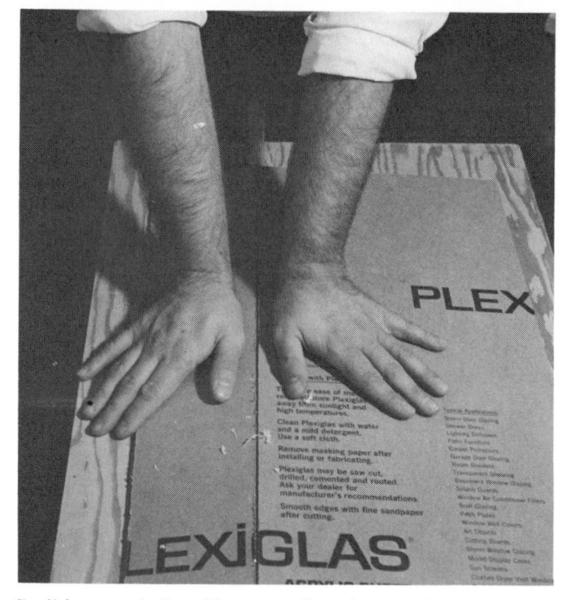

Scribing and breaking acrylic sheet with a special acrylic cutting tool. (*Courtesy, Rohm & Haas Company*)

TRIMMING FORMED PIECES

Formed ACRYLIC
Veneer Saw Blade
Shaper Table
Shaper Motor

(*Courtesy, American Cyanamid Co.*)

STOPPING CRACK

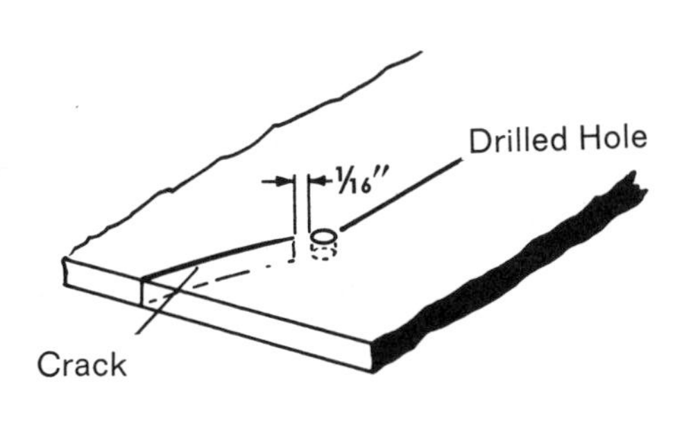

Drill small hole to stop crack

MACHINING DIFFICULTIES AND CAUSES—CHECK LIST

DRILLING

Difficulty	Common Causes
Tapered hole	1. Incorrectly sharpened drill 2. Insufficient clearance 3. Feed too heavy
Burnt or melted surface	1. Wrong type drill 2. Incorrectly sharpened drill 3. Feed too light 4. Drill dull
Chipping of surfaces	1. Feed too heavy 2. Clearance too great 3. Too much rake
Chatter	1. Too much clearance 2. Feed too light 3. Drill overhang too great 4. Too much rake (thin web, as described)
Feed marks or spiral lines on inside diameter	1. Feed too heavy 2. Drill not centered 3. Drill ground off-center
Oversize holes	1. Drill ground off-center 2. Insufficient clearance 3. Feed rate too heavy 4. Point angle too great
Undersize holes	1. Drill dull 2. Too much clearance 3. Point angle too small
Holes not concentric	1. Feed too heavy 2. Spindle speed too slow 3. Drill enters next piece too far 4. Cut-off tool leaves nib, which deflects drill 5. Drill speed too heavy at the start 6. Drill not mounted on center 7. Drill not sharpened correctly
Burr at cut-off	1. Cut-off tool dull 2. Drill does not pass completely through piece
Rapid dulling of drill	1. Feed too light 2. Spindle speed too fast 3. Insufficient lubrication from coolant

CUTTING OFF

Difficulty	Common Causes
Melted surface	1. Tool dull 2. Insufficient side clearance 3. Insufficient coolant supply
Rough finish	1. Feed too heavy 2. Tool improperly sharpened 3. Cutting edge not honed
Spiral marks	1. Tool rubs during its retreat 2. Burr on point of tool
Concave or convex surfaces	1. Point angle too great 2. Tool not perpendicular to spindle 3. Tool deflecting (use negative rake) 4. Feed too heavy 5. Tool mounted above or below center
Nibs or burrs at cut-off point	1. Point angle not great enough 2. Tool dull or not honed 3. Feed too heavy

TURNING & BORING

Difficulty	Common Causes
Melted surface	1. Tool dull or heel rubbing 2. Insufficient side clearance 3. Feed rate too slow 4. Spindle speed too fast
Rough finish	1. Feed too heavy 2. Incorrect clearance angles 3. Sharp point on tool (slight nose radius required) 4. Tool not mounted on center
Burrs at edge of cut	1. No chamfer provided at sharp corners 2. Tool dull 3. Insufficient side clearance 4. Lead angle not provided on tool (tool should ease out of cut gradually, not suddenly)
Cracking or chipping of corners	1. Too much positive rake on tool (use negative rake) 2. Tool not eased into cut (tool suddenly hits work) 3. Dull tool 4. Tool mounted below center 5. Sharp point on tool (slight nose radius required)

Difficulty	Common Causes
Chatter	1. Too much nose radius on tool 2. Tool not mounted solidly enough 3. Material not supported properly 4. Width of cut too wide (use two cuts)

Chart adapted. Courtesy the Polymer Corporation, Polypenco Division

STEPS IN MAKING A BODY ORNAMENT

Carolyn Kriegman plans design for an acrylic body ornament.

A table-mounted sabre saw is used to cut basic shapes.

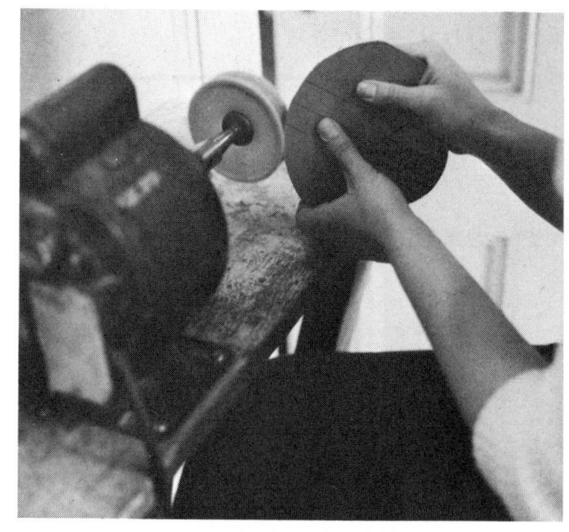

The shape is being buffed after sanding.

Piece is being heated in a Pyrex dish in a kitchen oven.

Carolyn Kriegman bends the heated part into shape.

Findings attach parts to one another.

CEMENTING

Cementing is relatively easy, but practice is required in order to avoid problems such as crazing and poor joint strength. Good transparent joints can be obtained.

If stresses were created by sawing, then edges have to be dressed to prevent crazing when the acrylic comes into contact with the cement. If a formed part has been subject to too much stress during forming, cutting, or drilling, it should be annealed.

Annealing is accomplished by placing the acrylic form, after all fabrication is completed, in a heat zone (oven) of 120°F to 175°F. for five to seven hours, and gradually cooling. Parts to be annealed must be clean, dry, and free of masking and coatings, and should be supported. The oven should have forced air circulation. Formed parts may deform at higher temperatures; therefore lower temperatures and a longer annealing time may be recommended, depending also on the size and thickness of the form.

All surfaces to be joined should fit accurately. If the joint is part of the original sheet surface, or if it is a smooth cut, it should be left as is and not be sanded or polished. Do not polish edges that are going to be cemented; this could produce a convex edge that will not bond well. Cementing should be done at room temperature, low in humidity.

Solvent cementing happens when the two surfaces to be joined are attacked and softened by cement or solvent to create a cushion. Only light pressure is needed to intermingle both cushions and to force out trapped air. This occurs within 20–30 seconds after the two pieces have been

An assortment of body ornaments by Carolyn Kriegman.

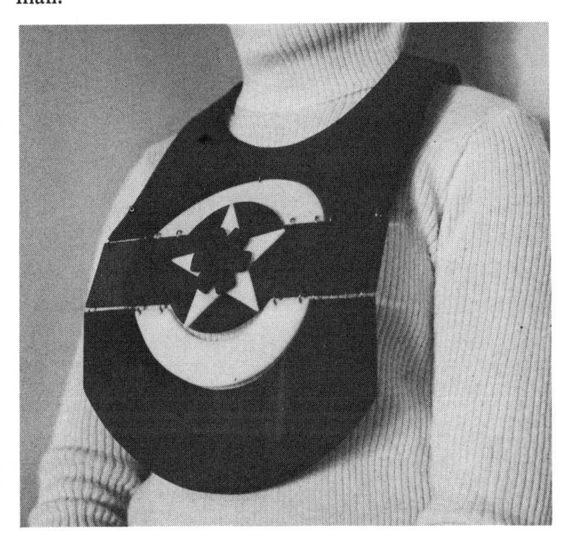

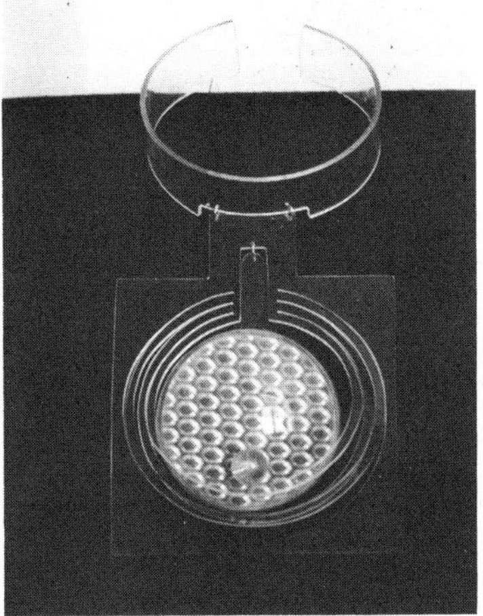

"Flowering" is caused by stress cracking in the drilled forms to the right and left. After the holes were filled with polyester resin, heating during curing accelerated the stress cracking, originally caused by too much heat during drilling. Annealing would have prevented it. The center form was drilled by using a coolant.

brought together. It is most important that no pressure be applied during this 20–30 second period. Ethylene dichloride and acetone are not recommended for cementing acrylic.

Capillary cementing by using a solvent applicator. (*Courtesy, Rohm & Haas Company*)

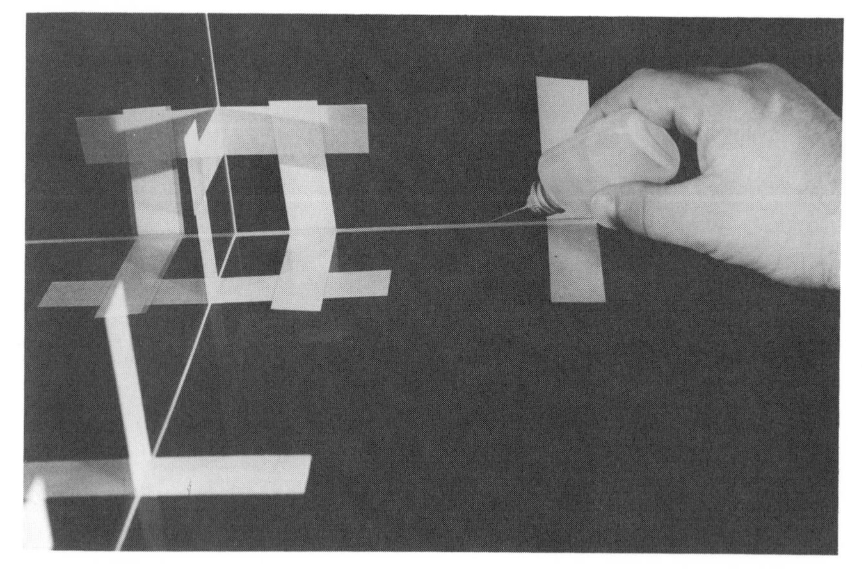

A caution: do not allow the cements to touch your skin. They may cause dermatitis. If cement comes in contact with your skin, wash it off immediately with soap and water. Clothing coming into contact with cement should be removed and washed before it is worn again. Overexposure to

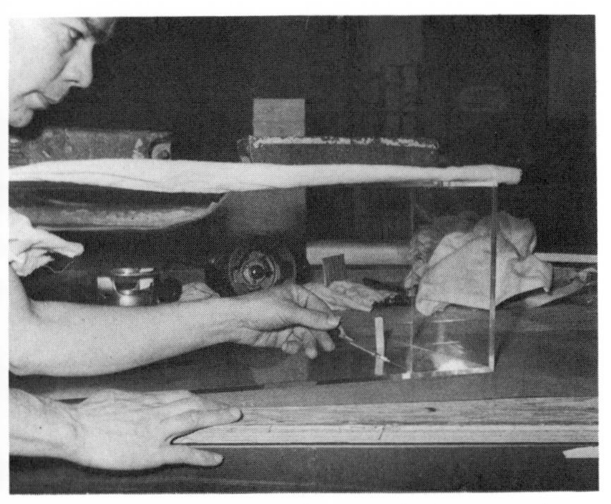

Capillary cementing at Just Plastics, Inc., using a syringe. Note flannel cover and weight on top of form,.

cement vapors can cause drowsiness, dizziness, and nausea. Adequate ventilation is essential. Do not smoke; these cements are volatile.

There are three cementing methods: dip or soak cementing, capillary cementing, and mortar joint cementing with viscous cements. Cement types are solvents and polymerizable cements. Unthickened cements are used indoors only; thickened cements may be used out-of-doors.

Dip or Soak Cementing

Start with a shallow aluminum, stainless-steel, or glass tray. Line up short pieces of wire or brads into the tray to keep the edge of the piece from touching the tray. Level the tray before pouring the solvent cement just to cover the brads or wires evenly. Stand the acrylic part on the brads upright (with supports) for 1–5 minutes. When the form is removed, stand it at a slight angle to allow excess cement to drain off. Quickly put soaked edge exactly in place on the other part. A steel square can be used for support. Hold parts together for 30 seconds without applying any pressure. After 30 seconds apply pressure gently to press out air bubbles. Take care, though, not to force out too much cement because, if you do, dry spots will form in the joint. A small bag with B.B. shot or small washers can be placed on the part to maintain firm contact for 10–30 minutes. The first 5 minutes are most critical. If the vertical part is allowed to slip, the bond will become a weak and unsightly one, even if it appears to stick. No machining or polishing should be done from 8–24 hours, until the joint fully cures. Cementing a multipart form, such as a box, requires a jig or another person to help.

CAPILLARY CEMENTING

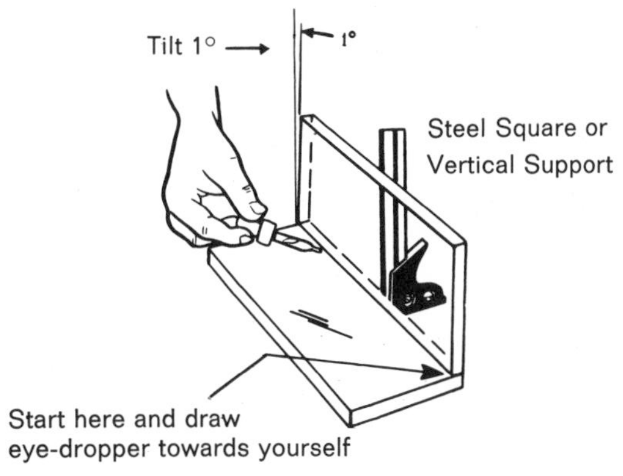

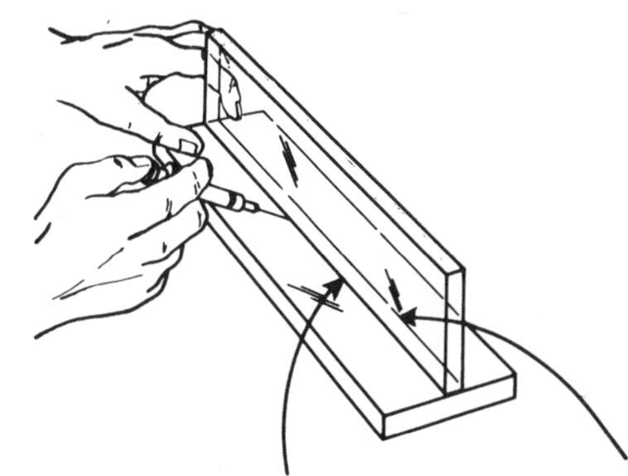

(Courtesy, American Cyanamid Co.)

SOAK CEMENTING

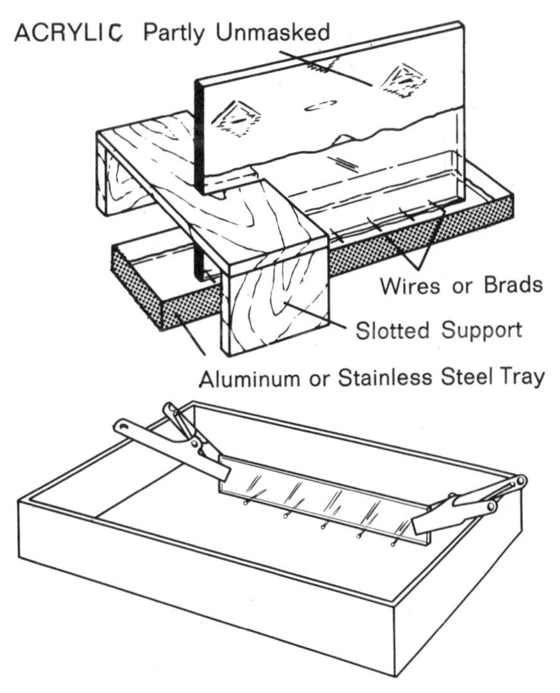

(Courtesy, Rohm & Haas Company and American Cyanamid Co.)

MAKING AN ACRYLIC MIRROR FORM

a. Parts are cut, assembled at the working area, and then cleaned.

b. Double-stick cloth tape is adhered to the mirror back.

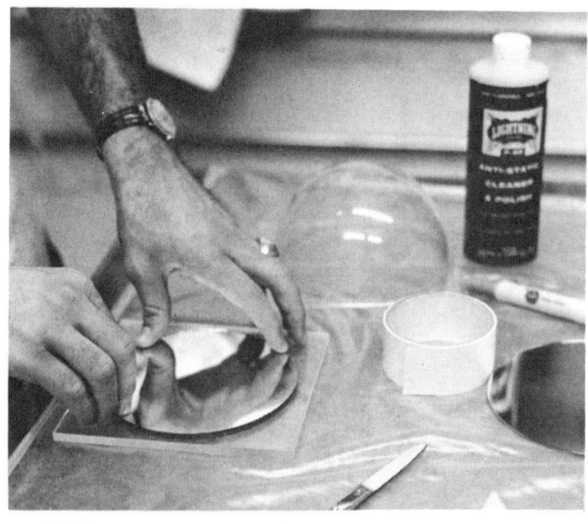

c. The mirror is set into place on both sides of the inside acrylic square.

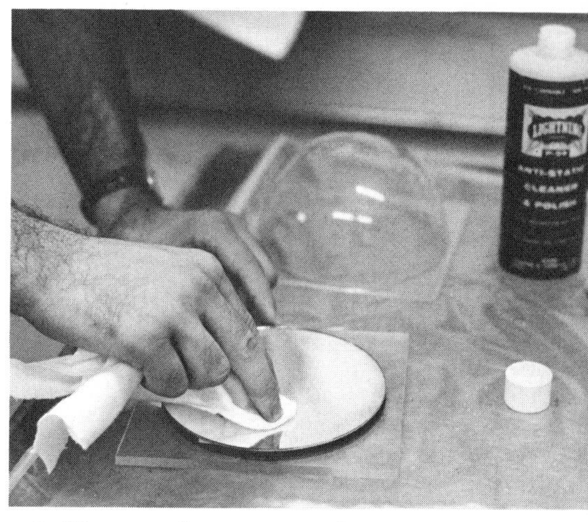

d. Finger marks are removed.

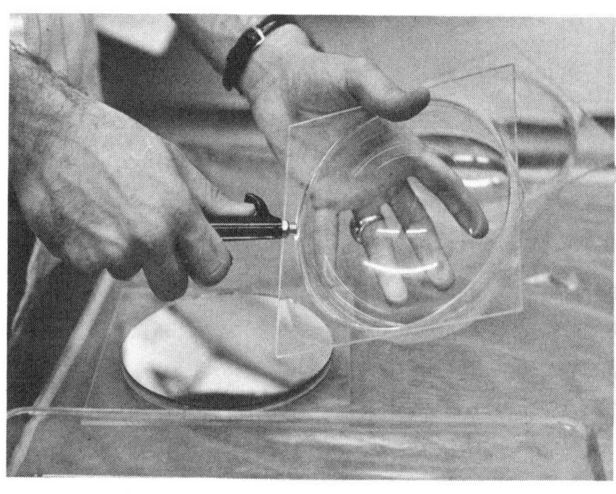

e. Dust is blown out of all the forms.

f. Blow molded forms are soak-cemented.

g. Then they are clamped in place with slight presure.

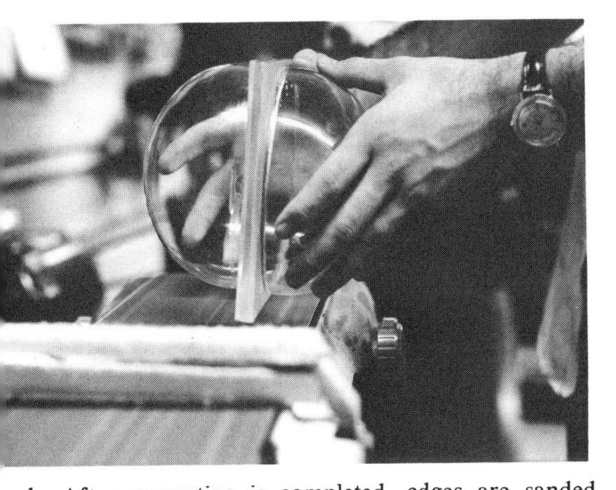

h. After cementing is completed, edges are sanded on a belt sander.

i. Then edges are scraped with a steel wood scraper.

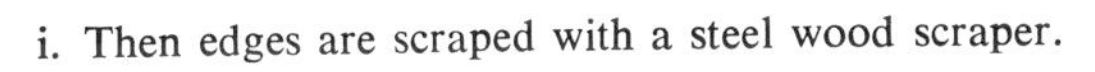

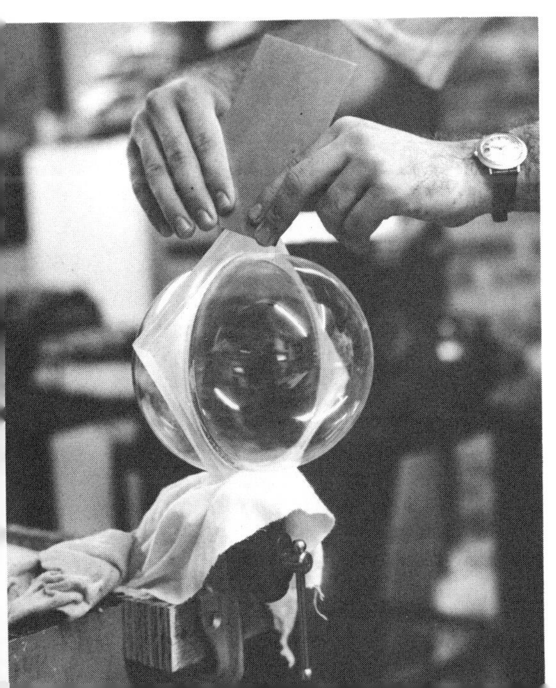

j. Finishing is completed by sanding with a fine-grit wet-or-dry sandpaper until a satin-smooth finish is obtained.

k&l. Two views of the completed mirror.

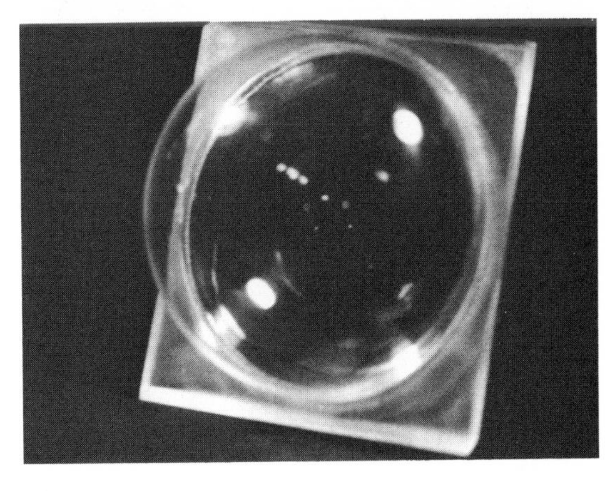

Capillary Method

Capillary action, the ability of solvent-type cements to spread between the crack of a joint area, can be used only if parts fit perfectly. To accomplish bonding with capillary cementing, set the parts in place and tilt the vertical part about one degree toward the outside so that the cement will flow. Then with a brush, eyedropper, or syringe draw a bead of cement along the inside and then repeat for the outside edge of the joint.

Mortar Joint Cementing with Viscous Cements

A viscous-type cement can be made by dissolving small clean chips of acrylic in the solvent and letting the solution stand overnight. Brush the cement on the edge of one piece and join the pieces, as in soak cementing.

PS18 and PS30 is best applied with a syringe. Since these cements shrink, excess cement must be applied. The cement must be mixed thoroughly in the correct order and proportions at room temperature, according to the manufacturer's instructions. Care should be taken not to mix air bubbles into the cement. After it is mixed, the pot life is only about 25 minutes; therefore the cement should be used quickly.

In using PS18 and PS30, close-fitting joints are not necessary. These cements fill gaps, but

Working with PS 30

a. PS 30 is measured into a clean jar. Forms to be glued are masked together.

b. After catalyst has been added, PS 30 is carefully mixed so that air bubbles do not become trapped in the viscous cement.

c. With a syringe, PS 30 is squeezed into the joints.

d. After curing, excess cement is routed off the form.

e. Wet sanding precedes buffing.

f. The completed form is wiped with a flannel cloth.

when polymerized they shrink and it is then necessary to refill the gap. Dams to hold the cement may be formed with masking tape, but care should be taken not to allow contact of the cement with the adhesive backing. Excess cement may be machined off after the cement has set.

Other Types

Epoxies may be used and applied in the technique used for viscous cements.

Applying of the cyanoacrylates requires special handling. First, both surfaces to be bonded must be thoroughly cleaned. (Most solvents such as acetone will mar the surface; use aliphatic naphtha.) Next, with a clean cloth or tissue apply a very spare amount of the surface activator (optional) to *one* of the surfaces. Allow a few seconds for the activator to evaporate. Then, to the other surface, sparingly apply one drop of the adhesive per square inch. (Do not brush or roll the adhesive onto the surface, because even slight pressure on the adhesive prior to bonding may initiate premature polymerization.) Immediately place the second surface in contact with the adhesive-coated surface and quickly rub the two surfaces together once or twice with light pressure to spread the adhesive and make the final alignment. Apply light manual pressure until the bond has begun to set. It takes two minutes for metals and plastics and 30 seconds for rubber surfaces. These adhesives are dated and should be stored in a refrigerator If you find a powdery residue clouding the bond, wash it away if possible. It was caused by the vapors escaping during polymerization. If curing can be done in a cold, dry room, and bonding held off as long as possible, then vapors will be less likely to collect. Almost any material can be bonded with the cyanoacrylates and the polyacrylates, except polyethylene and tetrafluoroethylene (TFE), paper and leather.

Other Methods

For opaque thermoplastic sheet—nylon, polycarbonates, polyethylene, acrylic, acetals, polysulphones, ABS, and PVC—electromagnetic bonding[3] can be used. Basically electromagnetic adhesive is a mixture of an electromagnetic, energy-absorbing material and a thermoplastic of the same composition as the parts to be bonded. These "adhesives" are available in ribbon or wire forms in various widths or diameters. A magnetic field is created by high-frequency induction-heating generators causing the parts to join. The strength of this kind of bond is excellent.

Ultrasonic Welding

Ultrasonic welding is a thermal bonding method that welds parts by means of frictional heat generated in the joint through mechanical vibrations of ultrasonic frequency. Ultrasonic welding equipment[4] consists of a stable power supply that converts 110-volt, 60-cycle electrical current to 20,000-cycle current, a transducer that converts this current to high-intensity me-

[3] EMA Bond, Inc., 147 Union Street, Northvale, N.J., 07647.

[4] Branson Sonic Power, 52 Miry Brook Road, Danbury, Conn., 06810.

CEMENTING WITH EASTMAN 910, A CYANOACRYLATE ADHESIVE

a. Application of surface activator to side one of a clean acrylic piece.

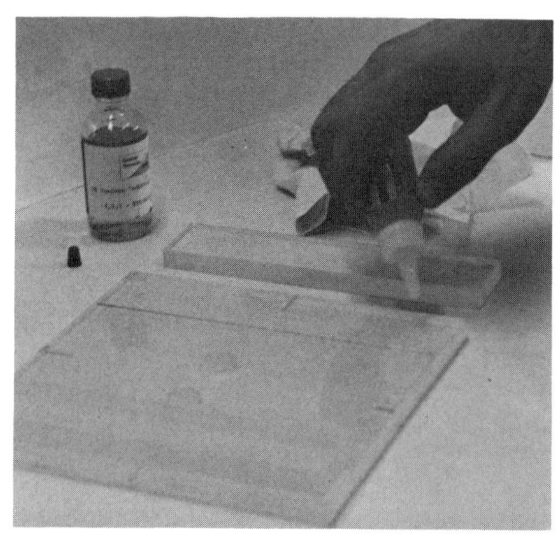

b. One drop of cement per square inch is applied to side two.

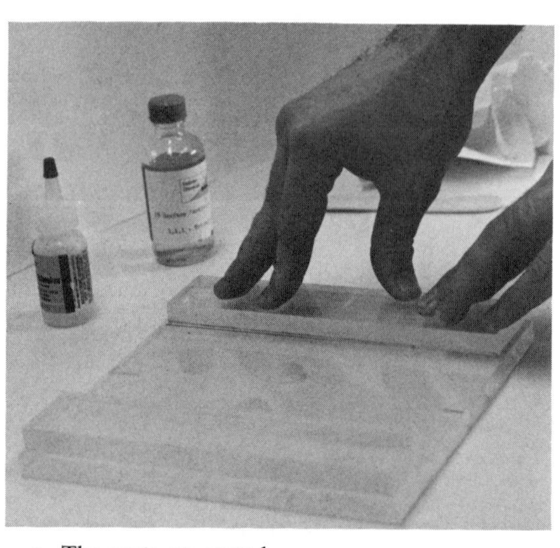

c. The parts are mated.

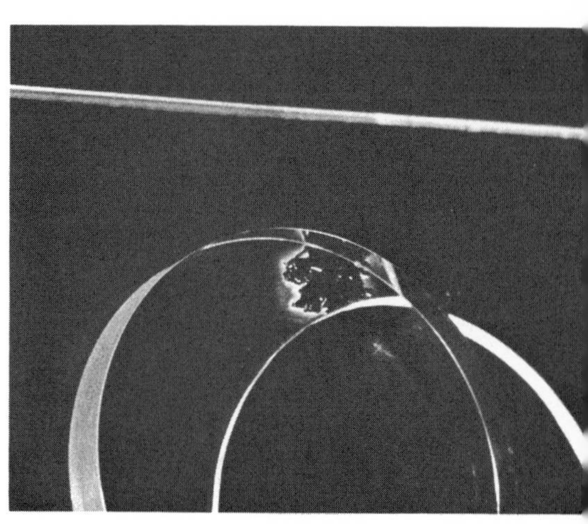

d. A powdery substance formed because of trappe vapors and too much humidity.

MAKING A SCULPTURE LAMP

a. A modified drill for acrylic is being "lubricated" with Ivory soap.

b. By using a flexible pad beneath the blow-molded dome to absorb shock, holes are formed.

c. Acrylic rod is cut and then placed into position in the holes.

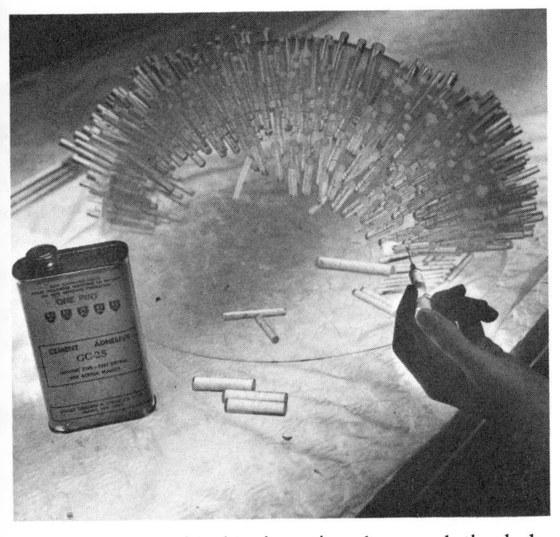

d. Methylene chloride is syringed around the holes to adhere the rods.

e. The two globe halves are taped together in preparation for cementing.

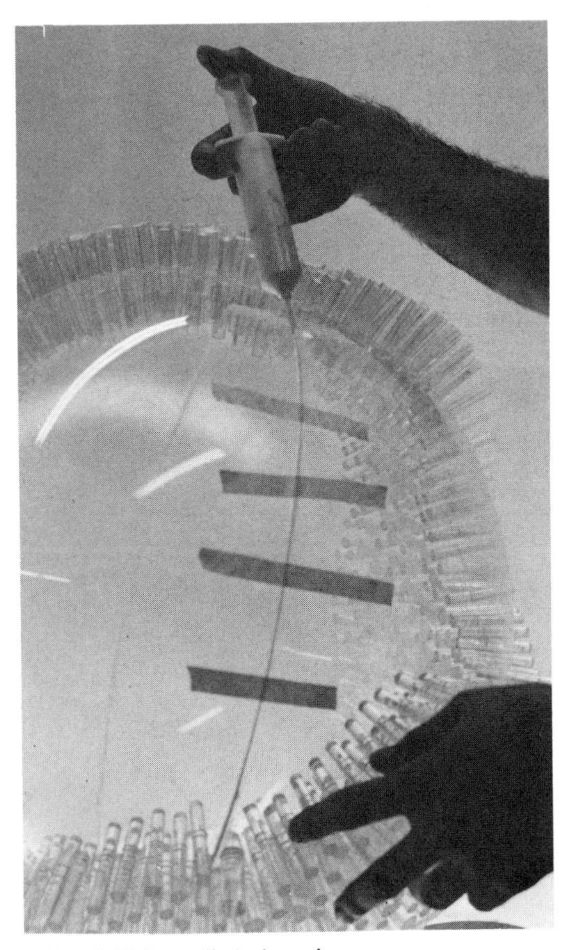

f. PS 30 is applied via syringe.

chanical vibrations of the same frequency, and a resonating metal section called a "horn" which conducts the ultrasonic mechanical energy to the plastic. Parts to be joined are held together under moderate pressure by the horn in a pneumatic holding fixture. When ultrasonic energy is applied, vibration of the part, in contact with the horn, generates intense frictional heat, and causes almost instantaneous fusing. Only the area to be bonded gets hot; the rest of the form remains cool. The entire process takes only seconds. No annealing is necessary for ultrasonic welding. Different joint designs can be used as well.

Cementing for ABS

Common solvents such as MEK, MIBK, or tetrahydrofuran can be used to cement ABS joints. These solvents can be used by themselves or bonding time can be stretched by making a 10-15% solution of ABS in the solvent.

It is best to apply the cement to the mating parts by means of a brush, sponge, knife, or gun. After the parts have been joined, hold contact in a jig for up to half an hour. MarBond [5] adhesives have been developed to cement ABS to a wide variety of substrates. They are based on epoxies and come in a variety of viscosities for brushing to putty knife and caulking gun applications.

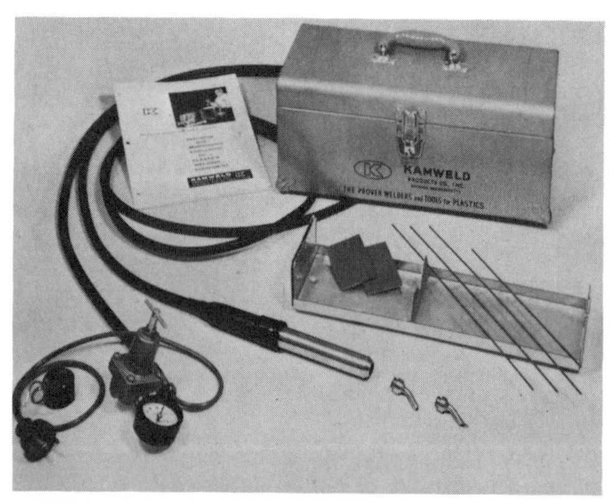

A portable welding kit for small-scale fabricating. (*Courtesy, Kamweld Products Co.*)

Cementing for Polycarbonates

Methylene chloride is generally used to cement polycarbonate to polycarbonate. Evaporation rate can be reduced, and therefore adhesion time extended, by adding up to 40% ethylene dichloride to the methylene chloride. When bonding polycarbonate to other materials such as metal, wood, rubber, glass, et cetera, use a solvent-free adhesive so that the surface of the polycarbonate does not embrittle. Surfaces should be treated by sanding lightly, followed by application of methanol wiped on with a soft cloth. Allow the areas to dry before bonding.

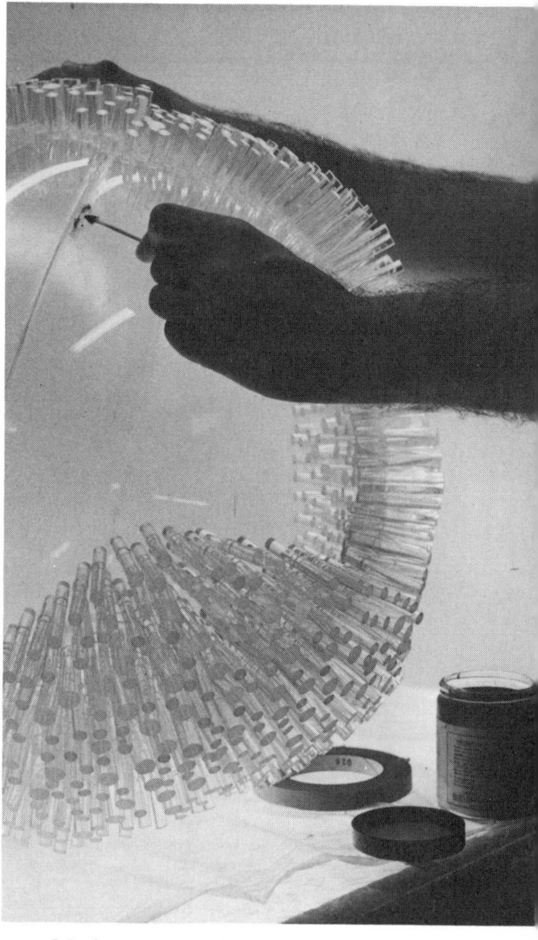

g. Marks and scratches are rubbed out with a rouge paste.

FORMING

Cold Forming

Within limits acrylic can be cold-formed at room temperature. The minimum radius of curvature must be at least 180 times the thickness of the sheet, e.g., a sheet of 1/16″ thickness can be bent to a curve of 11″ radius, 1/8″ to 22″, 1/2″ to 90″. Any tighter curves would cause stress

h. The completed sculpture lamp by Thelma R. Newman is 18″ in diameter, 26″ high. It rests on a stainless steel base containing a lamp. The acrylic rods effectively pipe light in a satin glow.

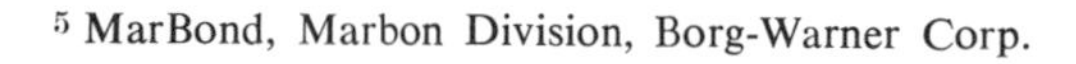

[5] MarBond, Marbon Division, Borg-Warner Corp.

crazing. Then the curved form needs to be attached by mechanical means.

Heat Forming

When heated to approximately 340°F, acrylic sheet becomes soft and pliable and then can be formed into almost any shape. Upon cooling, it will harden and retain its form. If there is an error, the form can be reheated, will return to its original shape (elastic memory), and can then be reshaped.

There are several basic methods to heat-form acrylic: strip heating for simple line bends, drape forming of two-dimensional curves, and stretch forming of complex shapes via manual methods, plug and ring forming, free blowing, vacuum forming, vacuum snap-back forming, and vacuum assist plug and ring forming, to name a few.

SIMPLE FORMING

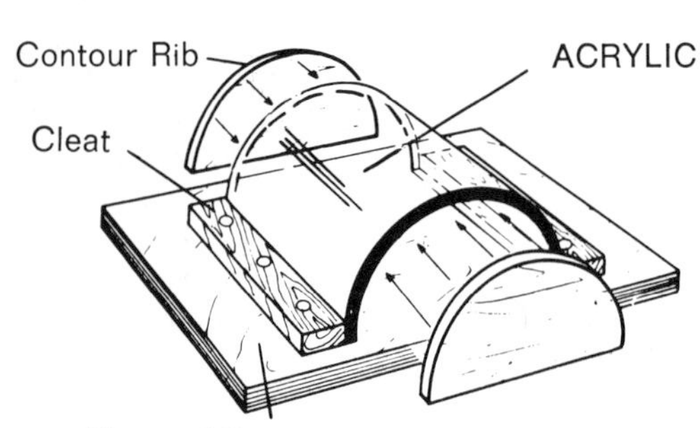

(*Courtesy, American Cyanamid Co.*)

SIMPLE STRIP HEATER

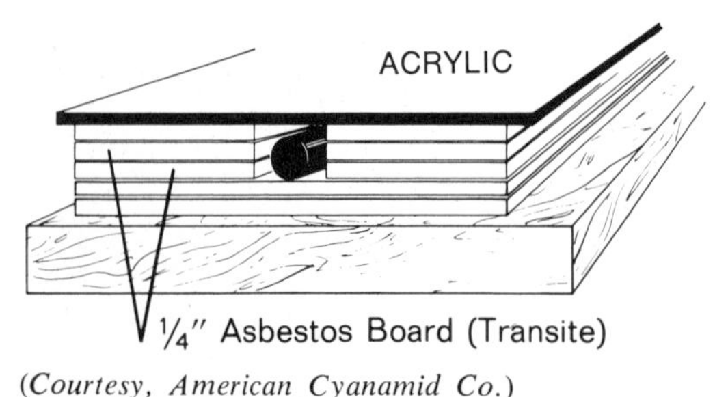

(*Courtesy, American Cyanamid Co.*)

LINE BENDS

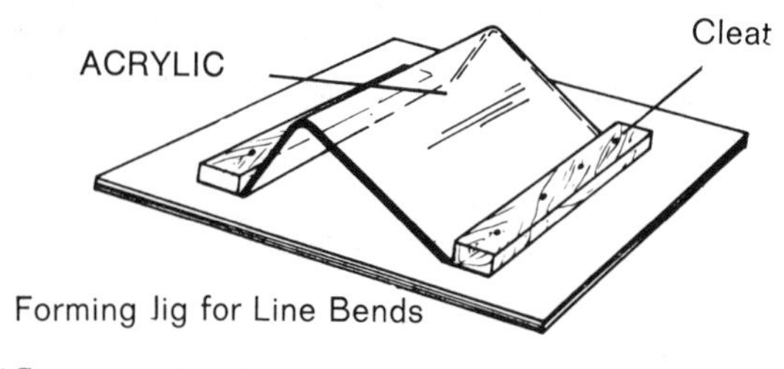

(*Courtesy, American Cyanamid Co.*)

OVEN

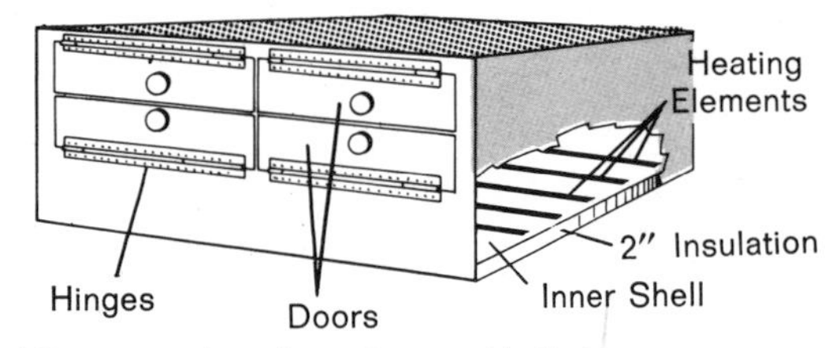

(*Courtesy, American Cyanamid Co.*)

DRAPE FORM

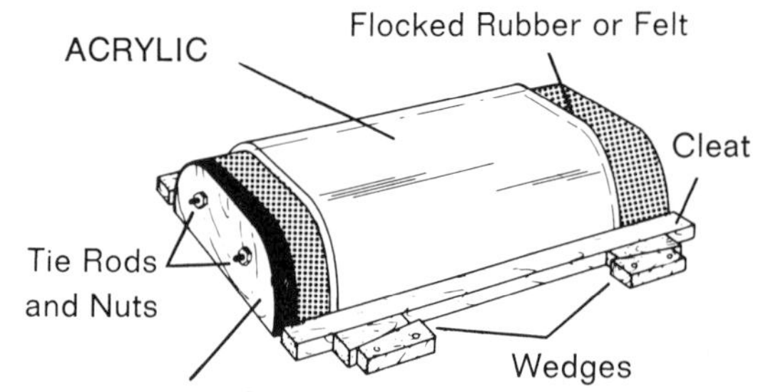

(*Courtesy, American Cyanamid Co.*)

Drape forming of a chair back and different stages in assembling. (*Courtesy, Cadillac Plastic, Louis E. Keyes*)

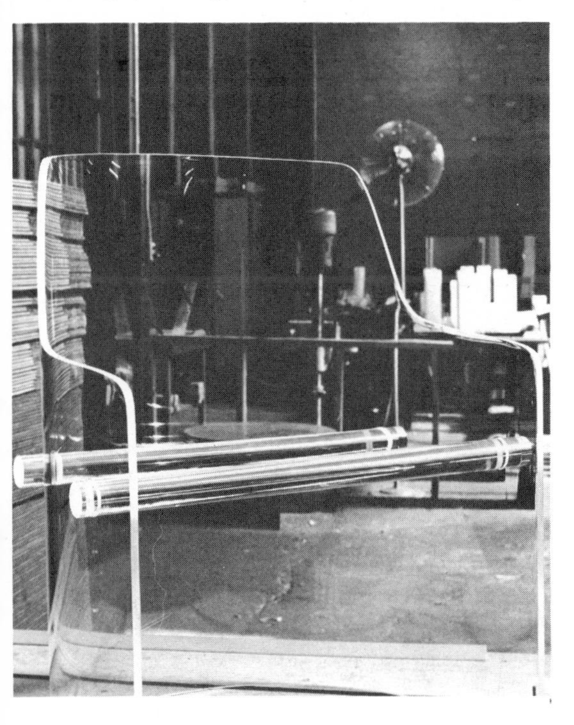

a. Flat sheet of acrylic is heated in a pizza oven on an asbestos sheet at Just Plastics, Inc.

b. It is quickly placed in the mold and the cover is closed.

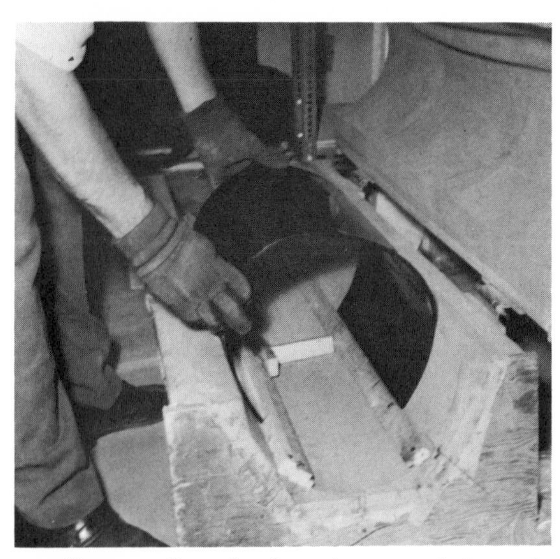

c. Upon cooling, the form is removed from the mold.

a. Wooden jig is being made for forming the acrylic rod.

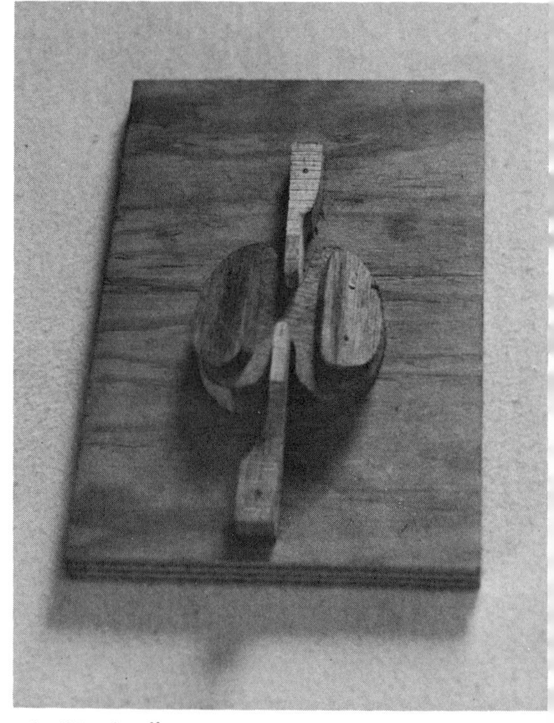

b. Wooden jig.

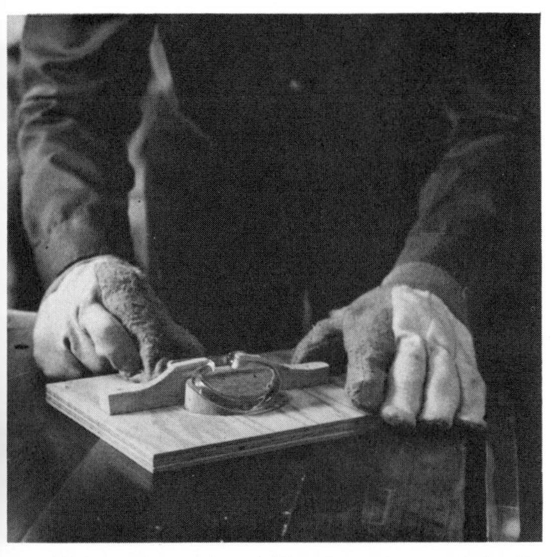

c. Hot acrylic rod is quickly wrapped around the jig.

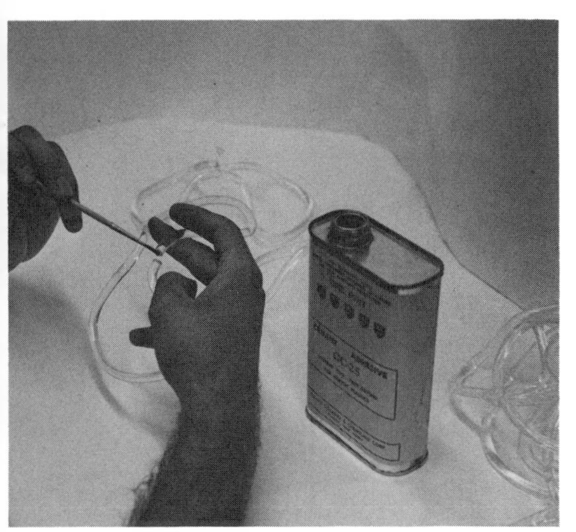

d. Ends of the rod are cemented.

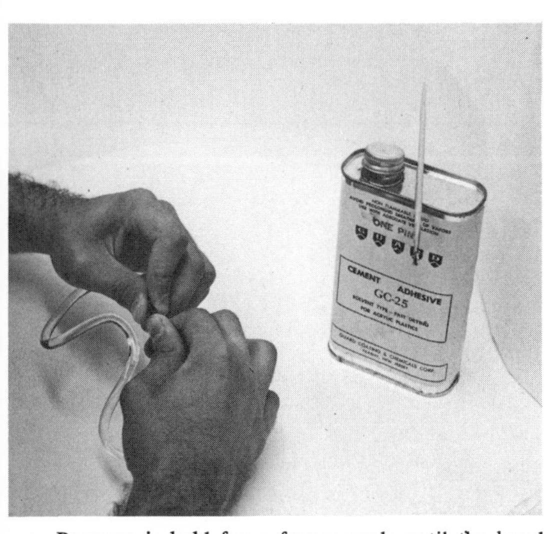

e. Pressure is held for a few seconds until the bond has taken.

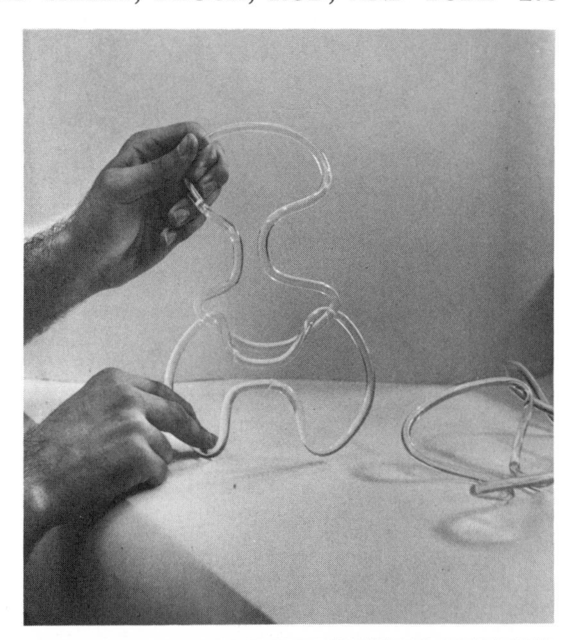

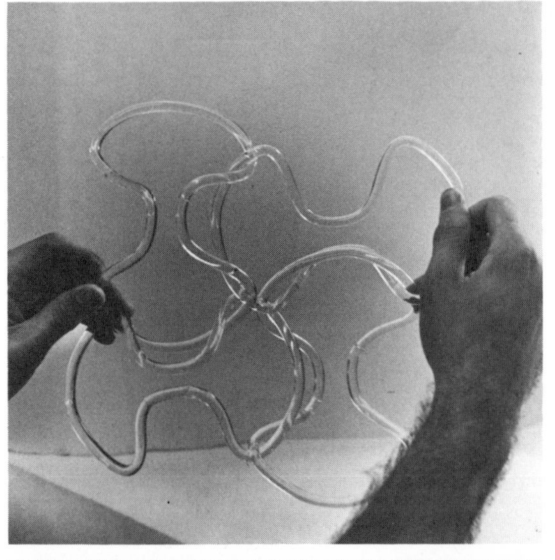

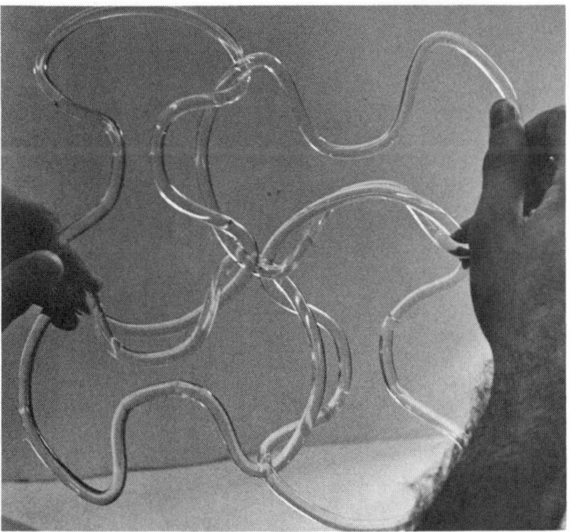

f,g,h. The parts are assembled.

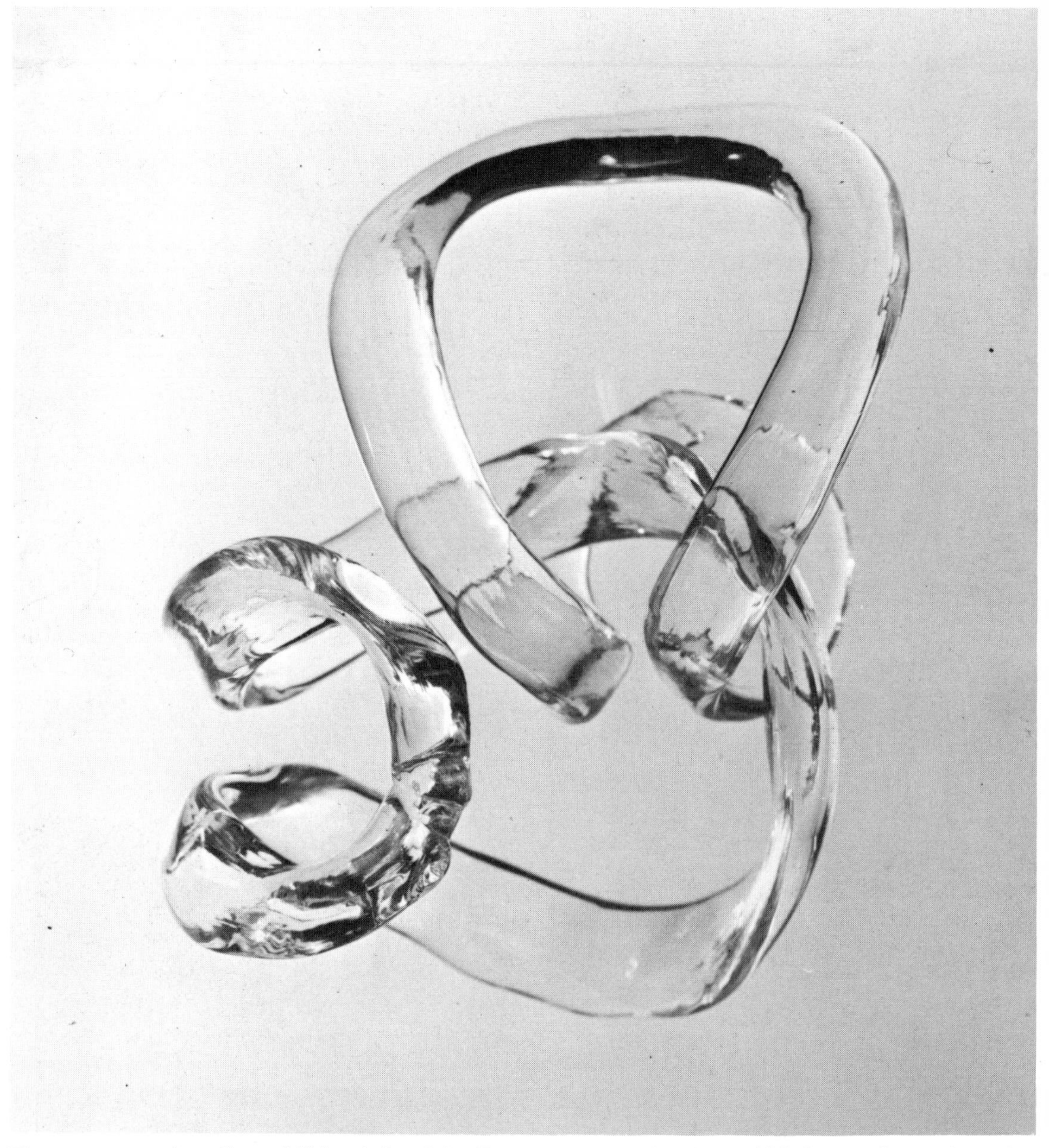

Three more curtain-wall possibilities designed by Dr. Angelo Mangiarotti, Milanese architect. These concepts were made in glass, which is very impractical when one considers safety, weight, and economic factors.

Strip Heating

When a form is composed of several straight line bends, the most expeditious way to shape it is with a strip heater or a tubular heater which will heat only the narrow area that has to be formed. Because the piece remains cold, except for the heated area, both sides of the angle will remain flat. Uneven heating, uneven cooling, and pieces longer than 24″ have a tendency to bow. For forming material over .250″ thick, two heaters should be arranged like a sandwich under and over the acrylic.

Strip heaters are very simple to put together. (See illustrations.) Rohm and Haas is marketing a heating element along with instruction as to how to mount it.

To heat the acrylic keep it at least ¼″ away from the heating element to prevent overheating of the surface. By varying the distance between supports, the width that is heated can be in-

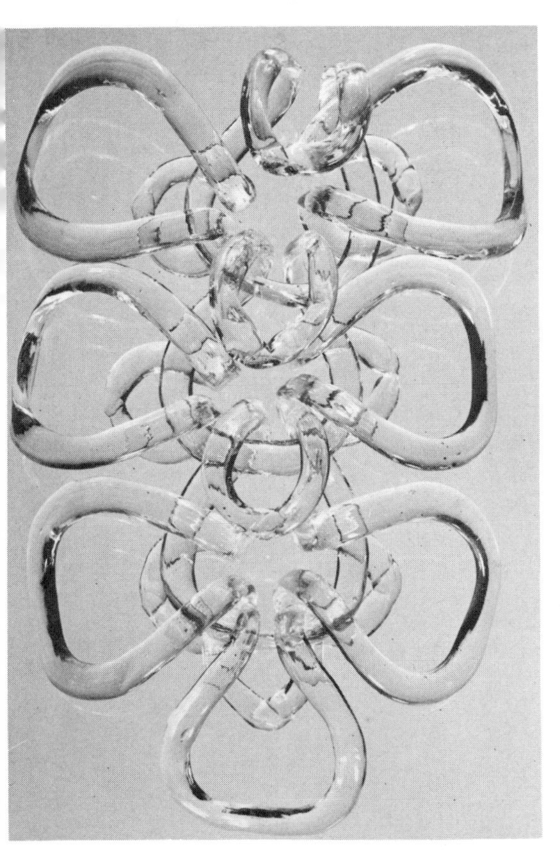

creased or decreased. For best results turn over the sheet so that it can be heated on both sides, particularly if the acrylic is thicker than 3/16″ (when you do not have two heaters). Overheating will cause surface blistering and burning. In order to estimate when the piece is soft enough to bend, test it (or use a test strip). You can then form it in a jig of some sort that will maintain the proper bend until the sheet has cooled and become rigid. Professional strip heaters come equipped with thermoregulators to prevent overheating.

Drape Forming

Simple two-dimensional curves can be formed by draping a heated sheet over a mold and permitting it to cool.

The best way to heat acrylic, or any sheet of plastic, is in a thermostatically controlled circulating-air oven. For experiments, however, infrared heaters or kitchen ovens will work. Some shops use a pizza oven and place the plastic horizontally on an asbestos sheet in the oven. Forced-air circulating ovens provide more even heating. Professional ovens are usually ver-

tical constructions carrying acrylic sheet on monorails.

Drape forming requires a mold. The simplest forms are two cleats that estimate the distance and contour supporting ribs to demarcate the shape and height of the curve. The more irregular the shape, the more support your mold has to provide. Epoxy and gypsum are sometimes used for more complex curves. To avoid "mark-off" (shape and texture of the support) cover the framework with thin sheet metal, Masonite,

Making a strip heater. (*Courtesy, Angelo Mangiarotti*)

a. Sheet asbestos is cut into strips; two narrower pieces are sandwiched between two wider pieces. Then holes are drilled to accommodate nuts and bolts.

b. Nuts and bolts are added to hold the structure.

c. One of two chromolux heating strips is added.

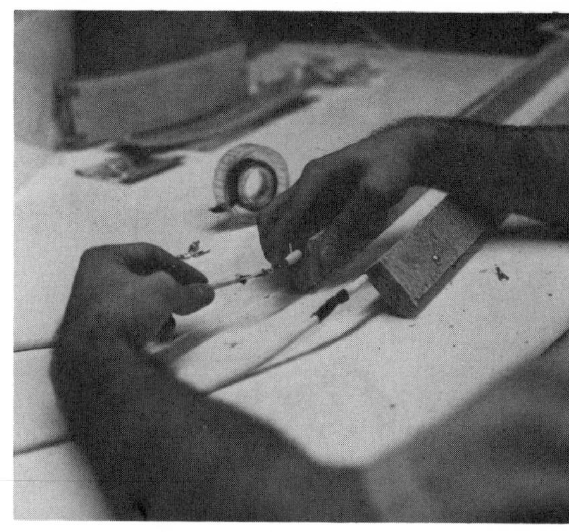

d. The two units are linked together and wired.

e. Acrylic shapes are braced and placed over the strip heater. They are now soft enough to bend.

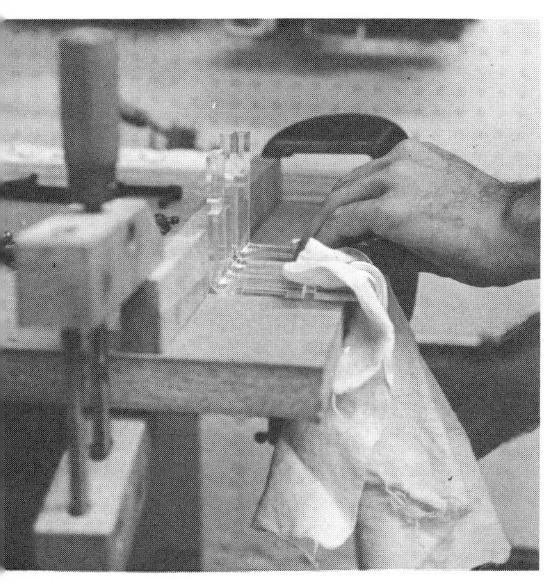

f. After bending, they are braced against a temporary jig until cool.

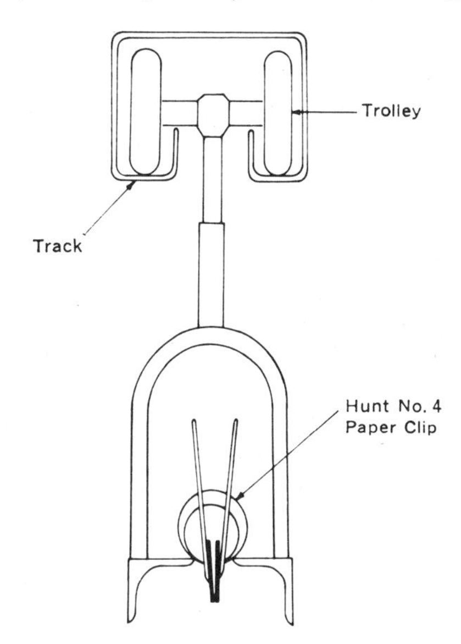

Cross-section of a monorail to hold acrylic sheet in a vertical oven for heating. (*Courtesy, Rohm & Haas Company*)

or heavy cardboard and flocked rubber, white felt, or heavy white cotton flannel. Cleats act as wedges holding the two ends of the sheet toward the mold. Hot sheet has a tendency to curl away at the edges.

In cutting your shape, expect about a 2% shrinkage. Remove masking paper, place the sheet in an oven, heat to softening, between 320°F. and 340°F. for about ten minutes for a ⅛″-thick piece. A sheet ready for forming will feel like soft rubber. Place the hot sheet on the mold, cover with a flannel blanket, and do not remove it until it has cooled to below 175°F.

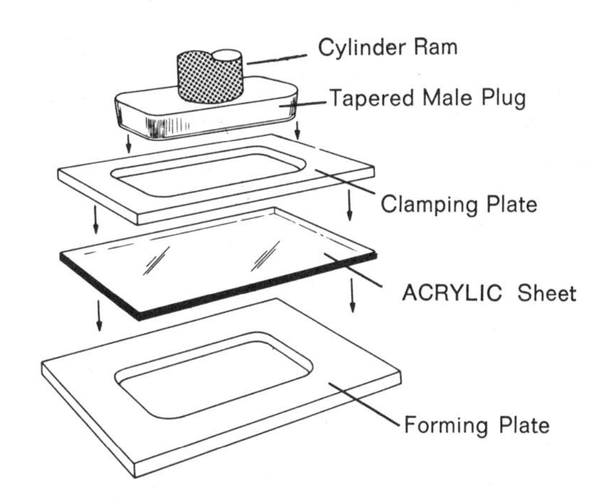

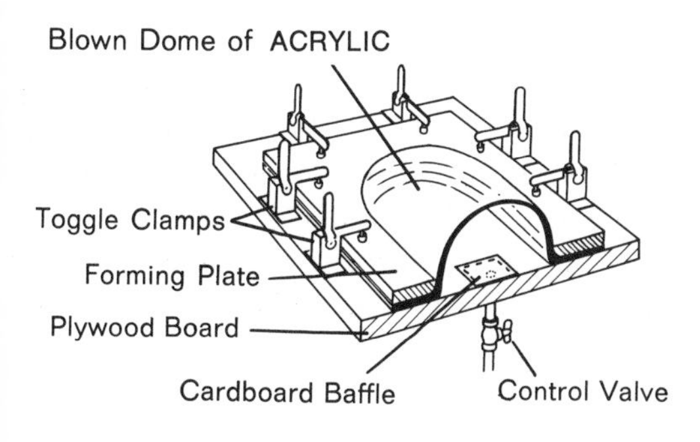

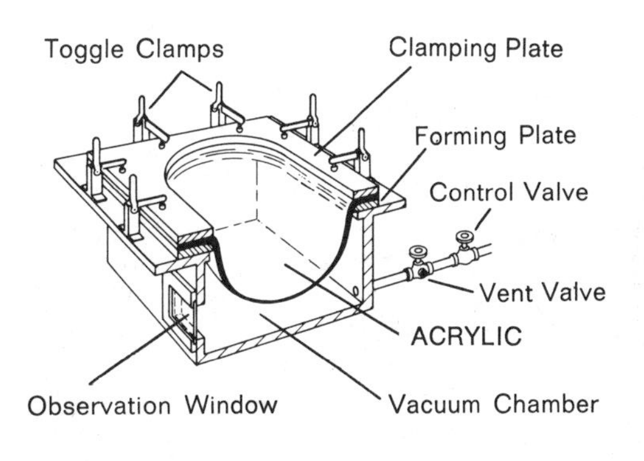

(*Courtesy, American Cyanamid Co.*)

RECOMMENDED SHEET TEMPERATURE FOR FORMING ACRYLIC

Type of Forming	Sheet Thickness ⅛″	¼″	½″
Two-Dimensional Drape Forming	265°F.	265°F.	265°F.
Free Blown	290°F.	285°F.	285°F.
Plug and Ring	300°F.	280°F.	280°F.
Vacuum Snap-Back	315°F.	295°F.	295°F.

Plug and Ring Forming

This method is useful for simple, shallow forms such as trays and sign faces. The mold is in three parts, consisting of a forming plate, a clamping plate, and a tapered male plug. The heated acrylic is sandwiched between the forming plate and the clamping plate. The tapered male plug, smaller than the inside hole of the forming plate, is forced by air-cylinder press into the hole. (A drill press can be used as a ram to force the male plug into both rings, forming and clamping plate to a predetermined depth.) The plates are made of Masonite, plywood, or metal, and the plug is made of hardwood or aluminum.

A small Di-Arco vacuum forming machine. (*Courtesy, O'Neil-Irwin Manufacturing Co.*)

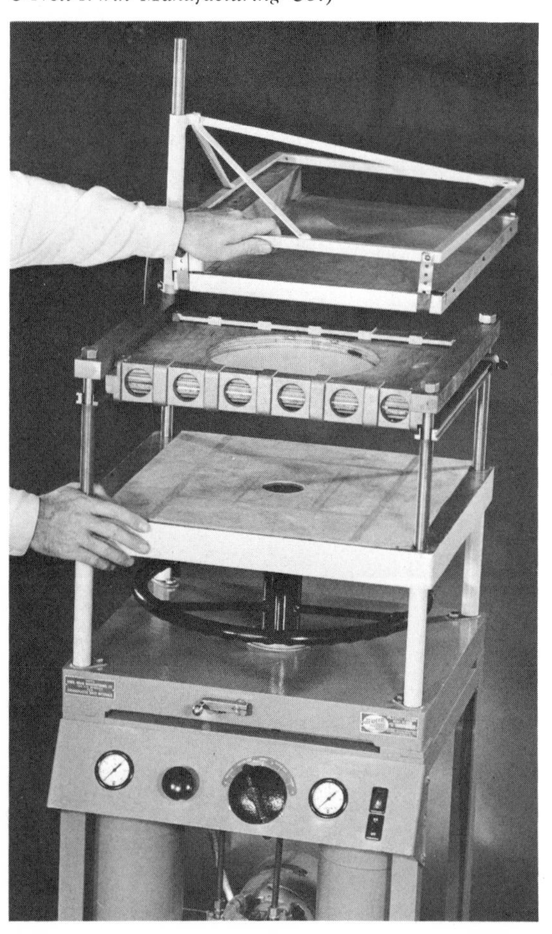

Free Blowing

Free blowing is one of the simplest forming methods. A heated sheet is clamped over a vacuum pot or pressure head and drawn or blown to shape. There is no possibility of mark-off because the sheet does not contact any form. The opening in the vacuum pot or cylinder head will determine the contour and, to an extent as well, the outside shape. The tendency is always toward a spherical shape. A simple example is attaching an air hose to the underside of a plywood board. Put flannel on top of the board, leaving space for the air-hose opening; place the heated sheet on this and a forming plate (ring) of Masonite on top; clamp with quick action toggle clamps or C-clamps if you have some help and run the compressor. Five to ten pounds of air pressure will form a 2′-diameter hemisphere (if it is clamped down properly). After the piece has been blown to the proper size, pressure is reduced to hold the piece at that height until it cools. Use the nomograph on page 00 to determine air presure for the diameter you want and thickness of the sheet.

Vacuum Forming

Similar to free blowing, in vacuum forming a vacuum is created in a chamber that draws the sheet through an opening in the forming plate. The vacuum chamber, usually made from welded steel plate, must be airtight. Masonite is used as the forming plate. As in free blowing, the cut-out in the forming place will determine the contour and the height (depth of draw). Control is by regulating the vacuum valve between the vacuum chamber and the vacuum storage tank.

Vacuum Snap-Back Forming

Like vacuum forming, snap-back forming is done in a vacuum pot. After the heated sheet is drawn into the pot, a male plug in the shape of the desired inside contour is lowered and locked inside the bubble formed by the vacuum approach. As the vacuum is gradually released, the acrylic snaps back slowly against the hardwood plug form, because of its elastic memory.

Free Blowing a Bowl

a. Hot acrylic sheet is placed on a blow molding stand at Cook's.

b. A "ring" with a top flat piece to form a flat base is quickly clamped in place.

c. Air is blown through a hole stretching the hot acrylic into a natural curve.

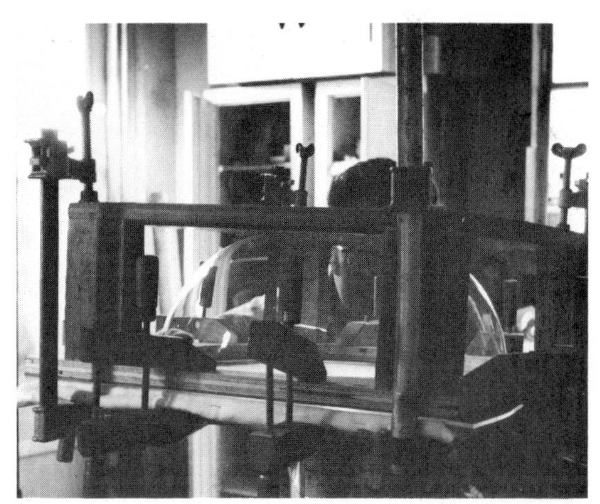

d. When the acrylic has reached the top limit, air pressure is reduced slightly until the form cools.

e. The bowl is removed from the blowing stand and the flange is cut off with a veneer blade in a shaper motor.

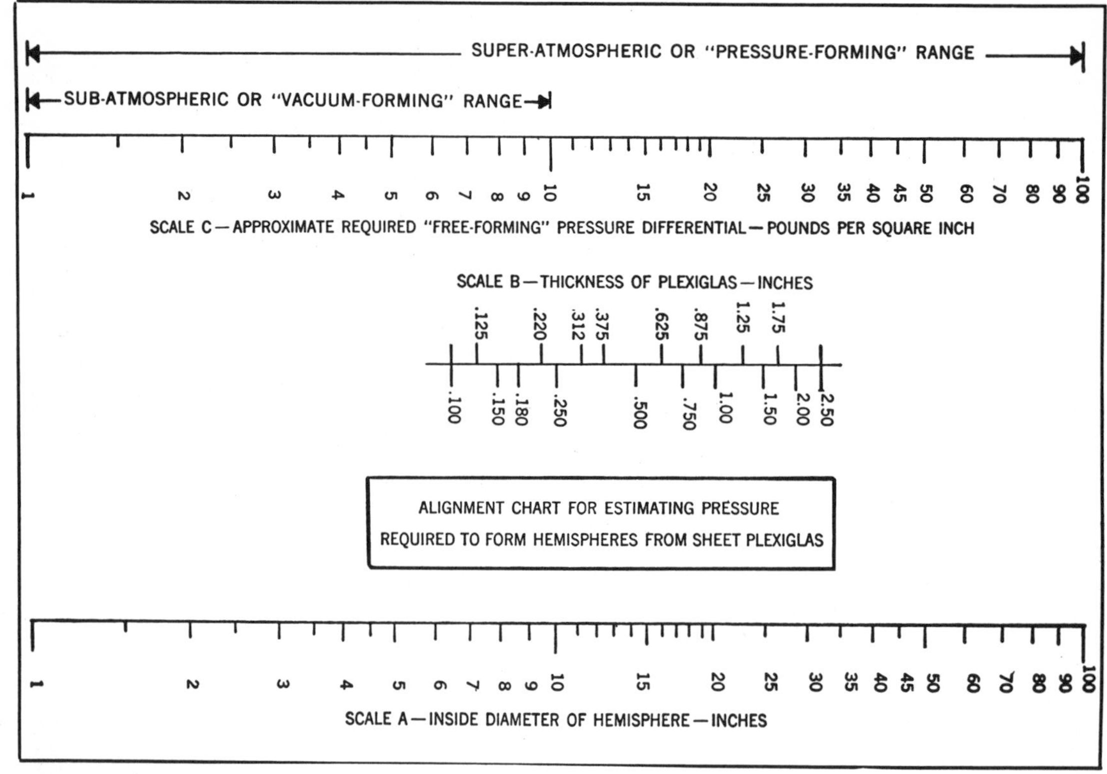

ESTIMATING PRESSURE REQUIRED FOR FREE BLOWING OR VACUUM FORMING
How to Use the Nomograph:
Problem:
Estimate the pressure required to form a 25″ inside-diameter hemisphere from .250″ thick acrylic sheet.
Solution:
Draw a straight line through 25″ diameter on Scale A and .250″ thickness on Scale B. Continue this straight line to intersect Scale C. Read off estimated pressure at the point of intersection of the line and Scale C—4.8 p.s.i. required.
The alignment chart can also be used to estimate the forces required for other methods of forming, other degrees of draw, and other than spherical shapes by estimating the diameter corresponding most closely to the desired required part.

(The plug has to have vent holes to prevent air from becoming trapped.)

Variations on these forming methods are blow-back forming, ridge forming, reverse blow forming, male and female forming.

FINISHING

For some designs it is best to finish parts at stages, and then complete the final touches after the piecę has been completely formed. For our purposes finishing implies scraping, sanding, ashing, buffing or polishing, flame and solvent polishing. Any of these steps can be skipped, depending on the quality of the surface and the purposes of the piece and "cosmetic" needs.

MAKING A CHAIR THROUGH HEAT FORMING

a. The form is designed. A pattern is made, traced onto plywood, and cut with a sabre saw.

b. The basic frame is put together.

c. Siding is put on wherever the Plexiglas will touch the mold.

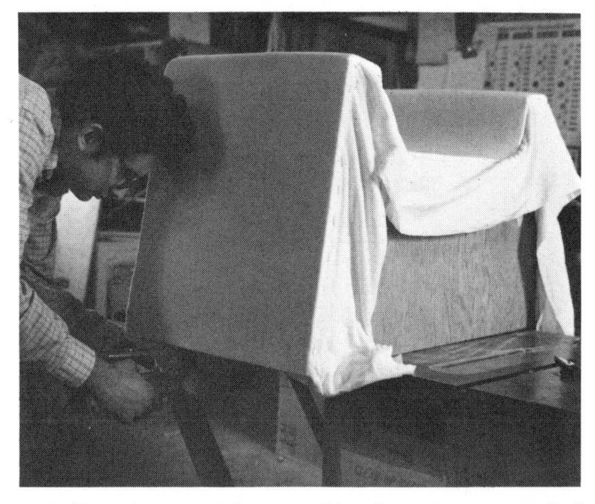

d. Two layers of heavy white flannel are stretched over the mold wherever the Plexiglas will touch.

e. The Plexiglas is cut to size and hung on a monorail, ready to be heated.

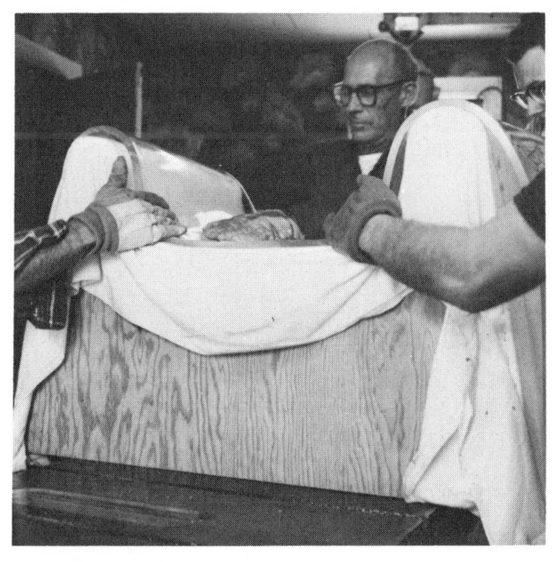

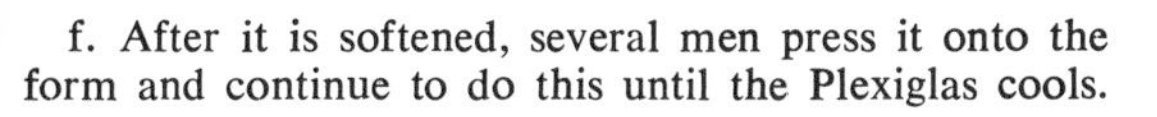

f. After it is softened, several men press it onto the form and continue to do this until the Plexiglas cools.

g. The bottom curve is simple enough to be strip heated.

h. After heating and bending, the form is braced until it cools.

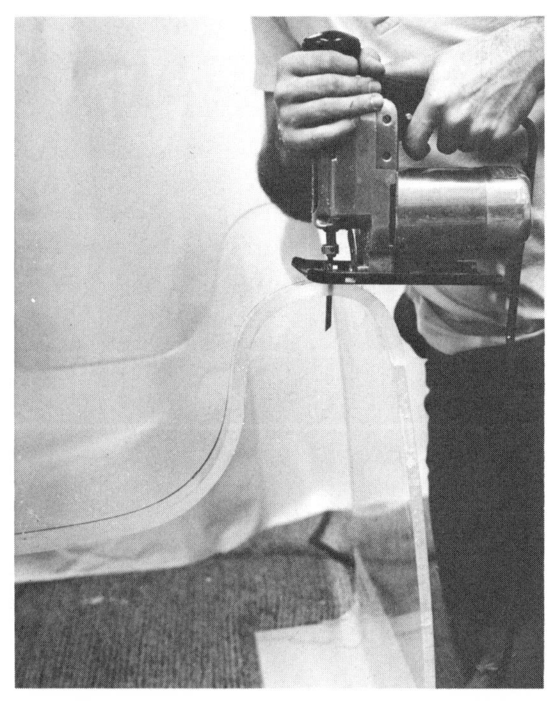

i. Mark-off of the monorail clips is trimmed off the chair with a sabre saw. The edges are then scraped and sanded. (Photos taken at Cook's.)

Scraping

Very rough saw marks on edges can be removed with a belt sander and/or by scraping with steel "wood" scrapers. Scraping is accomplished by scraping against the edge with a scraper edge in one direction from front to back or back to front. while the piece is held stationary in a vise. This step saves a great deal of time and laborious hand sanding.

Scraping acrylic with a steel wood scraper before sanding and polishing.

Sanding and Ashing

Both sanding and ashing are not necesary. Ashing consists of buffing with a slurry or paste (the thickness of mud) of pumice abrasive and water. It is usually used in production processes because it accomplishes the same purposes as sanding and polishing in less time. It requires special equipment—an ashing hood to keep the ash from flying out, a 2-horsepower motor, stitched cotton buffs 10″ to 12″ in diameter. In ashing, the part, held firmly against the wheel, is moved slowly until scratches are removed.

Sanding is used to bring the acrylic to a smooth matte or satin finish; it can be left this way (light will pipe through sanded edges) or from this stage can be polished to its original high lustre.

Disc sanders, belt sanders, drum sanders can be used. On these machines (which should be operated at slower speeds than for wood), aluminum oxide belts are recommended for dry sanding. Speeds higher than 2,000 r.m.p. will tend to cause overheating and gum up the belt. For wet sanding, wet-or-dry silicon carbide belts running at 2,800 surface feet per minute, with grit sizes varying from 220 to 280 for extra-smooth edges, should be used.

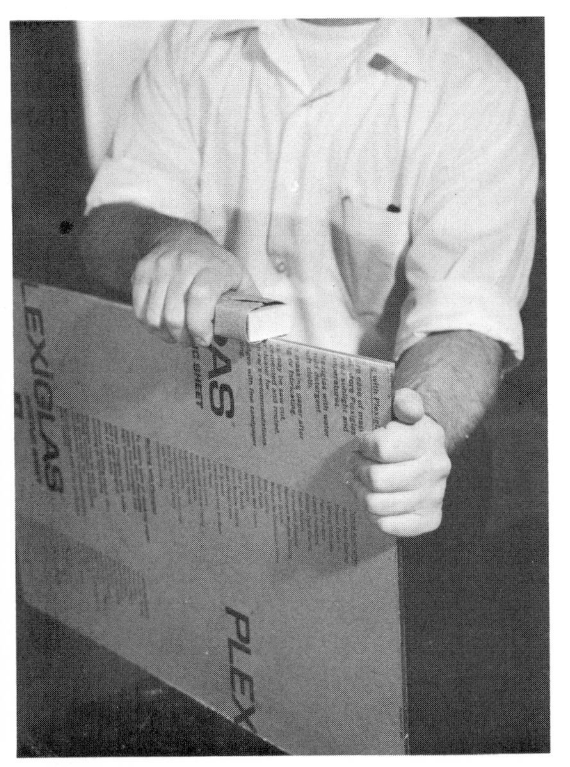

Sanding the edge of Plexiglas sheet. (*Courtesy, Rohm & Haas Company*)

Sanding by hand can follow scraping or belt sanding. If the scratches are very deep, use 240- or 320-grit wet-or-dry paper, sanding in one direction, followed by 400- or 500-grit paper, sanding in the other direction. It is essential that all deep scratches are removed before changing to the next finer grit. For extra-special work I use a series of grits from 400–600 of aluminum oxide Mylar-coated abrasive.[6] When the piece is satin smooth, dry the edges with a flannel cloth and you are ready to polish.

A pistol drill equipped with a buffing wheel can be used for polishing acrylic. (*Courtesy, Rohm & Haas Company*)

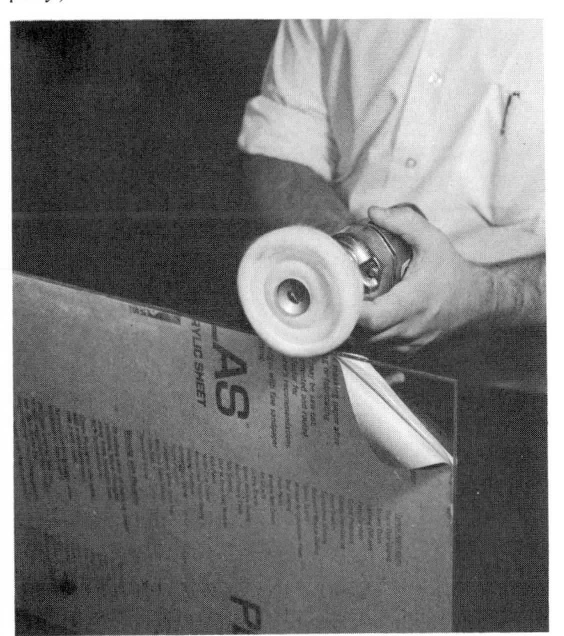

Polishing and Buffing

Wheels used for plastic should not be used to polish metals. For average use a one H.P. motor is adequate, but for production work a 1½–2 H.P. motor is necessary. The diameter of the wheel and the motor r.p.m. will determine surface speed. The following speeds are recommended by plastic manufacturers; however, I find that very slow speeds do not take out marks, while very fast speeds burn the acrylic. I use a 10″ wheel of unbleached muslin, 1½″ wide at 1,725 r.p.m.

6″ diameter	1,800 r.p.m.
8″ diameter	1,500 r.p.m.
10″ diameter	1,200 r.p.m.
12″ diameter	1,000 r.p.m.

Use a loose or packed buff (not a multistitched wheel); it permits wheel ventilation and thereby produces less heat. Apply polishing compound frequently but sparingly. Some compounds require a wheel dressing of tallow, at first, to hold the compound on the wheel. But there are new compounds that eliminate the need for frequent wheel cleaning (tallow and other greasy abrasives and polishes get hard and gummy after a while). XXX white diamond composition [7] eliminates the need for two to three wheels. (Very little tallow is necessary.) It polishes out surface scratches and brings acrylic to a high lustre with little effort. N42 chromium composition, 320 HR hard rubber composition, and 327 hard rubber composition also work effectively.

Rotation of the wheel is toward you. Therefore always keep the *upper edge* of your piece away from the wheel; otherwise the force of the wheel will tear the piece from your hands and send it flying. For edge polishing move the piece across the wheel from left to right. Do not try to polish the entire thickness at one time. When the bottom thickness has been polished, turn the piece around and repeat the operation so that polished areas overlap.

[6] Flex-i-Grit. See "Sources of Supply."
[7] The Matchless Metal Polish Co.
See "Sources of Supply."

In surface polishing keep the piece in motion at all times. Do not use too much pressure and do not start near the top. Stay away from the top edge so that the wheel does not "catch it." Start at the middle and work back and forth toward the bottom. Turn the piece upside down and repeat the operation.

A flexible shaft tool or a polishing wheel attached to a pistol drill can also be used. Use 4″–6″ muslin wheels, 1″–1½″ wide. Keep the wheel in motion, apply compound frequently, and use little pressure.

Some people complete the polishing operation by buffing the piece with a Domet flannel wheel, free of any compounds.

For hand polishing, fine scratches can be removed with Simoniz Cleaner and a soft flannel pad. When scratches have disappeared, wipe clean with a flannel cloth and apply Simoniz wax or Johnson's wax.

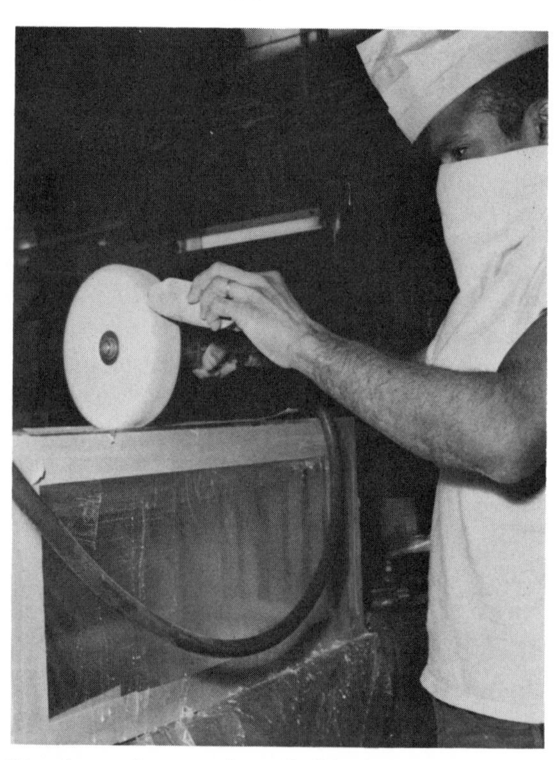

Very large pieces require a flexible-shaft buffing wheel. Compound is being applied to the wheel at the same time that polishing is done. (*Courtesy, Just Plastics, Inc.*)

EDGE POLISHING

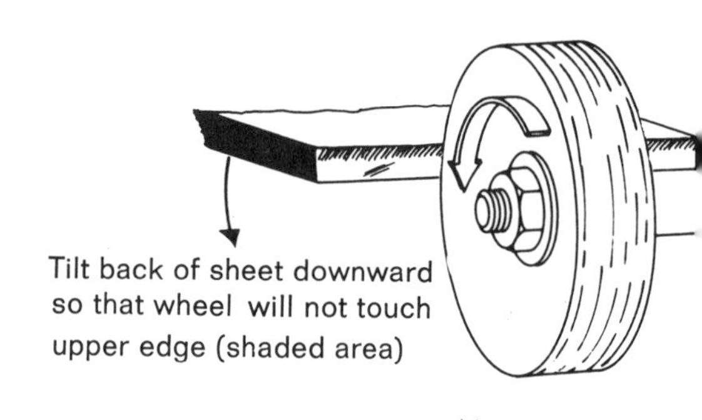

SURFACE POLISHING

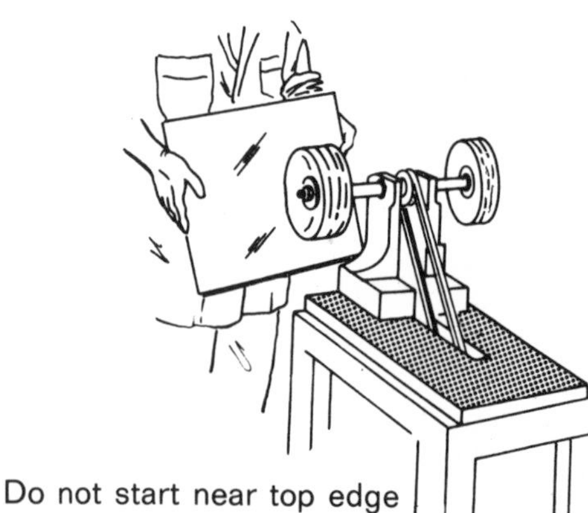

(*Courtesy, American Cyanamid Co.*)

Flame Polishing

This is a fast, economical process, except that it requires an oxy-hydrogen flame. Although I have used oxy-acetylene, it is not advisable since it deposits carbon. Use a welding torch with a #4 or #5 tip. Set the hydrogen pressure

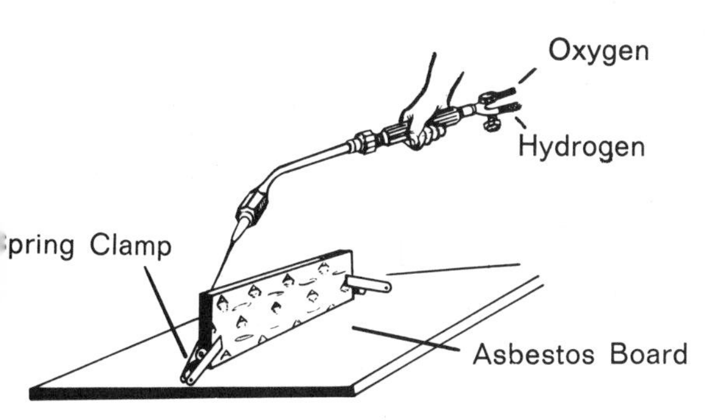

(*Courtesy, American Cyanamid Co.*)

at eight pounds, the oxygen pressure at five pounds. Ignite the hydrogen *first* and then turn the oxygen on and set the flame. The flame should be almost invisible and about four inches long. Move it along the surface at approximately four inches per second. A very slow speed will cause the surface to bubble; a speed that is too fast will not do the job. There is a tendency for flame polishing to cause crazing later. After flame polishing, annealing at the highest possible temperature will lessen the possibility of crazing.

Cleaning of Acrylic

To remove static use a 10% solution of ALL in water. To remove grease and dirt VM&P naphtha, kerosene, isopropyl alcohol, or household ammonia (diluted) and applied with a soft flannel cloth will accomplish the purpose. Soap and water will complete the job. Avoid benzene, gasoline, acetone, carbon tetrachloride, xylene, or denatured alcohol; they will cause eventual crazing.

DECORATING

Engraving, silk screening, spray painting, brush painting, dyeing, vacuum metalizing, and hot-stamping are some approaches to decorating the surface of acrylic.

Engraving

A flexible shaft tool with an assortment of "dental"-type bits does the best job of engraving. If the tool is operated by a variable-speed foot pedal, control is easier. Jigs, tape, and other supporting devices may be necessary to contain the tool within a described area. Various textures can be achieved through different types of bits, as well as by varying pressure and direction. Different depths catch piped light in varying amounts and glow. Three-dimensional carving effects can be achieved by internally carving shapes. (Remember those horrible internally carved roses?) Negative areas glow into positive forms. Control is difficult, however, and requires much practice. It is a good idea to experiment with your various drill bits on a clamped piece of acrylic in order to develop a vocabulary of possibility. Your dentist may give you his old bits, or you can find an assortment where they sell Dremel hobby equipment. A pantograph-controlled engraving machine with various-shaped cutters is used professionally to engrave acrylic.

Silk Screening

Silk screening is a multiple process. Translucent or opaque effects can be produced. Grip-Flex [8] hand-cut water-soluble or lacquer-proof film and photographic screens for greater detail, touche, and glue can be used to block the screen. For screen printing with a translucent material addition of Grip-Flex 19-210 clear is used as an extender. It also increases gloss, allows for deeper draw, and eases the screening operation. For opaque screening repeated runs may be necessary. Thinner reduces viscosity and slows the drying time and retarder keeps the screen sufficiently open while extending the working period.

Spray Painting

If stencil-masking tape or a spray mask is used, the form can be spray-painted with acrylic paint.

After cleaning the surface and removing static (static can cause patterns to form in the paint film), apply your masking. If a spray-masking film is used, build it to 10 to 15 mils (wet). Ammonia and water can be used for cleaning equipment. The following equipment is recommended for spraying:

Binks Co., Inc.:	PRESSURE	SIPHON
Spray Gun	18, 19, 21, 29	15, 26
Air Nozzle	63B	78SD
Fluid Nozzle	63A-SS	78SS
Atomizing Pressure	55PSI	40PSI
Air Supply	14CFM	4CFM
DeVilbiss Co.:		
Spray Gun	P-JGA	P-JGA
Air Nozzle	704	36
Fluid Nozzle	FX	EX
Atomizing Pressure	25PSI	40PSI
Air Supply	9.5CFM	6.5CFM
Regulator	H.Q.	HAA

[8] Grip-Flex, Wyandotte Paint Products. See "Sources of Supply."

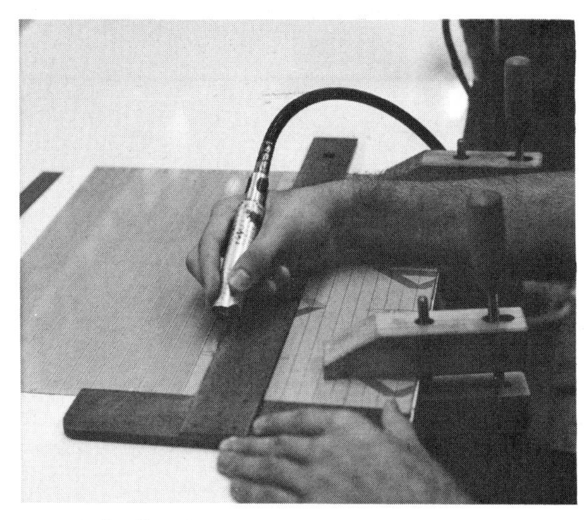

a. A flexible-shaft drill equipped with a dental burr is used to engrave straight lines in the acrylic shell.

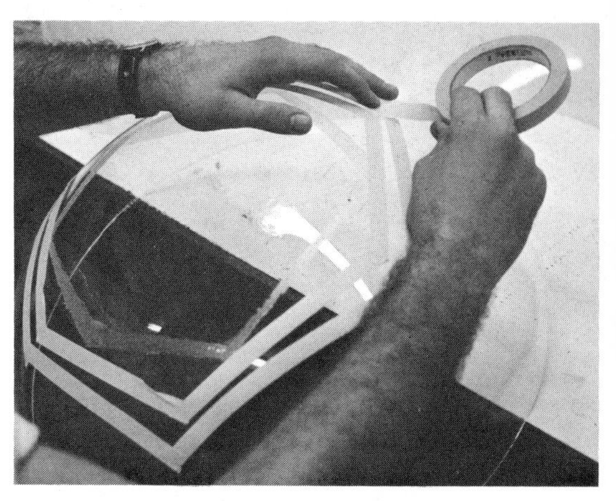

b. The free-blown bowl is masked with masking tape to help keep the tool within the described areas.

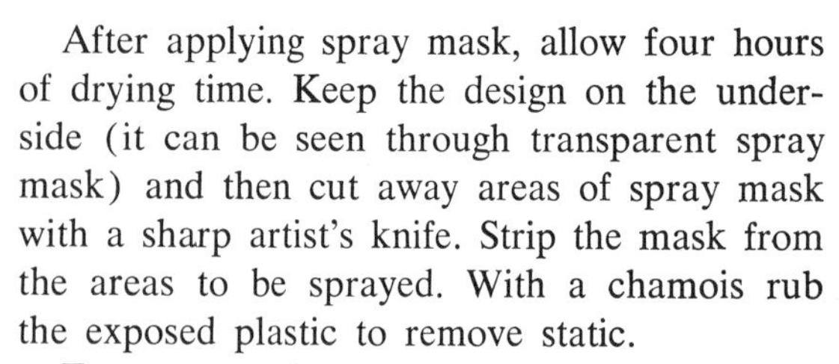

After applying spray mask, allow four hours of drying time. Keep the design on the underside (it can be seen through transparent spray mask) and then cut away areas of spray mask with a sharp artist's knife. Strip the mask from the areas to be sprayed. With a chamois rub the exposed plastic to remove static.

To spray, a back light in a well-ventilated spray booth is necessary, to determine whether the paint is being deposited uniformly. After the paint has been stirred, thinned, and placed into the siphon cup, turn off both valves of the gun and pull the trigger holding it open

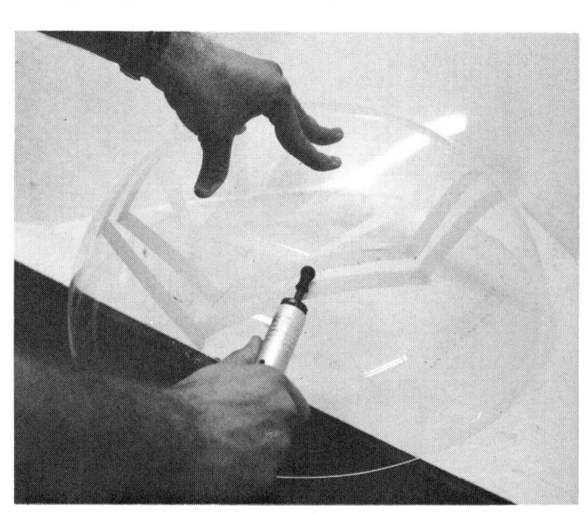

c. A grinding bit is used to create a pebbly texture on the bowl. Speed control is via a variable speed foot pedal.

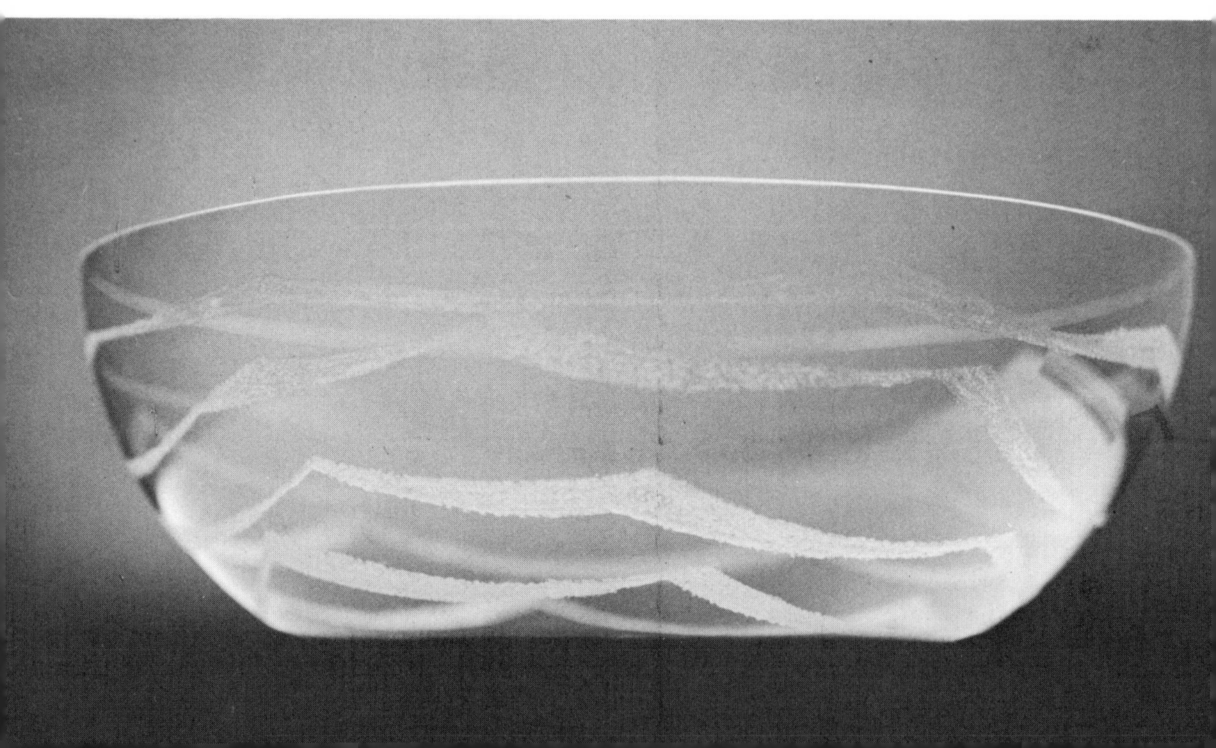

d. The completed punch bowl by Thelma R. Newman.

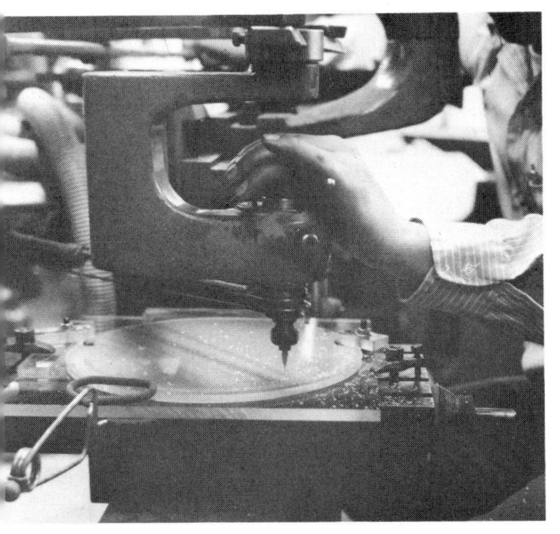

e. An engraving machine equipped with a pantograph is used for commercial engraving.

continuously while you start to turn on the air valve and watch the pressure gauge. Stop opening the air valve when the compressor can no longer maintain adequate pressure, and then crack (open) the paint valve slowly until the paint begins to appear. The final setting should be gauged by how much flow it takes to complete a coating in six or more light passes. Try not to manipulate the gun trigger except in the beginning and at the end. You will have to adjust your procedures for variations in each color, temperature, and humidity.

Brush Painting

Artist brushes can be used with acrylic paint. There are a great many brands of acrylic paint on the market; make certain, however, that the paint has an acrylic base.

Dyeing

Cellophane tape can be used to mask areas to be dyed. Immerse pieces in a mixture of ALL and water, or in Duponol[9] WA at 140°F. The pieces are then ready to be dipped in a dye solution prepared from acetamine[9] dyes.

Mix together:

Acetamine dye	0.25 grams
Merpentine	7.50 cc
Denatured alcohol	2.50 cc
Water	90.00 cc

Heat the dye solution to 180–200°F. (Lower temperatures are used for slow dyeing rates.)

[9] Duponol, acetamine dyes, and Merpentine are Du Pont products. See "Sources of Supply."

Keep the dye agitated for even dispersion of color. Depth of color is determined by the amount of time the piece is kept in the dye. After removal from the dye bath, wash the piece in cold water to fix the dye. A full range of transparent color is available.

Hot-Stamping

Hot-stamping is practical only for small imprints, under 4″ x 5″, because a considerable amount of time would be required to imprint a large area. A brass die is heated to approximately 350°F. and is brought down under pressure, which forces a coated foil to stick to the plastic. Colored, patterned, or metallic foil is transferred from a tape carrier by heat and pressure from a flat metal plate or silicone pad (for raised-area decorating). Pressure, heat, and dwelltime must be accurately controlled.

Vacuum Metallizing

Vacuum metallizing is a commercial technique applied to surfaces that will not be handled to a great extent. The process generally consists of applying a transparent lacquer base coat, disposition of a metal film (usually aluminum; sometimes copper alloys, silver, and gold) in a vacuum chamber, and then application of a lacquer top film. Surfaces larger than four square inches should be avoided. Application of vacuum metallizing on the second surface is more practical and more permanent.

Frosting

Sandblasting with sand, glass beads, metal, or fine powderized walnut shells, or roughening with various abrasive compounds to create a negative or frosted-surface finish is one method of frosting. Another is to spray the surface with frosting lacquer compositions.[10]

SPECIFIC DESIGN CONSIDERATIONS

1. The best way to mount acrylic sheet is in a channel system that allows for contraction and expansion.
2. Sharp 90° bends are bad. Any sharp change in direction will eventually crack at that point when the piece is under pressure.
3. Anything done to acrylic induces stress, *e.g.*, milling, hot-stamping, painting, drilling, et cetera. If a hole is drilled for a bolt and the same area is spray-painted, the piece will probably fail there.

[10] A Du Pont product.

4. Anneal your form to relieve internal stresses. Anneal the form after polishing and before painting, otherwise crazing will show in the painted area.
5. When drilling for a bolt in a large panel, elongate the shaft to allow for lineal movement. Cork, neoprene, or a nylon washer will prevent the bolt from slipping by providing cushioning.
6. Do not drill a hole near a corner; if it is necessary, open the hole out to the edge and clean the notch to avoid notch sensitivity.
7. Do not countersink a screw at the angle of the screw head, because compression will exert a pressure just as a wedge would and cracking will occur. Use a bolt if it is necessary to countersink the head, making certain that the opening is at right angles and not diagonal or wedge-shaped.

NEW VARIETIES OF ACRYLIC

Goodyear is developing a new transparent material, a rubber component added to acrylic to increase its physical properties, comparable to ABS and butyrate. It will be transparent and will be designed for conventional processing.

A cast acrylic/ABS composite sheet has been developed in opaque colors. Its weatherability permits outdoor applications with increased physical properties.

No doubt, tomorrow will find newer varieties and the potential of sheet plastic will be expanded to meet even more rigid specifications.

ACRYLIC CLINIC

DEFECT	CAUSE	REMEDY
Crazing	Clamps exerting excessive pressure	Use less force or use pads to distribute force
	Excessive stresses	Anneal parts
	Excessive solvent exposure	Change adhesive
	Inaccurate fit requiring forcing or flexing	Allow more leeway in design
Cloudy joint	Excessive humidity	Reduce humidity, if possible,
		or use slower evaporating solvent,
		or change adhesive
Ugly joint	Too many bubbles and dry spots	Improve fit of joint; apply solvent to both sides
	Cushion and/or solvent squeezed out of joint because of excessive pressure	Reduce pressure on joint or do not induce pressure until 20–30 seconds
	Temperature too low	Warm solvent or raise room temperature
Poor joint strength	Temperature too low	Heat solvent
	Not enough cushion	Use open joint
	Too many bubbles	Allow bubbles to rise to surface (PS 30) after mixing
	Foreign matter on surfaces being joined	Clean surfaces

DEFECT	CAUSE	REMEDY
Cement will not harden	Improper mixture	Check instructions
	Not thoroughly mixed	Mix for a longer period
	One component is stale (turns yellow or amber)	Throw away component; use fresh component
	Surface tension lost (cyanoacrylate)	Do not spread; just apply drops
	Inhibitors present	Clean surface; eliminate tape, stirring rod, et cetera
Parts will not bond	Improper cleaning	Reclean surfaces with solvent, resand, rewipe with solvent, and re-apply adhesive
	Poor surface fit	Resand or remachine to provide better contact
	Acidic surface (for cyanoacrylate cements)	For cyanoacrylate cements, use suface activator
Cracks around bolts	Fit too tight	Allow 1/16 inch per linear foot for expansion
Chips on edges	Teeth on blade too large	Use metal-cutting saw blades
	Feed too rapid	Slow down feed and increase tool speed
	Piece allowed to chatter	Hold down piece with more support when cutting
Dull surface	Attack by solvent	Polish piece and anneal; avoid cleaning with acetone, hydrocarbons, et cetera
	Use of abrasive materials, such as scouring compounds	Use household ammonia and water, All and water, or plain soap and water for cleaning
"Burned" sections and particles sticking to holes and edges	Use of wood drill	Adjust edge of drill
	Speed of motor and/or polishing wheel too great	Reduce motor RPM or size of polishing wheel; change to a loose muslin buff
	Dull tools	Use sharp tools; try carbon tools
	Too much frictional heat generated	Use coolant or air
	Tool too hot	Reduce time tool is used; allow it to cool
	Cut edges stick together because of too much frictional heat	Observe above
Specks in paint film	Foreign material trapped	Keep spray equipment clean
		Spray and paint in dust-free area
		Clean top of can before opening lid
	Failure to remove static charge from plastic	Wipe sheet with damp chamois just before painting

DEFECT	CAUSE	REMEDY
Webbing, feathering, or cottoning when paint is sprayed	Improper paint viscosity	Too heavy paint can be thinned with 5% dilutions
	Improper air pressure setting	Try higher air pressure, up to 80 PSI
Paint runs under mask	Too thin a paint	Use thicker material
	Film too thin	Use more layers
	Knife too dull; pulls up edge as it cuts	Use sharper knife blade
Bubbling on surface	Direct heat from flame	Flame-polish faster—4″–5″ per second
Binding and twisting of saw blade	Uneven feed	Feed slowly and evenly
	Tool dirty	Clean blade with alcohol, naphtha, or mineral spirits
	Improperly set saw fence	Allow a bit of leeway
	Masking-paper adhesive sticking	Fix masking paper, or remove it
	Tool too hot	Let tool cool; use coolant
Masking paper sticks	Very old material, exposed to heat, water, smoke	Soak with kerosene and peel away with a polyethylene or Teflon scraper

6

a gallery of designs with acrylic

Examples ranging from lamps, furniture, environments to jewelry and toys display a wide variety of interpretation in fabricating acrylic and other sheet into useful forms.

ENVIRONMENTS

The following pictured environments are meant to emancipate us from past conceptions of what an environment should be like. Certainly transparent acrylic can circumscribe an area invisibly and with curves, very different from post and lintel treatment of traditional materials.

Ugo LaPietra, architect-designer of Milan, Italy, had a hand in creating three of the different kinds of environments illustrated here.

Four aspects of a fashion shape designed by Ugo Lapietra. Acrylic cylinder—containers for dresses—move mechanically up to the ceiling. The ceiling becomes a showcase. A control board tele-selects which dress container will come down. Lapietra wanted to conceive of an interior in relationship to a fashion store and created a programmed boutique. A mirror wall doubles the room. (*Courtesy, Domus*)

Ugo Lapietra's environment is a big "container," large enough to enter. Sound and light interact by automatic controls and envelop the participant with a strange feeling, because lines catch light and demarcate the otherwise transparent exterior and at the same time appears to interact by vibrating with audio transmission. (*Courtesy, Ugo Lapietra*)

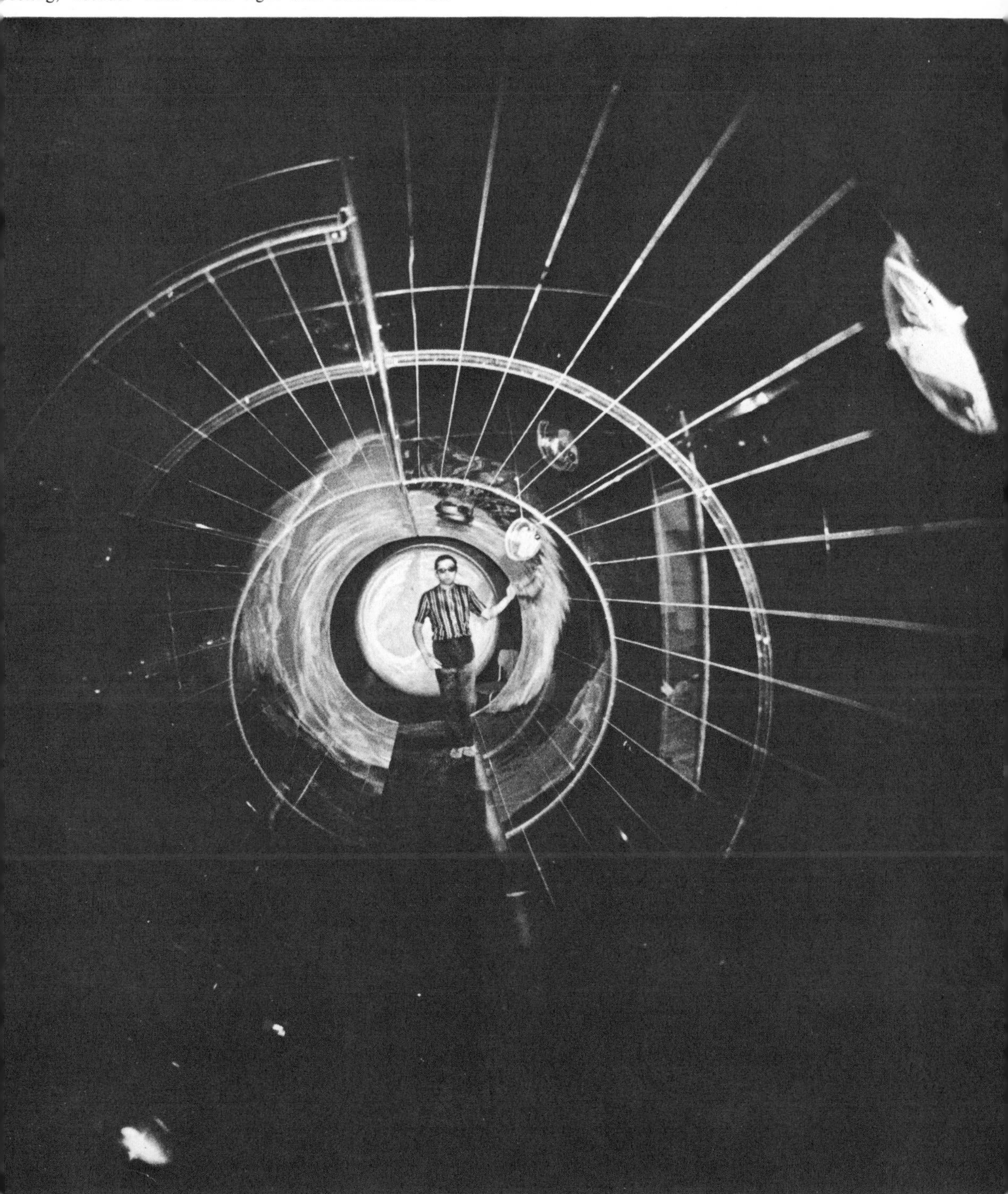

Another environment of clear acrylic by Lapietra. By pressing a button, on the left, audio-light transmission is controlled in the big environment.

The top unit of Lapietra's environment contain controls for transmitting sound and converting sound to light patterns. In the synaesthetic process the partici pant with his presence and intervention determines alters, and releases the interrelated events within th morphological model. (*Courtesy, Ugo Lapietra*)

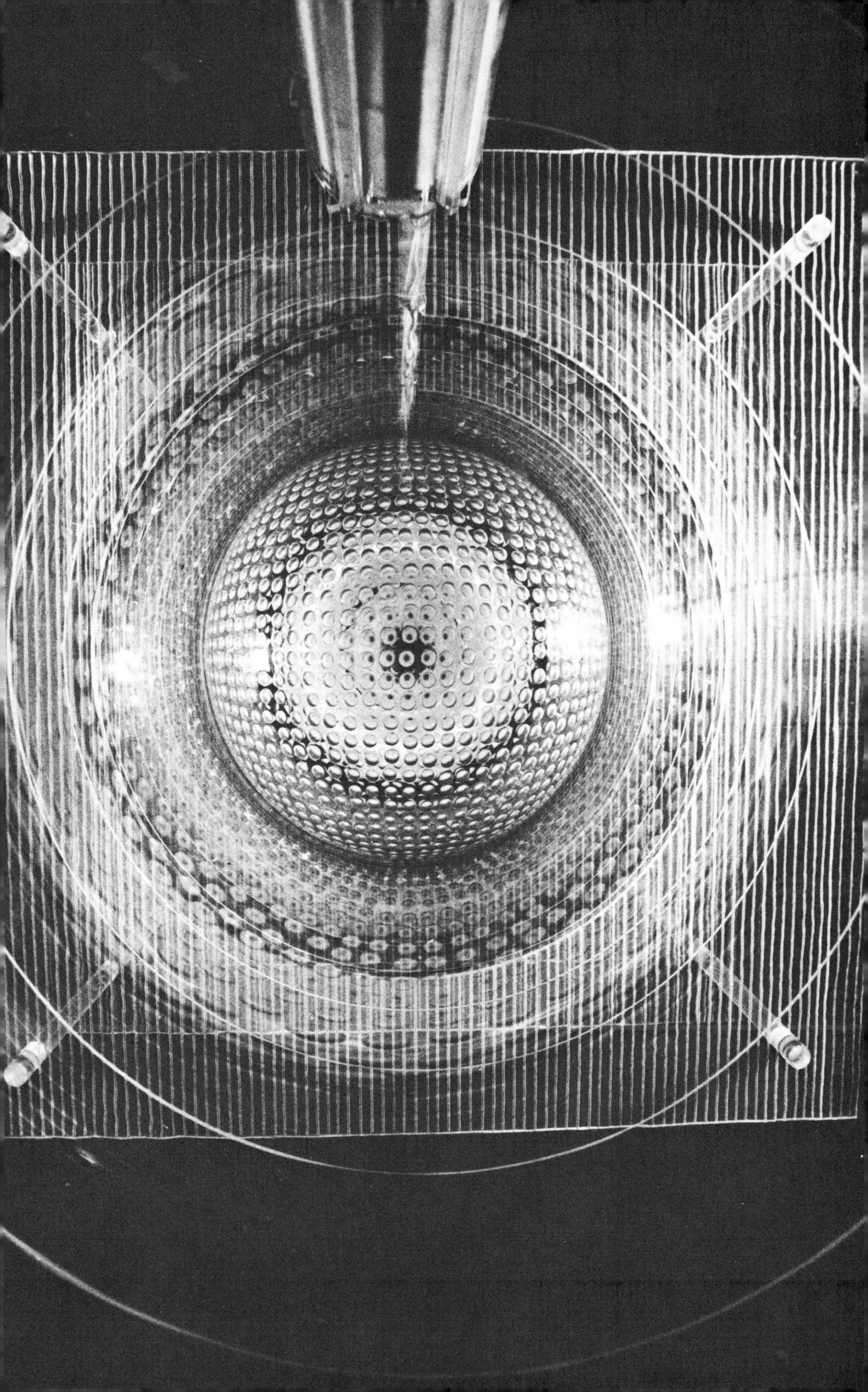

"A New Place" by Les Levine is a project for the Museum of Modern Art, made of free-blown Plexiglas connected with metal 40′ x 40′ x 7′. (*Courtesy, Le Levine*)

a. Lapietra collaborated with Lomazzi, D'Urbino, DePas, Beretta on this World's Fair environment designed to be made of blown PVC.

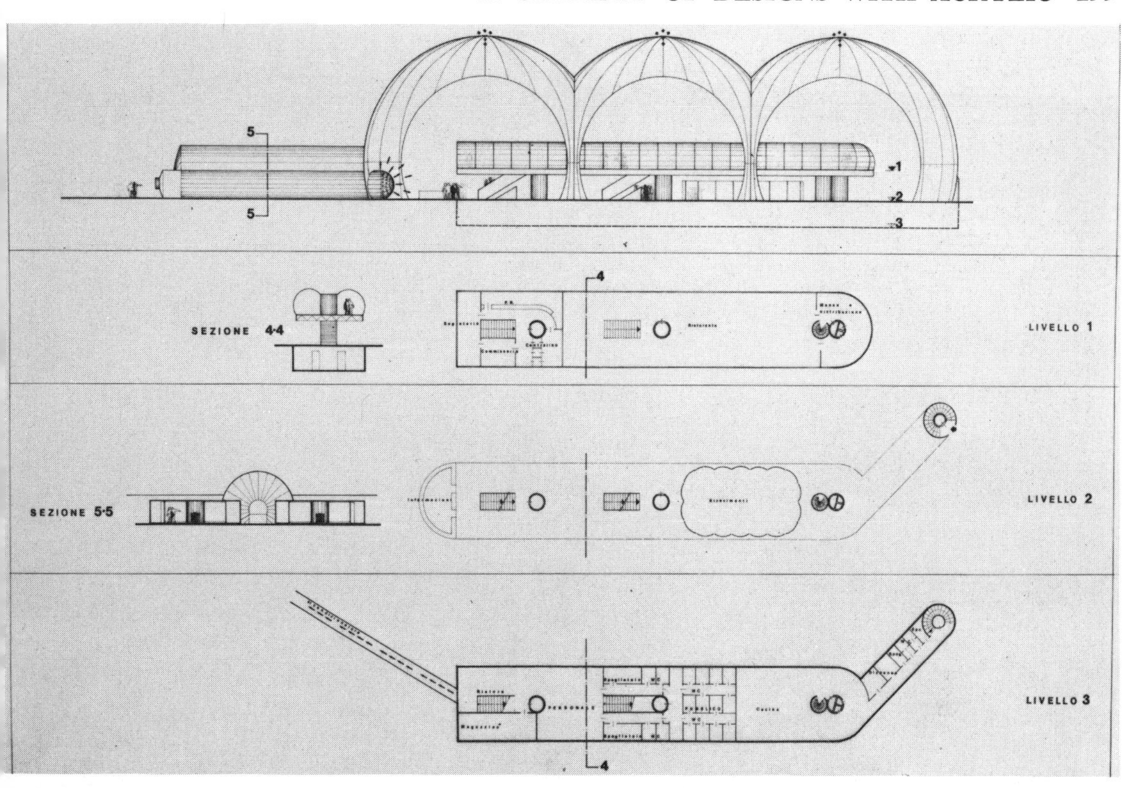

b. Plan and cross-section of the three levels contained under the blowup overhead.

c. "I.R." are two rotating entrances that bring visitors into a controlled central entrance. "G.C." are generators that control air conditioning and maintain air pressure within.

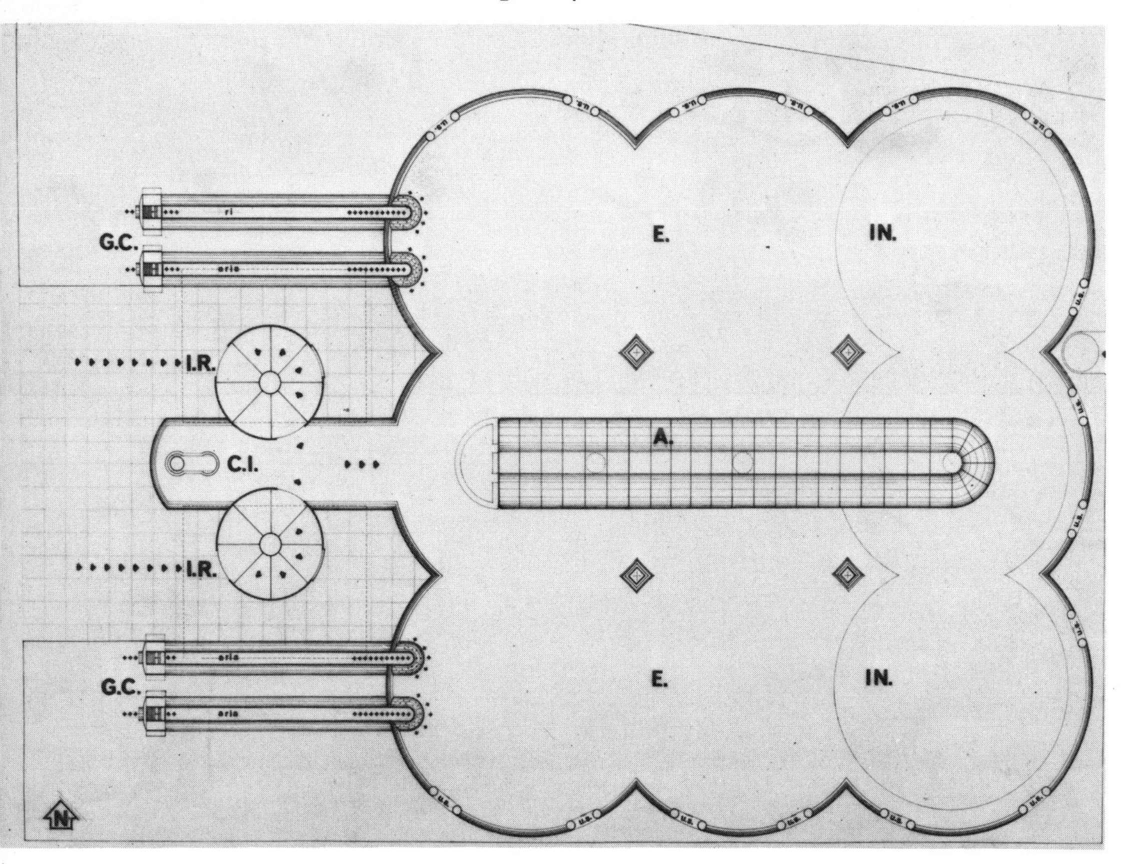

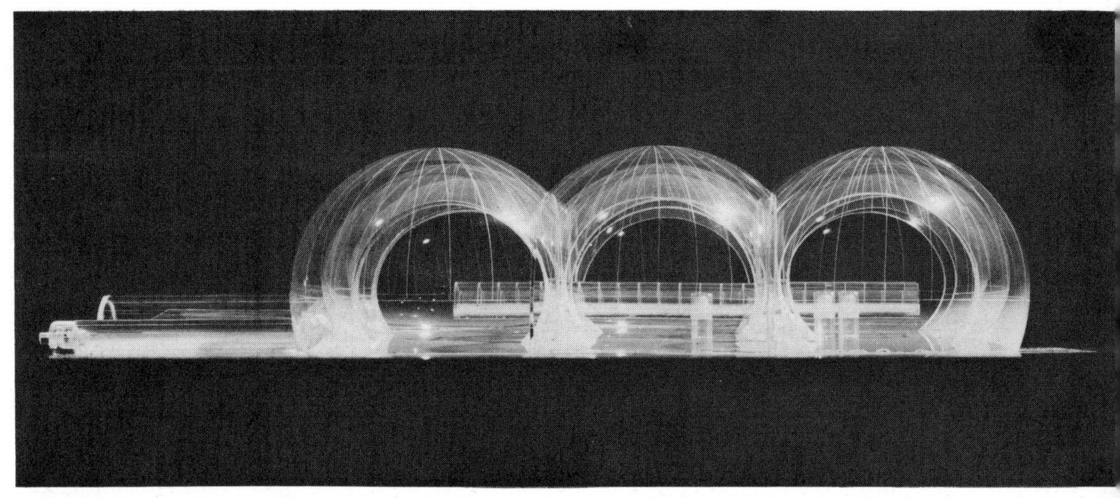

d. Side view of the model, executed in clear acrylic.

e. Close-up of model.

f. Units can be expanded almost to any length.
(Courtesy, Lapietra, Lomazzi, D'Urbino, DePas, Beretta)

WINDOWS

Epoxy and polyester-fiberglass windows have made the scene several years ago with some success. But the most remarkable technique made possible through plastics is a process called "Farbigem." Farbigem glass makes feasible

a. A Farbigem window at Riverside Park Methodist Church, Jacksonville, Florida.

b. Close-up of the Riverside Park Methodist Church window by Willet Studios, showing the Farbigem projections of layered glass. Colored glass underneath transmits color through the sculptured strata of plate glass.

c, d, e. Another treatment of the Farbigem process at Ohio State University, depicting details of a large glass mural showing the world history of medicine. Note that lead lines have given way to layering of glass projections.

whole walls, that project out at least a foot, in organic-looking carved strata of glass. Starting with a sandwich of acrylic (the "meat") and plate glass (the "bread"), areas of colored glass and/or masses of clear crystal can be laminated layer on layer on one side or both sides. Transparent acrylic glue is used to hold the whole thing together. Adequately describing this mar-

(*Courtesy, Willet Studios*)

f. An apprentice's panel at Willet Studios that illustrates the way colored glass in the "background" transmits through the projected layers of glass. Fractured edges of glass refract the light so that light does not obliterate edges of shapes as it could, if not controlled, in leaded glass. (*Courtesy, Willet Studios*)

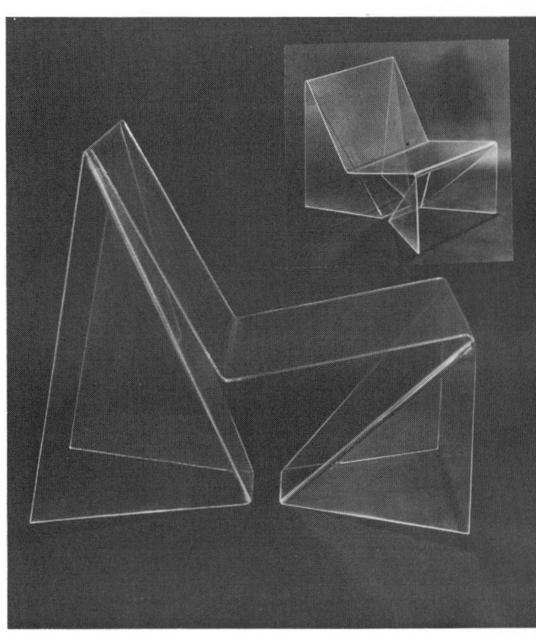

Neal Small's Plexiglas chair illustrates a "folding-back" solution to increasing strength of acrylic sheet. This shape probably was formed using a strip heater and jigs accurately to maintain the bends. 24″ wide, 30″ high, 30″ deep. (*Courtesy, Neal Small Designs, Inc.*)

velous three-dimensional effect is very difficult. It really has to be experienced. Pictures do not do it justice.

Willet Studios of Philadelphia have brought Farbigem to the United States from Holland.

FURNITURE

Take a piece of oak tag. Curve-bend-fold it and you have a possible model for acrylic furniture. Between the design model and production, though, are problem solutions as to how the design will hold up in use. If you take a thin, flat sheet and convolute it, or fold it back on itself, the sheet has been strengthened by dint of its new structure, much like corrugated cardboard is or a scallop shell. Furniture, therefore, need not be made of very thick stock (¾″ or more), if these structural concepts are followed.

A piece that is to absorb heavy weight, such as a chair, should not have right angles, but rather curves—preferably, curves that fold back on themselves. If verticals are necessary, then thick stock must be used.

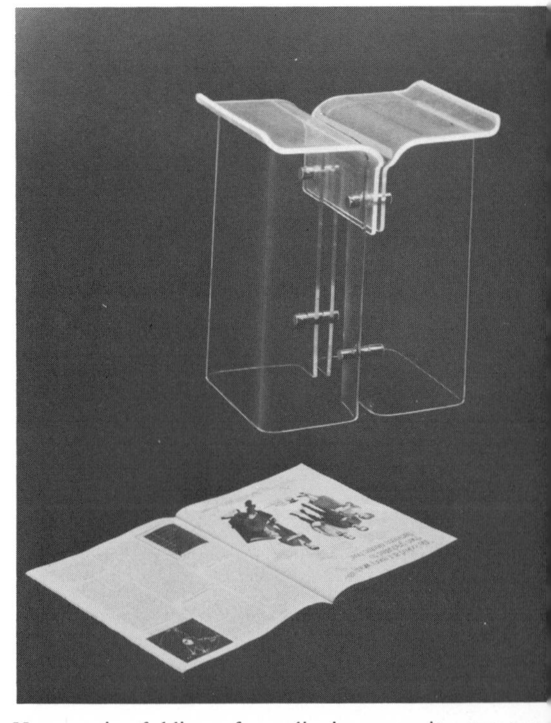

Here again, folding of acrylic increases its structural strength. The base is of clear polished Plexiglas and the seat is either smoke or fluorescent hot pink Plexiglas. Connectors are of polished chrome. Designer, Neal Small. 16″ wide, 13″ deep, 16″ high. (*Courtesy, Neal Small Designs, Inc.*)

Two pieces of ¾″ clear Plexiglas form this chair. They are joined with polished chrome fittings. Seat area is 16½″ deep, 24″ wide; over-all height, 24¼″; over-all depth, 23″. Blosser Foyer Lounge Chair by Design Group. (*Courtesy, Design Group*)

Cocktail table. A cylinder of cast acrylic is topped with an engraved acrylic sheet. Designer, Thelma R. Newman.

Acrylic cut and glued into a lounge chair by Spiro Zakas. Upholstery consists of covered cushions of urethane. (*Courtesy, The Spiro Zakas Association*)

Footstool, bench of opaque acrylic is made of folded ¾" stock. Design by Spiro Zakas. (*Courtesy, The Spiro Zakas Association*)

End/magazine table of red or white Plexiglas by Superstudio. 32" x 16½" x 12" high. (*Courtesy, Stendig, Inc.*)

The chaise is thermoformed acrylic. One-inch thickness was necessary because of the span. The upholstery is metallic-vinyl. The tray-table also was made of thermoformed acrylic with an epoxy finish for the top. (*Courtesy, Project One*)

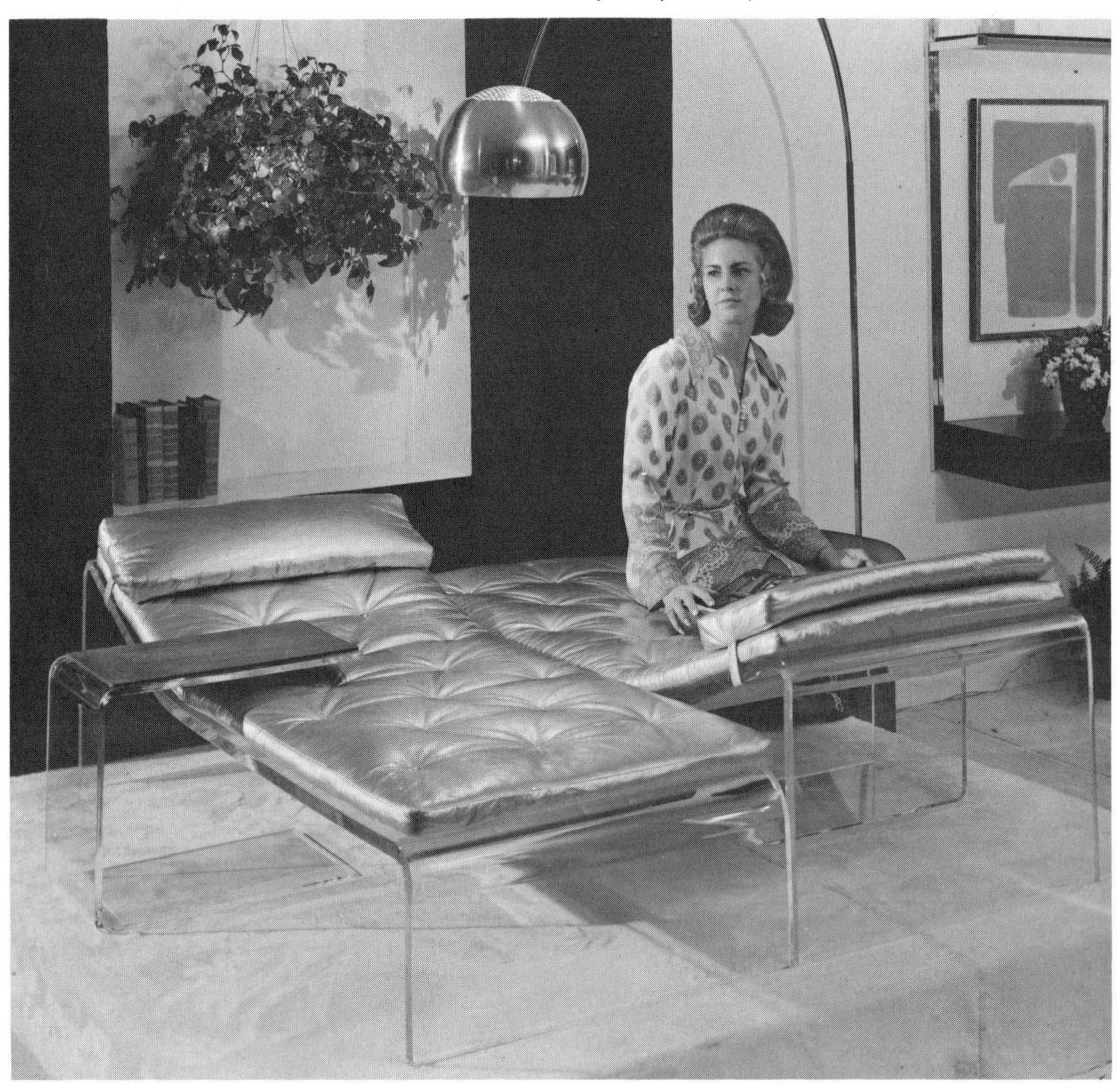

Ingenious slices and four bends makes a table by Neal Small 36″ x 36″ x 15½″. (*Courtesy, Neal Small Designs, Inc.*)

LAMPS

Acrylic's unique light-piping quality can intensify light wherever the acrylic surface is cut, incised, sanded. If utilized intelligently, light piping can be used to transmit and diffuse light as well as to mask the light source and create textures and patterns—all possibilities in the same design.

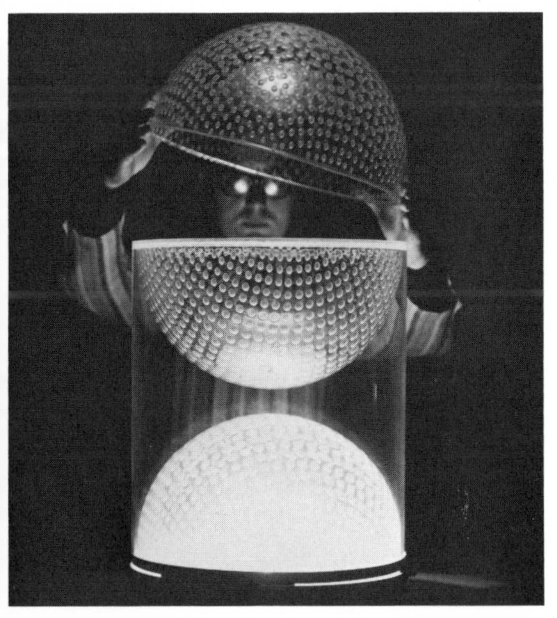

Ugo Lapietra's lamp uses three acrylic half spheres; two are connected to a cylinder. Routed circle indentation, flame polished and later free blown, stretch the dot patterns. (*Courtesy, Ugo Lapietra*)

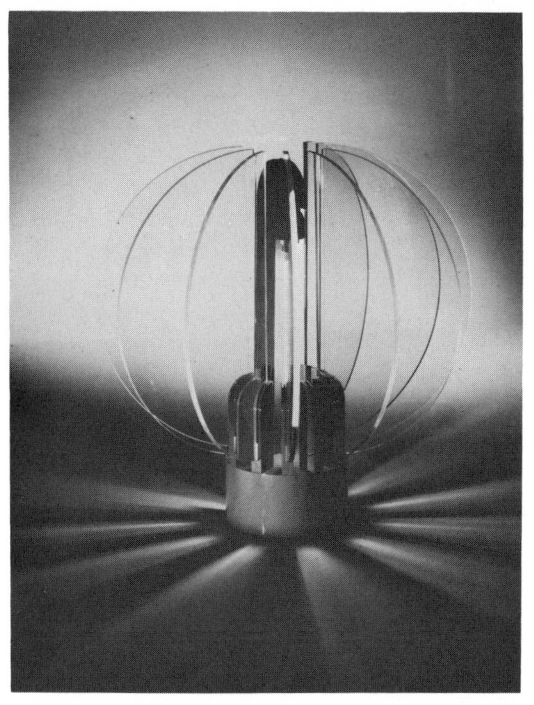

Gae Aulenti's lamp of acrylic fins on an aluminum base for Kartell projects radiating light patterns. (*Courtesy, Gae Aulenti*)

Wihelm Vest of Austria designed this opened artichoke lamp of acrylic for George Kovacs Lighting. Because of open spaces between the patterns of opal acrylic, allowing heat to ventilate, large bulbs can be used. (*Courtesy, George Kovacs Lighting, Inc.*)

Designer Filippo Panseca created this light-mirror in an acrylic globe. (*Courtesy, Kartell-Binasco*)

Another desk lamp by Neal Small is built of red, black, yellow, or white opaque Plexiglas with white translucent Plexiglas sides to diffuse the light of a 40-watt bulb. (*Courtesy, Neal Small Designs, Inc.*)

Neal Small's award-winning desk lamp is made of satin black Plexiglas with a mirror mounted in the top to reflect a light beam without glare. It uses a 50-watt reflector bulb. (*Courtesy, Neal Small Designs, Inc.*)

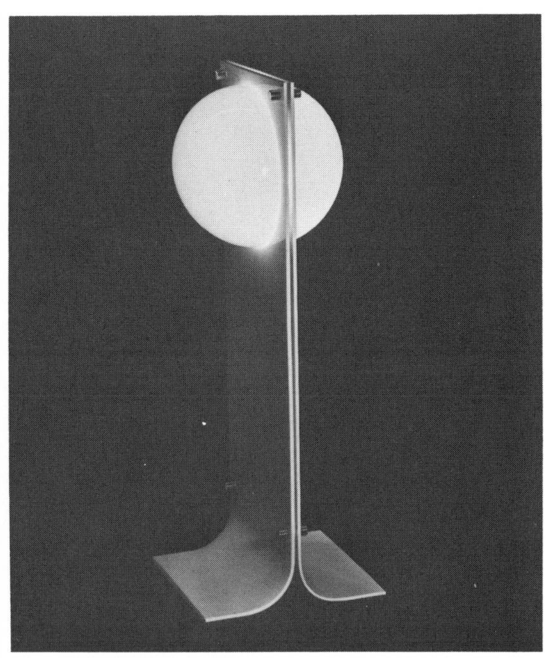

Neal Small's floor lamp has seen its imitators. White Plexiglas has been used but clear acrylic also should be very effective for this design. 40″ high x 14″ wide. (*Courtesy, Neal Small Designs, Inc.*)

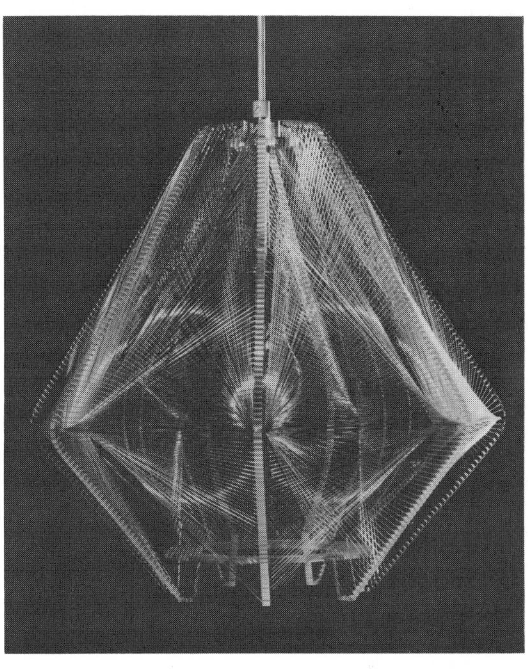

"Plexima" is made of slotted clear acrylic, strung with clear nylon string. The design owes its inspiration to the sculpture of Naum Gabo. (*Courtesy, Koch and Lowy*)

A far cry from the traditional lamp and shade, these area lamps are of blow molded acrylic. The top half is red, black, or white opaque acrylic and hides the bulb, and the bottom half is patterned light diffusing clear acrylic. There is a vent slit at the top. (*Courtesy, Neal Small Designs, Inc.*)

CONTAINERS

Vases for flowers, wet or dry, pitchers, magazine racks, umbrella stands, display containers are but a few uses for acrylics. Tubes and sheet fabricated by bending, slicing, blowing, and gluing are all that is necessary for a simple processing repertoire.

A creative idea, these double vases by Neal Small probably were made by strip heating and bending. PS 30 would be a recommended cement for applications, such as these, where the joint is exposed to water. (*Courtesy, Neal Small Designs, Inc.*)

These pitchers are merely tubes that are sliced; the area to be formed is heated and a jig is used to force out the lips of the pitcher. Sizes vary from 5″ high to 9″ high. Designer, Neal Small. (*Courtesy, Neal Small Designs, Inc.*)

This umbrella stand is made of an extruded tube of opaque white acrylic. The central stem has half circles drilled at the bottom to allow water that drips off umbrellas to overflow into the base. The base also serves as a balance to keep the stand from tipping. Designer, Thelma R. Newman.

Another Neal Small design that utilizes bending and folding to strengthen the structure. Record or magazine rack, 12″ wide x 12″ deep x 12″ high. (*Courtesy, Neal Small Designs, Inc.*)

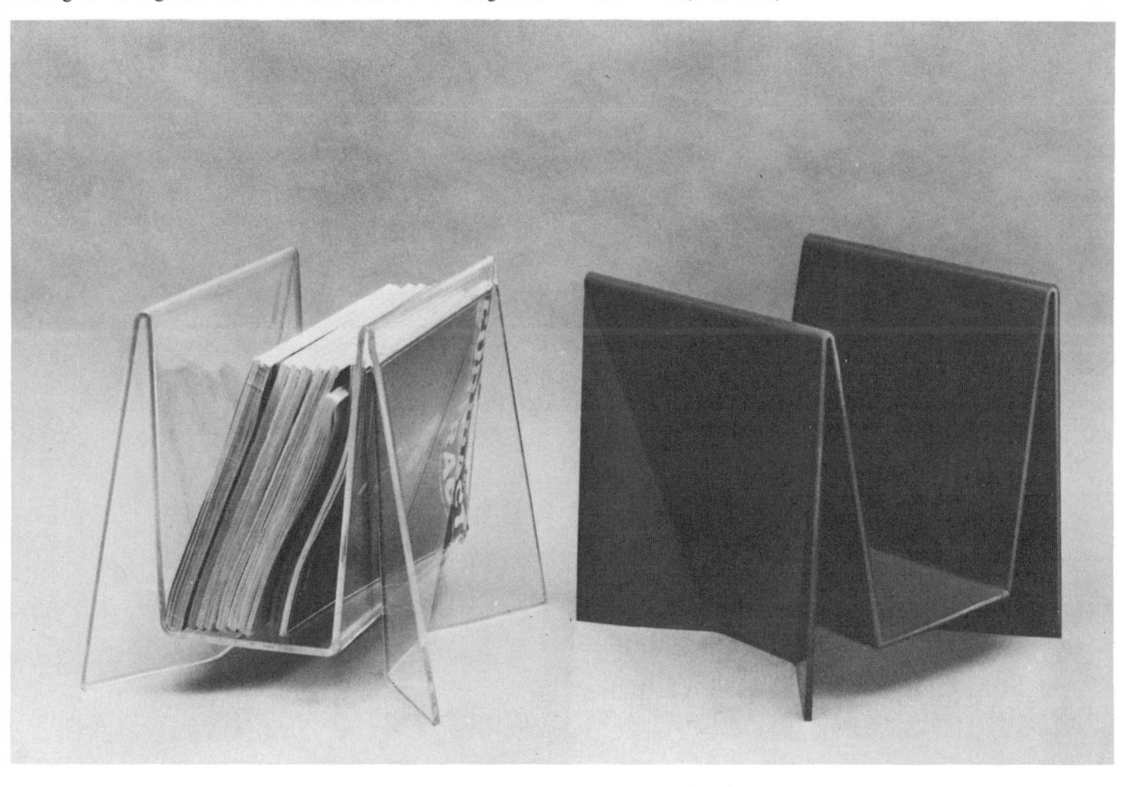

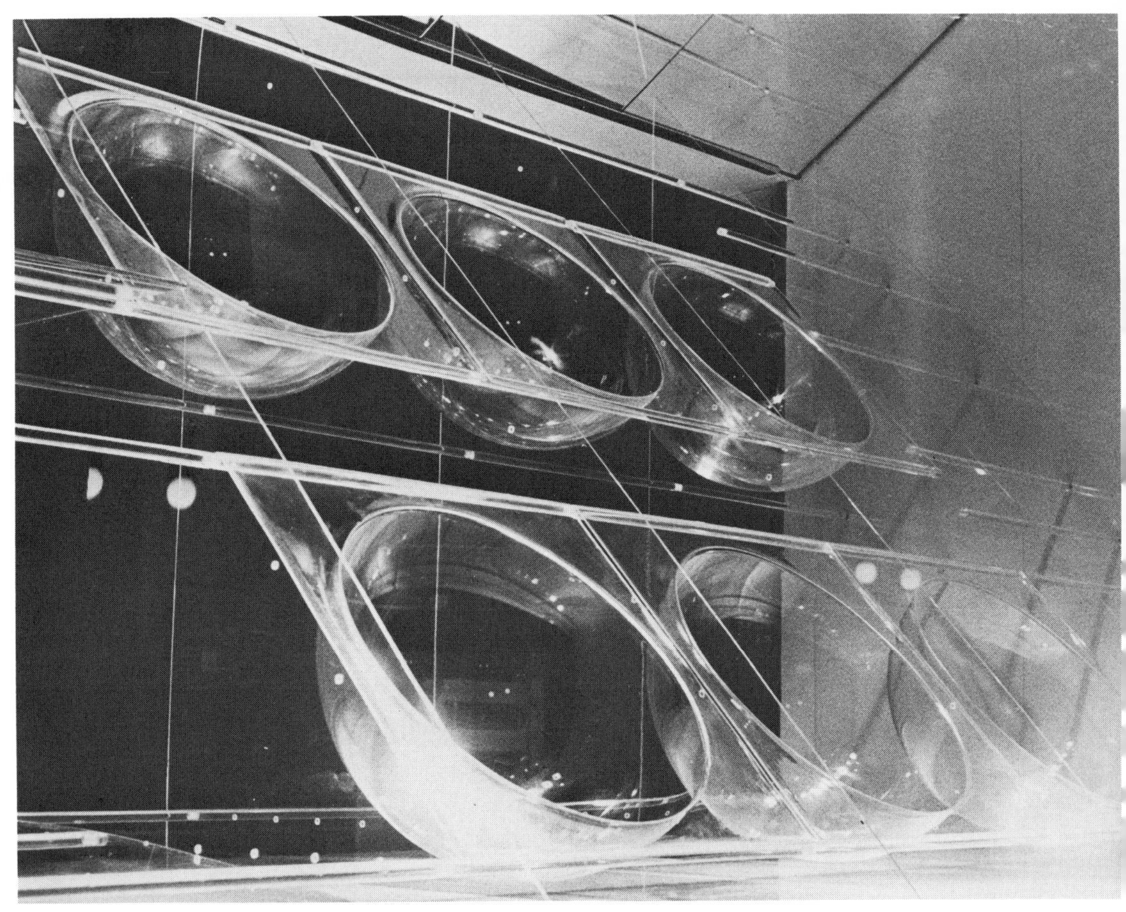

Blow molded acrylic half spheres complete with flange are used as store window display containers. It is an advantageous structure because light is not blocked and the merchandise can be seen from both sides. (*Courtesy, Ugo Lapietra*)

ORNAMENTS

Clear acrylic has a gemlike quality. Its pleasing surface works well and harmonizes with other materials by dint of incorporating the textures and colors of wall, skin, table, floor—whatever—in its own form.

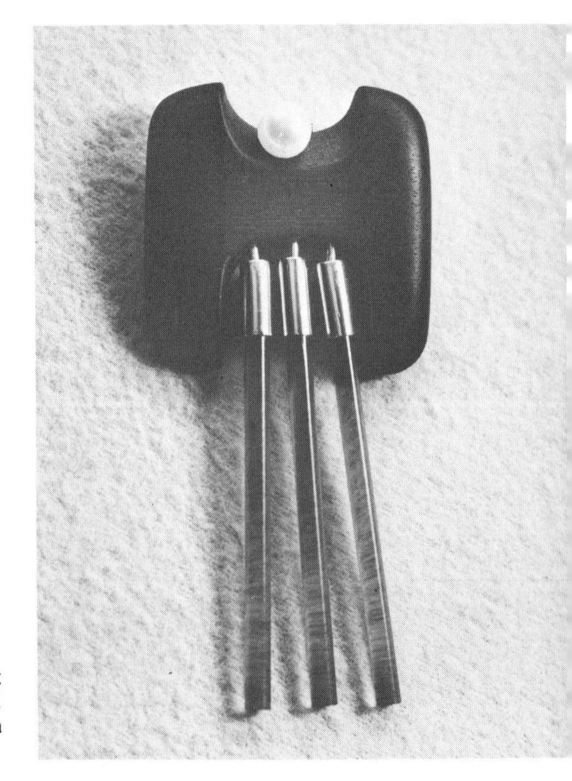

This pin is made of ebony with sterling ferrules holding red Plexiglas rods. A 6¼ mm cultured pearl is nested in the ebony concavity. 3½" x 1½". Designer, Alan Landis. (*Courtesy, Alan Landis*)

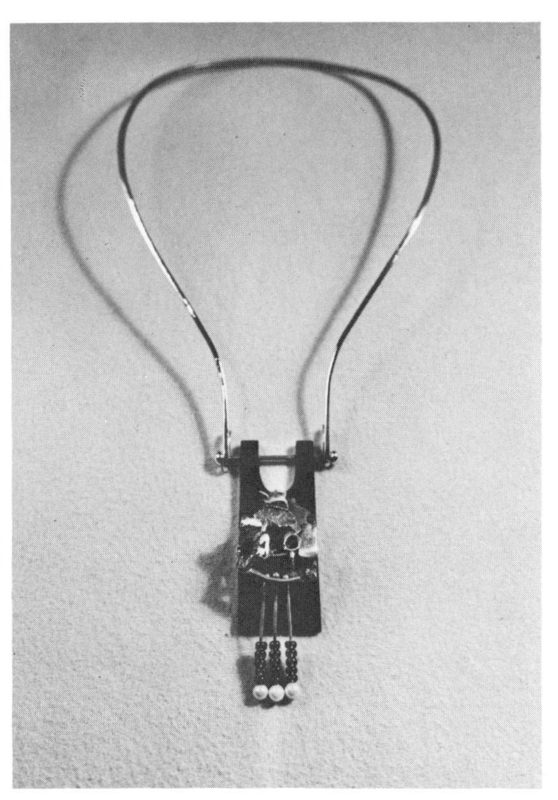

Alan Landis' necklace is made of sterling silver and green Plexiglas. The beads are cultured pearls and glass. Acrylic is treated as importantly as "precious" materials. 10″ x 5″. (*Courtesy, Alan Landis*)

Bix Lye's clear Plexiglas rings were exhibited in the Museum of Contemporary Crafts' Plastic as Plastic exhibit. Colors are cemented, holes are meticulously polished, surfaces are carved and pierced to catch light. (*Courtesy, Bix Lye*)

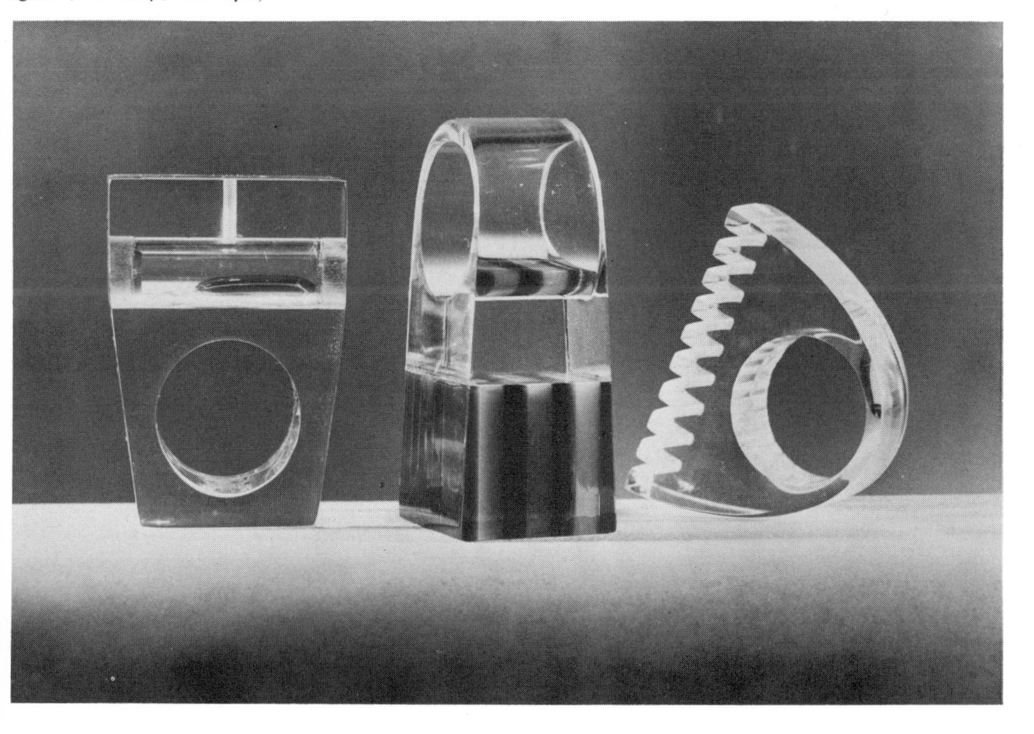

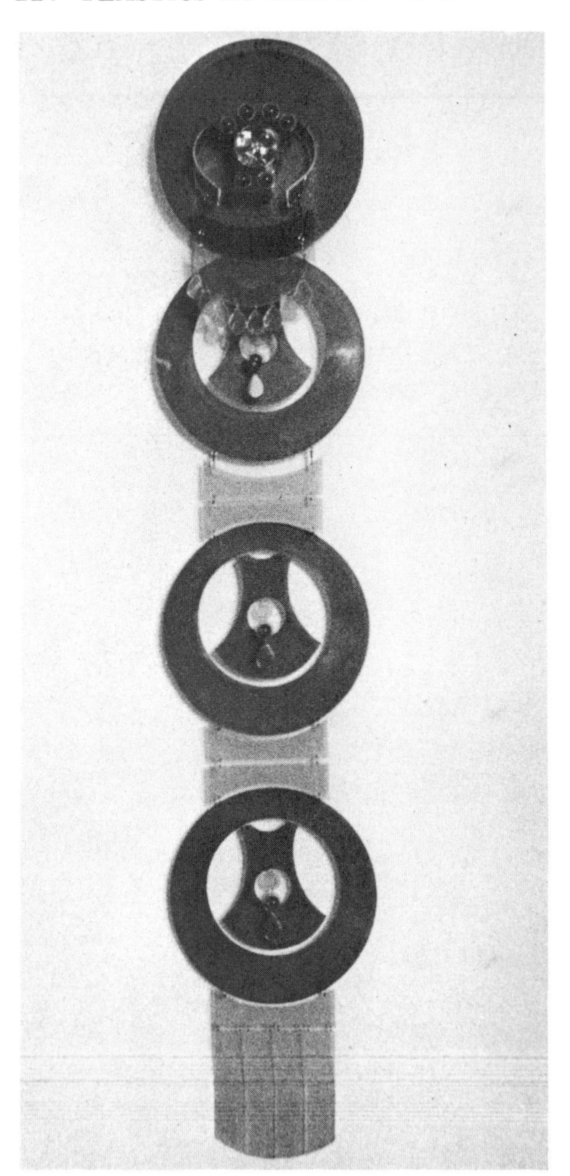

Carolyn Kriegman's humorous design can decorate either wall or body. Acrylic parts are linked together to become a flexible mobile ornament.

Nancy Thompson's bracelet is made of three transparent Plexiglas shapes that are connected with two-way swivels and brass beads. The design is an alternation of open and closed relationships. (*Courtesy, Nancy Thompson*)

Yellow and orange acrylic rods are assembled with gilt screweyes and jump rings. The necklace, by Nancy Thompson, is based on a modular relationship. (*Courtesy, Nancy Thompson*)

PLAY FORMS AND OTHER THINGS

Small quantities of a creative idea can be tested because manufacture (by hand) is simple. If an idea succeeds, then large quantities that look the same as the hand-crafted ones can be produced through injection molding mainly, which will bring costs down after the costly mold is amortized.

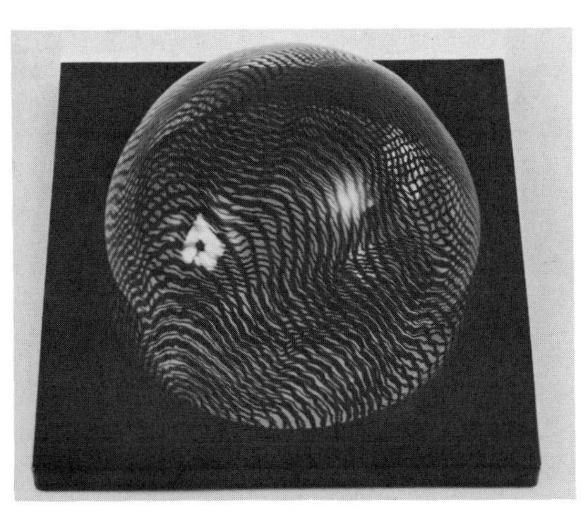

Henry Pearson's "Moirasphere" is a blow molded dome silk-screened with abstract curvilinear lines that are reflected and repeated in a moire effect by a mirror in the base. The form is manufactured now out of a less expensive plastic, perhaps acetate. Involvement is a concept in most of the transparent adult "toys" on the market today. (*Courtesy, Artmongers Manufactory, Inc.*)

Two acrylic cut-outs can be rearranged for an interplay of light and pattern in Walter Stein's "Spring Wind." (*Courtesy, Artmongers Manufactory, Inc.*)

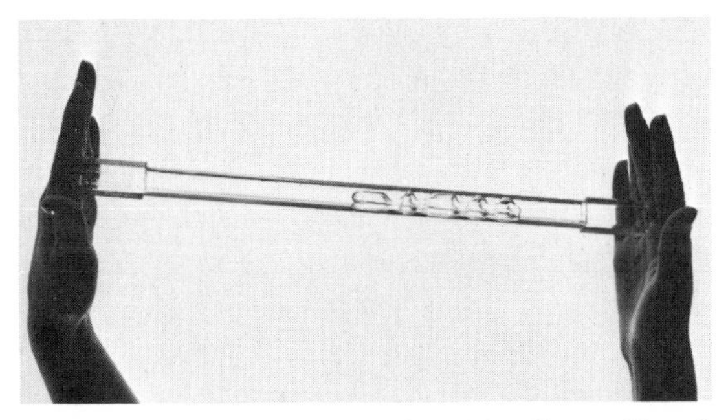

"Bloop, Bloop, Bloop" was designed by Casper Henselmann. The clear acrylic tube encases small floating disks that move at flick of the wrist. (*Courtesy, Artmongers Manufactory, Inc.*)

"Space Moon" by Merle Steir is adult entertainment. Three clear, blue or gray acrylic triangles are interlocked with a solid 2″ acrylic ball. The sphere acts as a lens to reverse and magnify changing arrangements of the triangles. (*Courtesy, Merle Steir*)

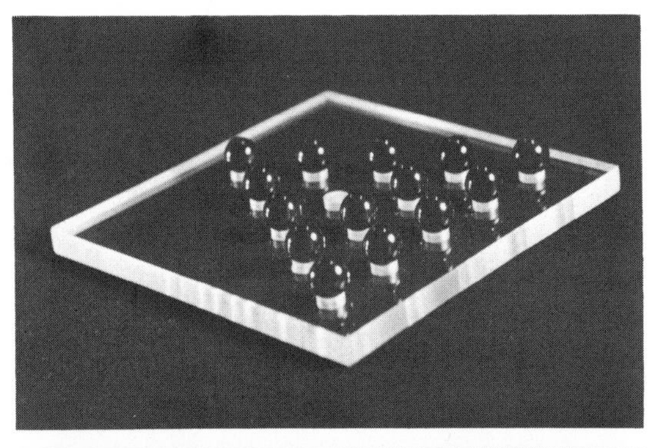

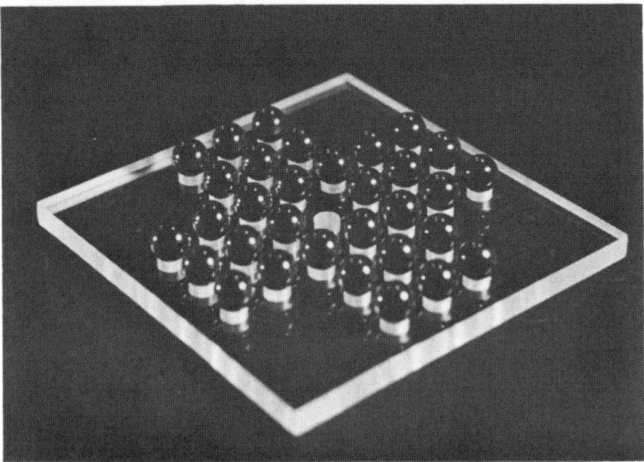

Jay and Lee Newman designed these acrylic solitaire games. The base is of clear ½″ acrylic and the balls are of transparent colored acrylic. (*Courtesy, Jayly Innovations*)

You could call it a sculpture. It is a "Hard Pillow" by Neal Small that is molded of rigid acrylic filled with down. 21″ x 27″ x 8″ high. (*Courtesy, Neal Small Designs, Inc.*)

Whether letter opener or cheese knife, this satin-polished Plexiglas rod, by Neal Small, is simple and functional—one of those "Why-didn't-I-think-of-it?" ideas. (*Courtesy, Neal Small Designs, Inc.*)

They are called sunglasses, but they could be identified by future civilizations as a "twentieth-century acrylic face ornament." (*Courtesy, Bernard Kayman, Ltd.*)

7

designing with foam, film, and fiber

Why consider foam, film, and fiber in one chapter? Is there a relationship? Some qualities correlate; others complement. Foam, film, and fiber each can be worked with directly without the intermediary of complex equipment (except in direct foaming and foam furniture forming from molds). These plastics are immensely flexible, both versatile and pliable (but for rigid foams). Film and fiber can be used to cover foam, but each group of materials can exist because of its own attributes. With creative designing each material can express singularly different possibilities. The uniqueness of these plastics—foam, film, or fiber—provides the potential for break-through in furniture, clothing, and structures, whether the material services a pace-setting design idea or whether the potential of the material is extended.

Flexible urethane foams that have their own skins, and integral covering, have been perfected recently. Old concepts of upholstery and furniture structure no longer are necessary with these materials. Dense and less dense foams can be designed into the same piece—again, as a homogeneous material—with no gluing or fastening necessary. There also are exceedingly tough rigid urethane foams that can be as dense and hard as wood or as soft as a cotton candy. Rigid and flexible foams also can be layered or combined in the same piece. These are long-lasting, versatile materials. Imagination can go soaring here because there are few limitations. One does not have to worry about size, gluing and fastening, weight, hardness-softness-strength, et cetera. Almost infinite combinations are possible. One imaginative possibility is to stretch burlap or a similar woven support between cables, spray about three inches of foam on the burlap, and then finish it off with a urethane finish or a thin layer of fiberglass. Organic surfaces with a grottolike appearance that defies wall-ceiling constrictions will become one of tomorrow's environments.

Free rising of rigid polyurethane foam. The amount of time it takes foam to rise is called "cream time." (*Courtesy, Reichhold Chemical Company*)

DATA CHART FOR POLYURETHANES

Appearance: Clear and colorless as a "raw material"; white as a foam

Physical Properties: Excellent adhesive properties; excellent abrasion and tear resistance; excellent resistance to impact and shock; good dimensional stability

Moisture: Low water absorption; dimensional stability and electrical properties less affected than nylon

Heat: Softens above the temperature of boiling water; begins to deform at 200–300°F.; burns slowly; special flame-resistant varieties available; foams more heat-resistant

Solvents—Chemical: Good resistance to most common solvents and chemicals; attacked by oxidizing agents

Light and Environments: Light has a negligible effect on its properties, although UV radiation will quickly discolor the surface of white foam; inert to fungus, bacteria, and animal and rodent attack; degrades when exposed to ionization; long-lasting—age has little effect

Available Forms: Film, tube, rod, coatings, molding powders for injection, compression molding and extrusion, adhesives, and foams—rigid and flexible

POLYURETHANE

Polyurethanes are thermosplastics, but their foams are thermosetting. They are produced in the form of film, tube, rod, molding powders for injection, and compression molding and extrusion as well as in foams and coatings. Emphasis in this chapter will describe the application of foams to design form.

Development of isocyanate polymer (polyurethane) first began in 1937 in Germany by Otto Bayer. It was introduced into the United States in 1953 as a replacement for foam rubber. Polyurethane, as foam, is an open-cellular material that is produced by reacting a polyisocyanate and a polyhydroxyl material under controlled conditions. Blowing agents and catalysts control reaction rates, and surfactants control cell structure. Other additives may include flame retardants, fillers, pigments, and the like. In mixing, an excess of isocyanate reacts with water to form carbon dioxide. The carbon dioxide that results expands by the heat generated in the reaction, entrapping the carbon dioxide to form a cellular structure. By controlling all of these elements chemists can "tailor-make" urethane

foams to almost any resiliency and density. Some rigid urethane foam systems produce a self-skinning surface that picks up every detail from the mold.

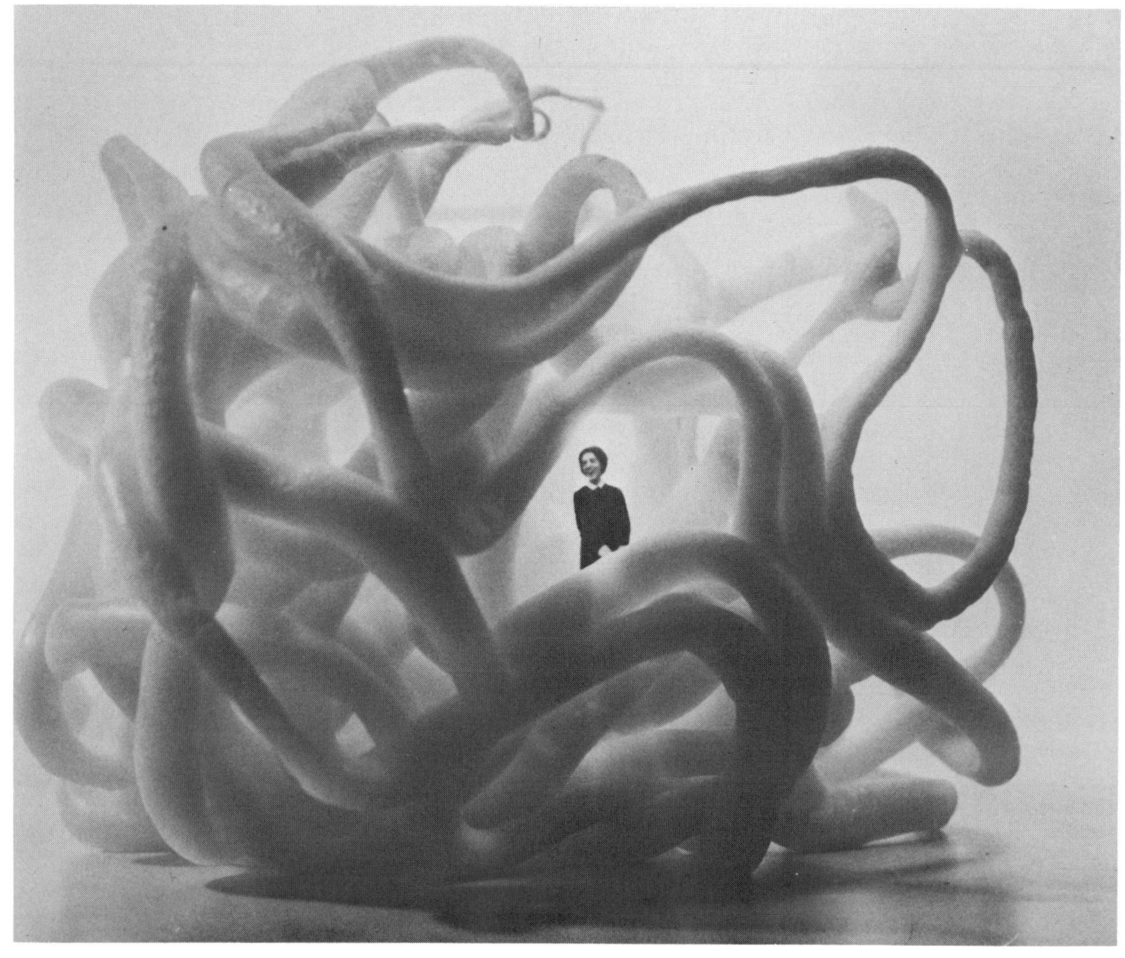

Extruded rigid urethane foam that foams and sets instantly like shaving cream from an aerosol can, if it were rigid, could create this labyrinth of an environment. (*Courtesy, Chemstrand Co. and Michael Lax*)

RIGID FOAMS

Urethane foaming is generally quite rapid and hence requires specialized equipment for metering and mixing components when it is used in production. Urethanes can be foamed, however, in a well-ventilated area, after carefully weighing components and then mixing them with a mechanical paint-mixing tool and pistol drill. Sometimes components need to be heated or cooled. There are three general application methods and many formulations.

Pour-in-Place Rigid Foams

Foam components are measured, mixed for a prescribed time, and then poured as a liquid stream into a cavity where they react and expand to fill the void. The mold form should be sufficiently reinforced to absorb the slight-to-moderate pressure generated during the foaming operation It usually takes from 5-30 minutes for complete setting, depending upon system formulation, room temperature and humidity. Eliminating volatile fumes can be a problem; excellent ventilation is essential. Rubber gloves, goggles, and even a face shield should be used, as well as clean clothing that completely covers exposed skin. If these precautions are observed, there is no reason not to work with polyurethanes.

Frothing Rigid Foams

Frothing is a variation of the pour-in-place method. A lower-boiling blowing agent converts to a gas as its pressure drops; foams are dispensed in a partially expanded state, much like aerosol shaving cream. Because of pre-expansion prior to the chemical reaction, a more spherical cell structure emerges, which results in a stronger material but slightly lower in density. This approach requires dispensing equipment. Just as in frothing shaving cream, air can be enfolded and trapped, leaving unfilled voids.

Spraying Rigid Foams

Special spray guns meter, mix, and atomize [1] the materials. Some guns also preheat the materials before dispensing the foam. This type of equipment is usually employed for filling large expanses such as roof areas and hulls of ships, for packaging large forms, for covering frozen cargoes, and for insulating tank cars and buildings.

Rigid urethane foam can be purchased in slabs or blocks in a wide variety of sizes. It can be cut with a band saw, knife, or heated wire and glued with polyvinyl acetate or chloride (Elmer's) or with a rubber cement.

Hand Casting Rigid Urethane Parts

A good variable-speed drill (1,500–3,000 RPM), a clean container made of paper, metal, or polyethylene large enough to hold components, and a paint-mixing blade are all that is needed to do a good mixing job. Selecting the proper urethane system is important. A fluorocarbon-blown foam (Freon) tends to yield thicker skins, be of a lower density and more sensitive to mold temperature than water-blown systems. Density is also important. For forms that have to hold nails or screws, water-blown systems are recommended. Water systems have less density gradation. In rigid foams #2 density is low and #15 is high.

Volume is the next determination. How much material do you need to mix to fill a certain area? Here is a short cut. Fill your mold with water. Weigh the water in pounds and divide by 62.4. This will give you the volume of your mold in parts of a cubic foot. Meanwhile, a mold has been made and readied. A silicone rubber mold (needs no release agent) or a mold of another material such as urethane elastomer, fiberglass, epoxy, vacuum-formed polyethylene sheet, latex, or metal sprayed with a release agent/barrier coat (usually an acrylic base) for urethane is most regularly used. These coatings provide both release from the mold and a barrier coating for subsequent finishing operations. (A 5% Ivory Flakes solution in methanol will work as a release agent.) Since polyurethane foam casting requires closed molds, venting must be provided to remove entrapped air. This is achieved by placing a piece of breathable-type paper [2] over the top of the mold after the material has been poured, but before the top cover (or platen) has been put on.

Weigh out the isocyanate component "A"; then weigh component "B." Combine both components in the container and thoroughly mix them for 30 seconds (or as the manufacturer recommends). Then quickly pour the material into the mold, being certain in the process to wet all the mold surfaces so that no dry spots or other surface defects will result.

The cream time (time it takes for the foam to rise) at room temperature in a room-temperature mold will usually take 45–90 seconds, depending upon formulations. (Heating of mold or component "A" will reduce cream time.) The processing time and end properties, therefore, can be influenced by controlling the temperatures of the mold and the isocyanate component of the resin system. Demolding usually takes 15–20 minutes. Strip the mold from the part and remove the part before it reaches complete exotherm to increase mold life. Removal of a piece as soon as the cover plate is lifted prevents warping.

Flexible Urethane Foams

These foams are produced in gigantic, continuous buns or slabs, which are then cut into standard sizes and shapes. Foams are also directly foamed into shape in a closed mold, and are also produced as a sandwich composite with skins of paper, aluminum foil, et cetera.

Working with Flexible Foam Slabs

Slabs of urethane can come in ice cream colors and various densities. Heavier density foams are longer-wearing because they have a closer cell structure. Lighter density foam has more air in it. Generally foams are purchased according to density (weight per cubic foot) and indent load (compression). A soft, nondense urethane foam may be indicated by 1.15 ± 0.05 density and 18–24 indent load. A medium density urethane foam may be 1.25 ± 0.05 density and 22–26 indent load. A very dense piece would be 2.25 ± 0.15 density, 27–32 indent load, and a hard, firm piece would be 1.50 ± 0.05 density and

[1] Equipment: Martin Sweets Co. Mod Mods 123 and 456, 3131 Market Street, Louisville, Ky. 40201.

[2] "Part Wick," made by Paper Corp. of America, or Wallhide 626, made by Brown Paper Co. See "Sources of Supply."

Al Vrana's design for "Obelisk" is a massive structure consisting of eight large units ranging in size from 12′ x 20′ in length and 1,000-4,000 lbs. each in weight. Dyplast, an expanded polyester foam, is used for the interior structure. Then the ferro cement process (developed by Luigi Nervi), a construction method of using multiple layers of fine steel mesh embedded in Portland cement mortar, is applied over the foam core. It creates great strength with light weight. For example, concrete weighs 43 lbs. per cubic foot, but this method weighs 26 lbs. per cubic foot.

An element for "Obelisk" is being measured in preparation for cutting. Then parts are cut and adhered in place before coating with ferro cement.

The final photo shows an element of "Obelisk" by Al Vrana. (*Courtesy, Al Vrana*)

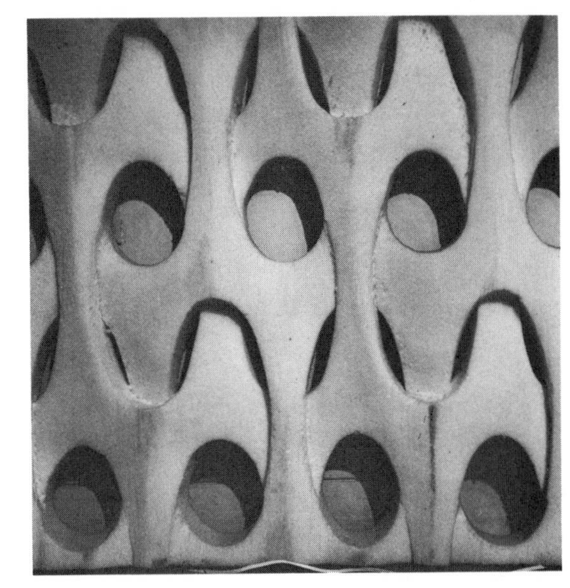

Carved styrofoam maquettes for a room divider by Jack Marshall.

a. Air was blown into the mold to clear out dirt.

STEPS IN MAKING A RIGID URETHANE FOAM CHESSBOARD

(This procedure took place in a laboratory of the Reichhold Chemical Company. Dom Andrisani was the expert.)

It was decided that a free rise density of 10-12 lbs. and a molded density of 20-25 lbs. per cubic foot would be adequate for the chessboard. (The lower the density, the softer the material. Foam can approximate the weight and density of wood, which averages 32-40 lbs. per cubic foot.)

Measurements of the mold cavity were made:

Length, 18.5″ x width, 18.5″ x depth, .75″ = 256.73. If you take cubic inches and divide by 1,728, you have computed cubic feet.

Therefore:

$$\frac{256.7 \text{ cubic inches}}{1{,}728} = 0.148 \text{ ft.}^3$$

Multiply by the desired density, which in this case, is 20 lbs.

$0.148 \text{ ft.}^3 \times 20 \text{ lbs.} = 2.96$ lbs. of foam

Converting to grams:

Foam 2.96 lbs. x 454 = 1,344 grams.

700 grams of component A

= 1,400 grams

700 grams of component B

This provides a bit extra for loss in the container. Reichhold's Polylite 34-718 urethane resin and Polylite 34-842 isocyanate were used.

b. A lacquer barrier coat was sprayed into the mold to act as a release agent to extend the life of the mold, and to facilitate decorating later. Barrier coat was sprayed onto the release paper as well.

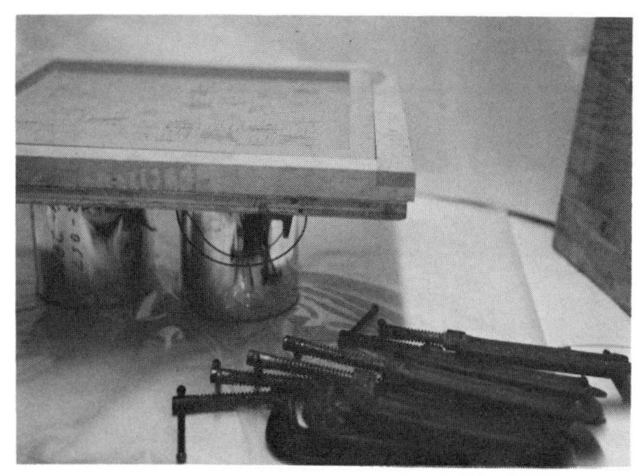

c. Clamps were adjusted and readied for rapid clamping.

d,e. The resin and isocyanate were weighed out and set aside. Then the mold was heated to 90°F.

f, g. Ingredients were combined and mixed for 30 seconds at a mix speed of 2,500 r.p.m., with the components at room temperature.

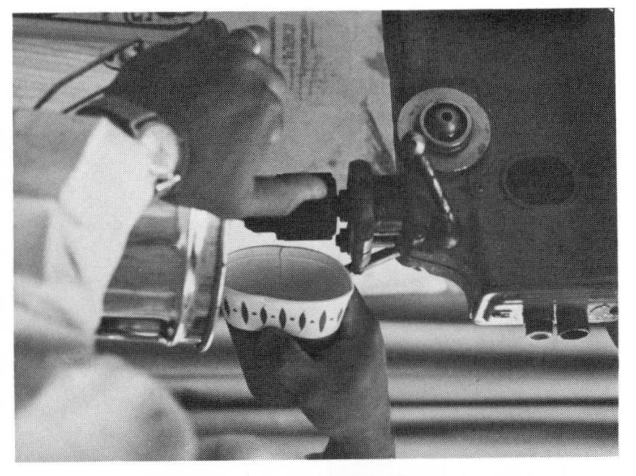

h. The mixture was poured into the mold.

i. Release paper, which had been sprayed previously with barrier coat, was placed over the filled mold to allow the mold to breathe, and then the mold cover was put on top.

j. Clamps were quickly attached.

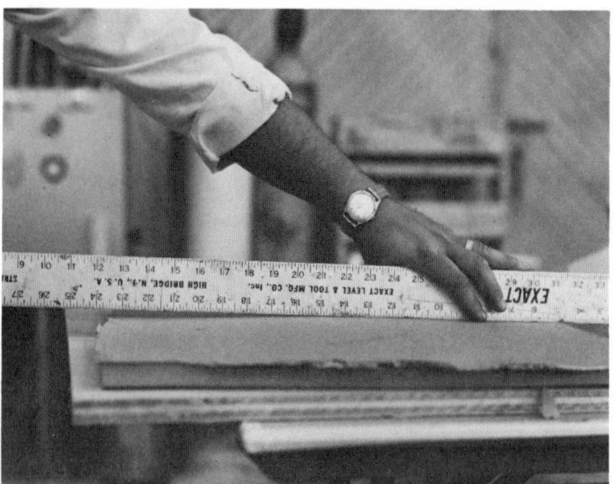

k, l. In a first attempt, steel reinforcing bars were not used on the mold and warping resulted because of foaming pressure.

m. In 10-12 minutes the form was ready for demolding. Cover plate was removed; release paper (Part Wick) was lifted off.

n. The inside frame was lifted away from the mold and the awaited moment was at hand.

o. The mold was lifted away from the chessboard.

p. And excess foam, or flash, was trimmed away from the part.

q. The completed chessboard, before the finishing coat was applied.

r. If this chessboard were to be mass-produced, the plant would look something like this. A mold press is in the foreground, to the right is the assembly line, and in the background is foam metering equipment. (*Courtesy, Flexible Products Company*)

s. A close-up of a mold being transferred to the mold press for curing. (*Courtesy, Flexible Products Company*)

t. A close-up of foam metering equipment. (*Courtesy, Flexible Products Company*)

Kenneth Winebrenner's rigid urethane mirror was made by foaming directly over a wood-framed mirror and then painting the foam. (*Courtesy, The Virginia Museum of Fine Arts*)

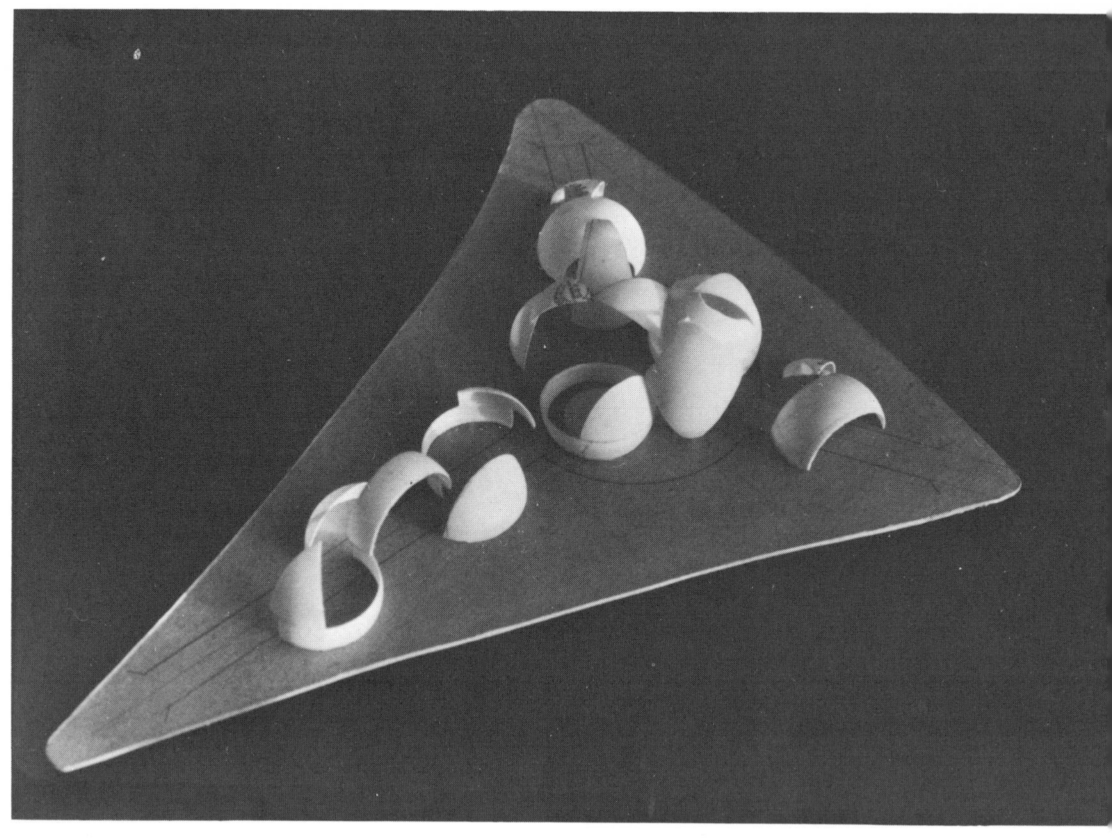

Felix Drury's "Three Rivers" Project
Model of the "Three Rivers" environment made of pin, pong balls.

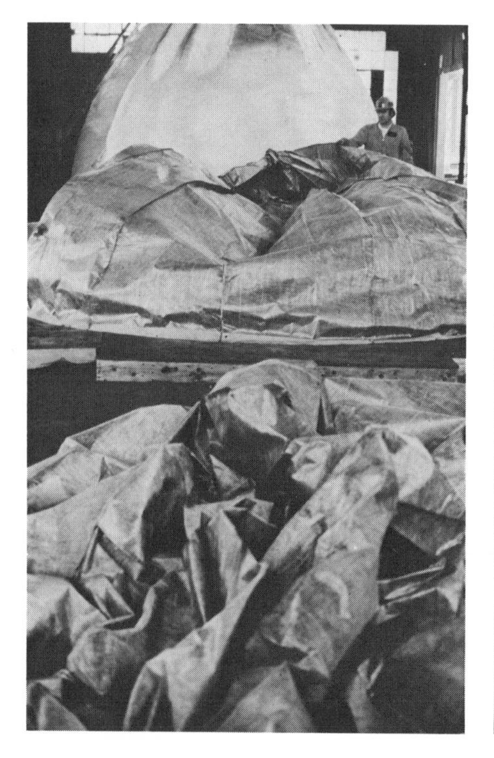

a,b,c,d. The initial forms were sprayed over "bal loons," first with barrier coat and then with urethan at a warehouse. It took three weeks to complete the job

e. Parts were cut to size with a saw. Openings were cut into the shell after air was released from the "balloons."

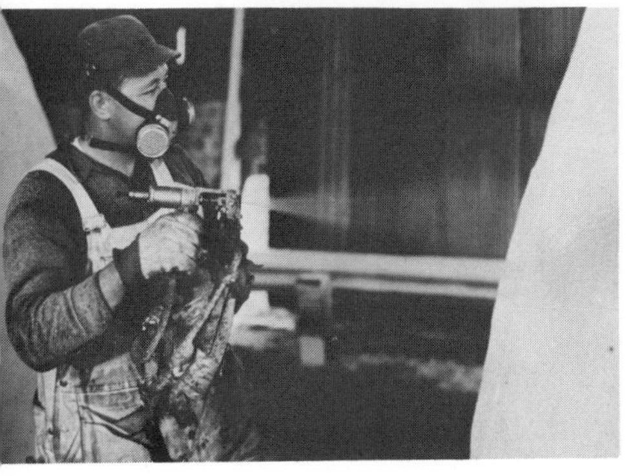

f. The forms were loaded on a river barge and floated down the Allegheny River to the downtown dock.

g,h,i. The shapes were unloaded from the barge at Gateway Center and brought to a small triangular park for erection.

j,k. Workmen assembled the parts and, with portable spray units, gave the environment a final coating.

m,n,o,p. Five viewpoints of Felix Drury's environment for "The Artist Looks at Industry" exhibit. The glass beads in a window are glass marbles used in the manufacture of fiberglass yarn. PPG Industries financed the project and provided some technical help.
(*Courtesy, Edward C. MacEwen of PPG Industries*)

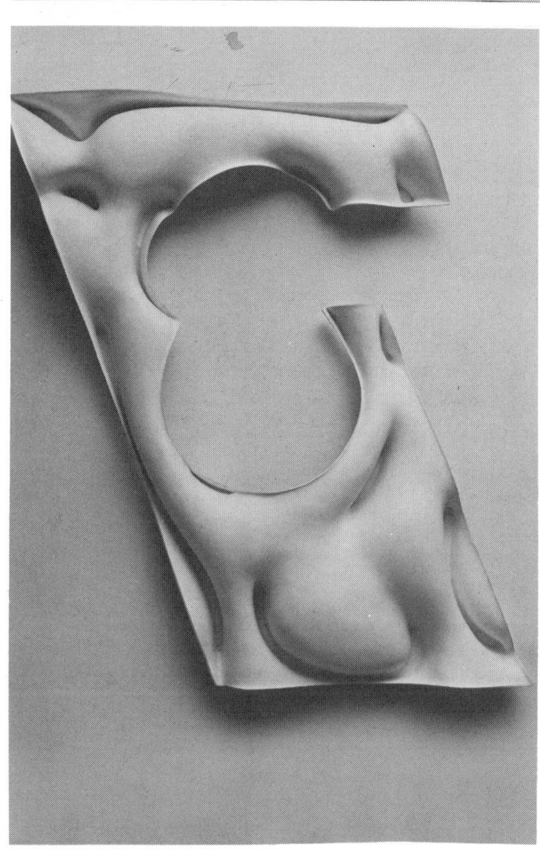

Three views of a model for a projected recreation area in a park conceived by Architect Angelo Mangiarotti. (*Courtesy, Casabella and Angelo Mangiarotti*)

"Ensculptic" is an environmental sculpture in plasti or, if you will, a home in Minnetrista, Minnesota. Th house starts with a mast that is a steel reinforced cor crete hollow column that also serves as a fireplac chimney. Nylon cables are attached near the top c the mast and then anchored in the ground at predete mined points. (*Courtesy, James L. Littlejohn an Ensculptic, Inc.*)

Another model for a park concept by Angelo Mangiarotti. (*Courtesy, Casabella and Angelo Mangiarotti*)

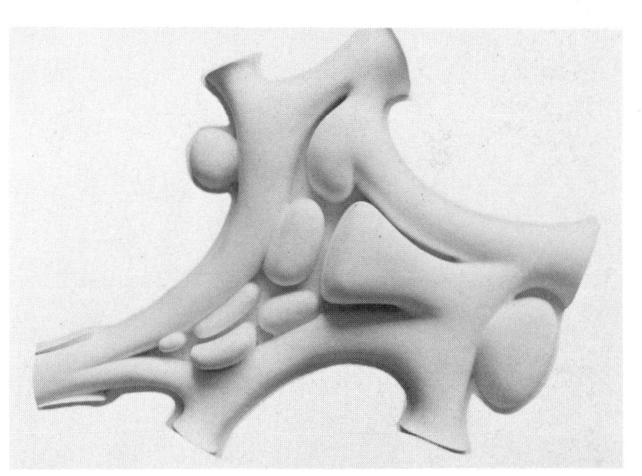

Large precut sections of inexpensive burlap are sewn between the cables, and a "webbing" is ready for the first application of rigid polyurethane foam. Foam is sprayed on the outside of the burlap tent. The roof thus formed is sprayed with fiberglass reinforced polyester.

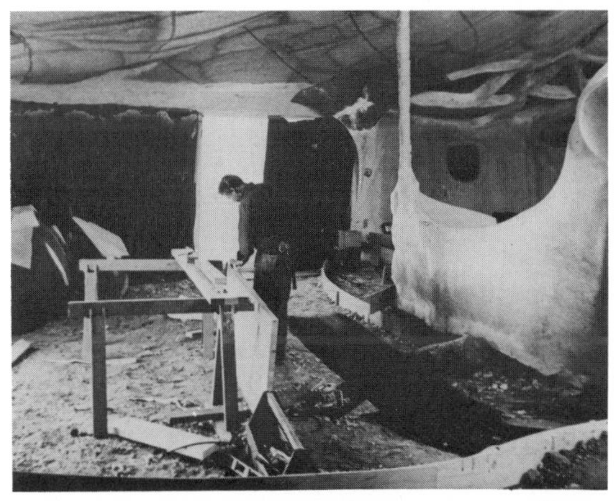

Walls are formed underground with concrete blocks; where cabinets and shelves need to be, walls are of plywood; the other walls are formed over extended steel mesh with enough rigid uprights to hold the mesh to the desired curves between floor and ceiling. At this point, wiring and plastic plumbing pipes are installed on the wall forms and underground. Then the walls are sprayed with foam, thus sealing in the wiring, plumbing, door and window frames, and other built-ins. The floor is reinforced concrete, finished with a poured epoxy covering.

Some views showing the sculptured interior. Urethane foam spray-up is utilized for what it can uniquely do and is not forced to assume a traditional post and lintel structure best suited for other materials that have joints and seams. Ensculptic, Inc., has been formed to advise on building these houses. This house belongs to James L. Littlejohn.

48 indent load.[3] Of course, the price would be higher, the heavier the foam.

Cutting

Cutting of flexible foam is done on a vertical band saw. Tremendous saws with 40″ possible cutting heights and a 1.1 HP motor with a table that would accommodate a cutting length of 89–121″ are used by industry. Some of these saws are adjustable so that they can be angled.[4] The ordinary band saw used for cutting wood will also do a good job in cutting smaller pieces. Because of lack of resistance of the material, hand tools are very difficult to control. There are, however, hand tools resembling an electric knife or hedge cutter that can be used as a hand-held cutter. The versatile hot wire cutter, applicable for cutting all kinds of foams, performs well when heat can be adjusted for varying densities.

Ian Walker developed instructions for making the following hot wire tool:

Extend a single strand of #10 nichrome wire at a right angle to two pivoting members. The tension on the nichrome wire is adjusted by a turnbuckle at the other end of the pivoting members. In appearance this tool looks like the old-fashioned buck saw or the contemporary Danish Saw. An electrical conductor is attached to both ends of the nichrome wire, which thus completes the electrical circuit. The electrical lead is then attached to a Veriac. This device controls the ampere flow through the nichrome wire. By controlling the amperes it is then possible to control the temperature of the nichrome cutting wire. To achieve a uniform cutting tem-

[3] Indicating system of Stauffer Chemical Co. See "Sources of Supply."
[4] LeBrecht Machine Co., Inc. See "Sources of Supply."

Planks are laid over four quadrants in staggered joints. H-clips assist in twisting the foam shell into double curvature segments of the hyperbolic paraboloid. (*Courtesy, Modern Plastics*)

Cement mortar using the ferro process is brushed on top of the foam shell to provide stiffening.

Completed roof structure.

perature: 4.2 to 4.4 amperes are required to cut polyurethane foam of 1.2 density. As the density of the foam increases, more amperes are required. If there is too much heat, small nodules of melted foam are deposited along the cut and a scorched look is evident. It is also a good idea to develop some kind of jig to help guide the cutting tool. With a jig thicknesses of up to 18″ can be accurately controlled, but if the thickness goes beyond that, the cutting wire tends to bow and produce an uneven, concave cut. The best bet is to have larger pieces cut commercially where precision cuts are possible.

Coating and Covering Foams

High density skins and lower density cores with coloring built in eliminates the need for any kind of surfacing material. Integral skin foaming has recently been perfected; furniture of this material has been designed mainly for public places.

Upholstery is still the primary means of covering foams. Foams can be nailed and stapled without fasteners pulling away. Sometimes wooden supports are molded into the foam, if

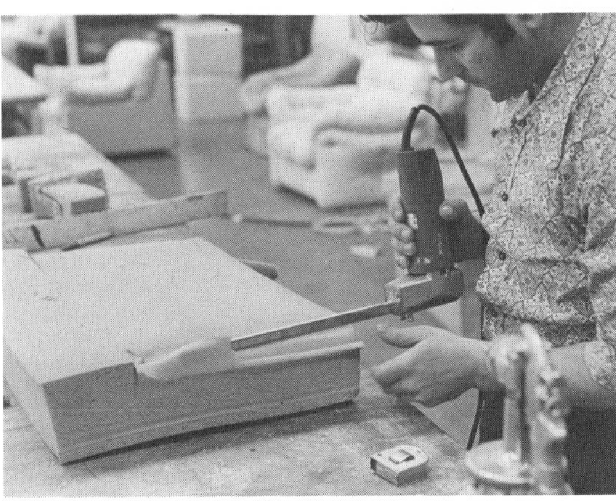

a,b. Carving flexible urethane to pad an FRP chair base at Cassina Center in Italy, using a Bosch electric knife.

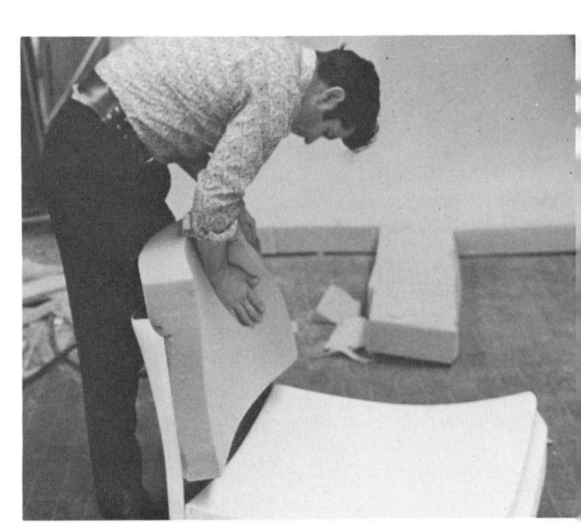

d. The cushion is pressed into place.

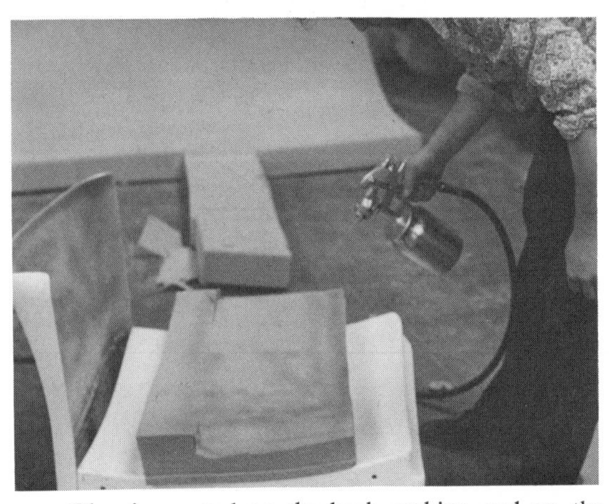

c. Glue is sprayed on the back cushion and on the back of the chair.

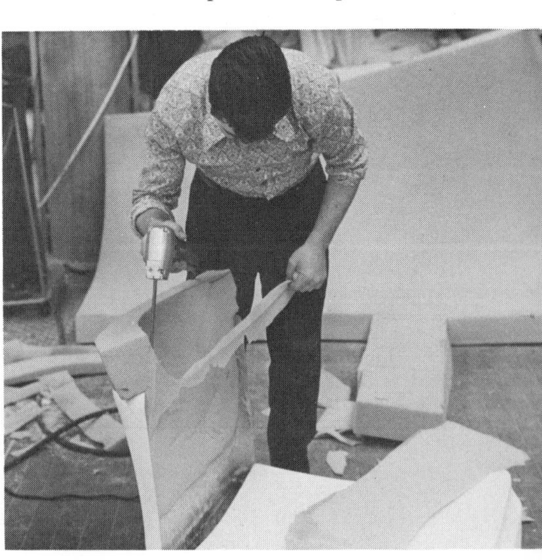

e. Final trimming to follow the designer's sketch.

finishing requires a special kind of support or attachment. Vinyl has been used as a popular "textile" covering. Coating the foam is another possibility.

In making a choice as to what finishes to use consider:

1. Type of cure of foam—air dried or elevated temperatures
2. Type of foam
3. Surface preparation, if any
4. Number of coats to be applied

a. Flexible urethane parts are cut for this three-position chair. The lounge position is shown here.

Ian Walker's "Orpheus and Charlie Brown" is made of flexible polyurethane foam tailored to cover a speaker system. Afterward the foam was painted. (*Courtesy, Ian Walker*)

b. Stretch fabric attached with two zippers hinge the foam parts and allow the sections to open into a lounging position as shown here, or to open fully into a stretched-out position—a contoured bed. (*Courtesy, DePas, D'Urbino, Lomazzi*)

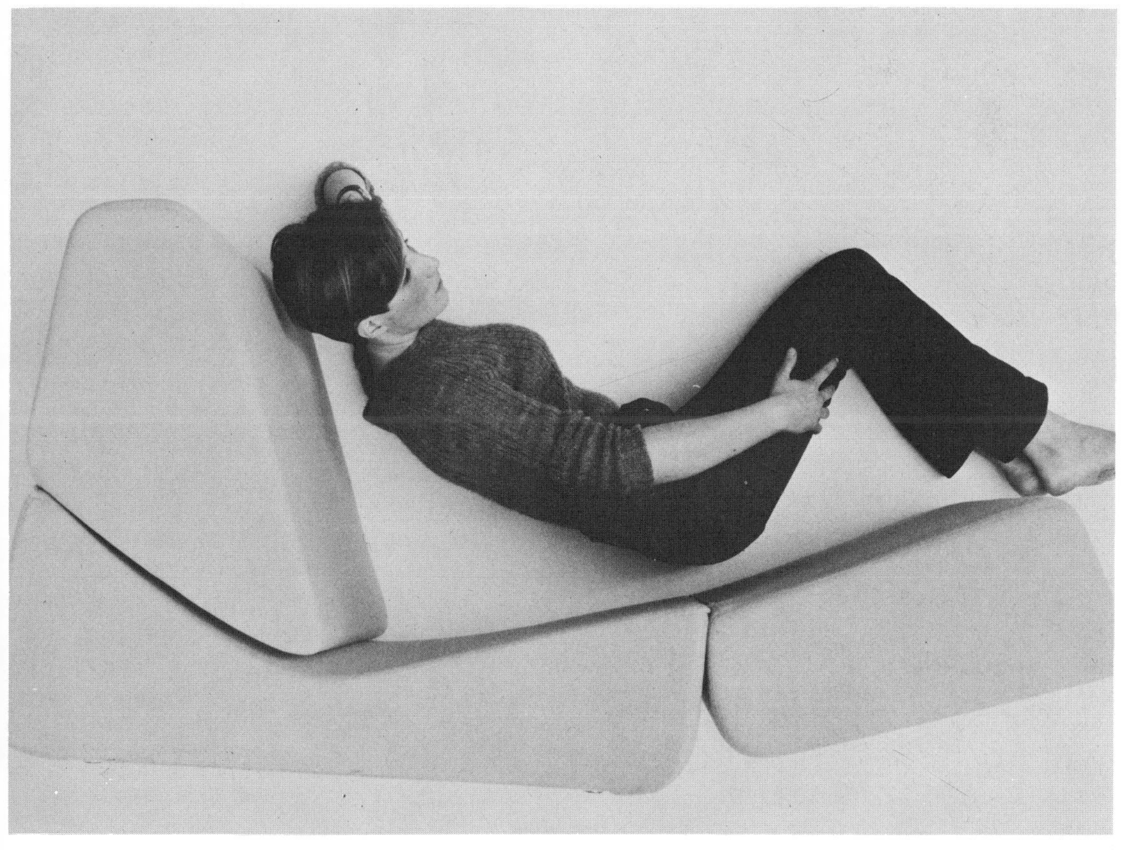

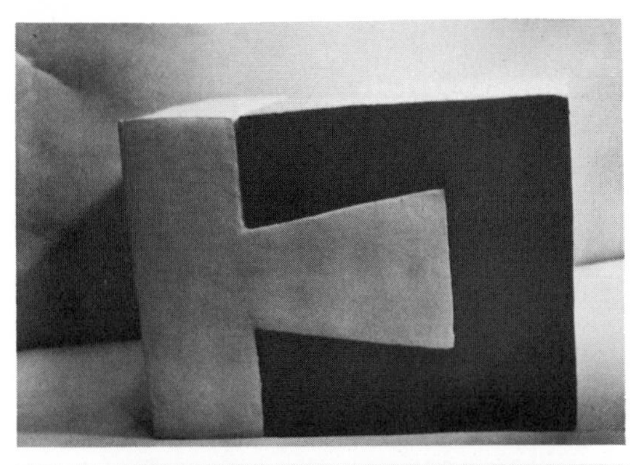

"Dovetail" is a chair and table or a bench—whatever you want it to be. It is made of painted, flexible polyurethane foam. Designer, Ian Walker. (*Courtesy, Ian Walker*)

5. Length of time between coats
6. Type of application, e.g., spray, dip, roller, brush, et cetera.

Commercially coating is a three-step operation: first, a base or ground coat is applied (not necessary if a base coat was used as a release and barrier coat); second, toning and novelty effects are added; and, third, protective gloss, semigloss, or matte coats complete the operation. Care in selection of the proper finish is important; hydrocarbons in some finishes such as lacquers will attack polystyrene foam but will not effect filled polyester and urethane foam pieces. Use the Scotch tape test to determine effectiveness of your coating. If the Scotch tape pulls off any coating, it is not good.

One company [5] has developed an elastomeric thermosetting "enamel" that has high elongation, weatherability, resistance to abrasion and extreme climatic conditions. It can be a one-coat high-gloss system for finishing urethane foam of any description. Some urethanes may require a special primer, though. Because of its properties, Durethane has been recommended for coating foam automotive bumpers. For use as a spray, it is necessary to thin with 1 part Cellosolve acetate and 6 parts toluol. After application, the form has to be baked for 30 minutes at 250°F. A 1.5 mil film thickness is recommended. Metallics, pastels, reds, blacks, and whites are available. Durethane 600, also an elastomeric lacquer, requires no baking but will dry at room temperature in one hour. It is not recommended, however, for exterior use.

Polystyrene foam can be colored with water or alcohol-soluble dyes and pigments. Inks, paints, lacquers, and other coating materials must adhere to the foamed polystyrene without attacking it to an undesirable extent. These materials achieve permanent adhesion through a carefully controlled solvent action which dissolves the foamed surface very slightly. "Controlled action" means that the evaporation rate and drying conditions have to be designed specifically for polystyrene foam.

Bonding Foams

When foams are used, it is often necessary to bond them to themselves and to other surfaces. Unfortunately, there is no one material that will do the job. In selecting the proper adhesive, consider whether the foam is open or closed cell, has porous or nonporous surface, is flexible or rigid.

Adhesives for bonding foams are available in water-based and solvent-based formulations, as mastics and hot melts, as wet stick or contact bond adhesives. They also are supplied in various viscosities, for spraying to viscosities applicable to roller coating, troweling, brushing, et cetera. When an adhesive is applied to two pieces of foam that have closed cells and were bonded when wet, the bond will be poor until the adhesive has dried. If the solvent cannot permeate through the foam, the adhesive can remain wet

[5] Automotive Finishes of PPG Industries Inc.—Durethane. See "Sources of Supply."

for weeks. A hot melt or contact adhesive would be more practical, then, for this application. If a solvent-type adhesive is used, apply the adhesive to the foam surfaces, allow the solvent to evaporate, and then join the surfaces.

Adhesive Types—Advantages and Disadvantages

Water-based adhesives, latex or resin types, are easy to apply, inexpensive, give better mileage, are nonflammable, take longer to dry.

Solvent-based adhesives are fast-drying, easy to apply, more water-resistant, tend to attack some types of foam, are difficult to spray (can be brushed on), and present fire and health hazards.

Hot melt adhesives are easy to apply but require application of heat and heating equipment to melt the adhesive prior to use. They bond instantly.

Mastic-type adhesives are very good for filling in voids and gaps and are thixotropic. Application is slow and the adhesives are more expensive.

100% adhesives solid, such as epoxies and urethanes, bond very well, have the best water and solvent resistance. They are more expensive than most adhesives and usually are available as two-component systems requiring mixing prior to use.

Safety Factors in Working with Urethanes [6]

It must be stressed that urethanes are great to work with and safe if precautions are followed. Component "A" containing free isocyanate is irritating if it is inhaled, is ingested, or contacts the skin or tissues. Adequate ventilation throughout the work area is extremely important, together with eye protection. Gloves and skin-covering clean clothing is also desirable. An organic vapor respirator should be available for emergency use, as well as an easily accessible place to wash with plenty of water if isocyanate contaminates the skin. Spills should be covered with soda ash or ammoniated saw dust and removed promptly. Spraying operations require a full mask supplied with uncontaminated fresh air.

Component "B," the cross-linking agents, often contains amine products. These are very irritating to the skin and mucous membranes. If

Welding a frame for the outer shape of a chair for urethane casting at Cassina Center, Italy.

skin contact does occur, the affected area should be flushed with large volumes of fresh water. Component "B" also contains a blowing agent such as fluorocarbon II, also dangerous for the

[6] Courtesy, Flexible Products Co., 1225 Industrial Park Drive, Marietta, Ga. 30061.

A completed chair mold ready for the injection of foam. (*Courtesy, Oppenheimer of Italy*)

Finishing the interior of a mold to be used for urethane casting requires fine sanding and waxing because the surface is accurately translated by the urethane. (*Courtesy, Cassina Center*)

Making a flexible urethane "poster" at Cassina Center

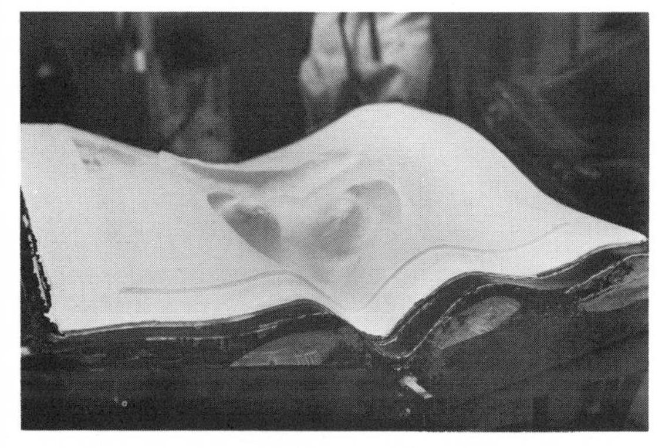

a. The mold is cleaned, waxed, and readied for pouring.

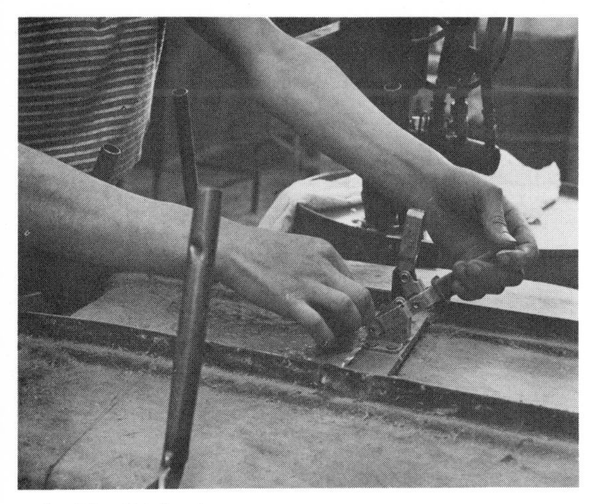

b. The lid is clamped on.

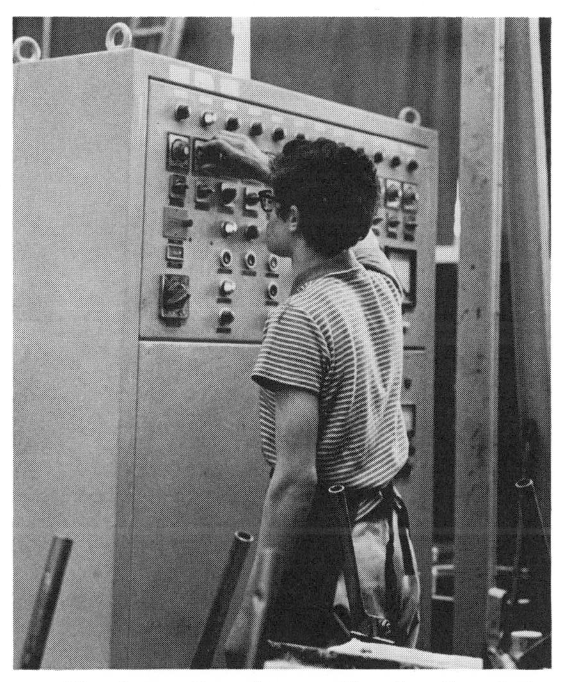

c. The foam dispensing machine is adjusted for proper urethane foam density and metering.

d,e. The machine nozzle is placed on the mold opening and the urethane mixture is pumped into the mold cavity.

f. As foaming is completed, it forces air out of vent-holes in the mold and the foam eventually rises out of the holes.

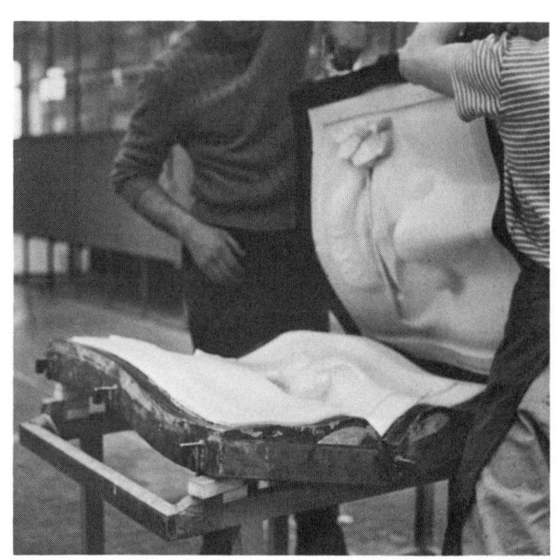

g. After curing time is completed, the mold lid is unclamped and the mold is opened.

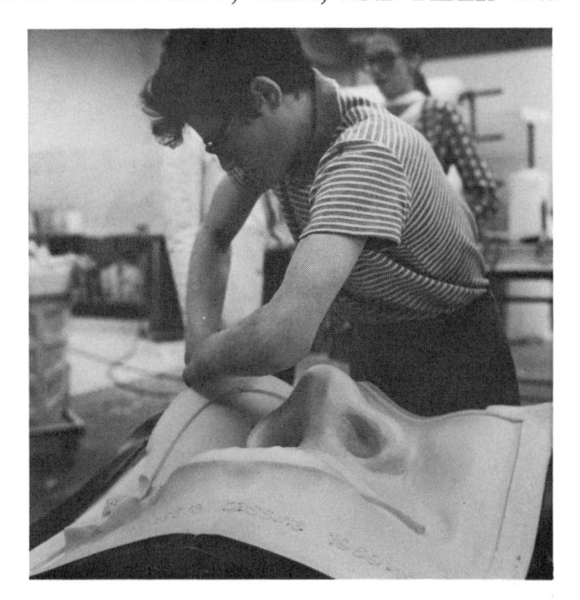

h,i. The "poster" is carefully released from the mold and removed.

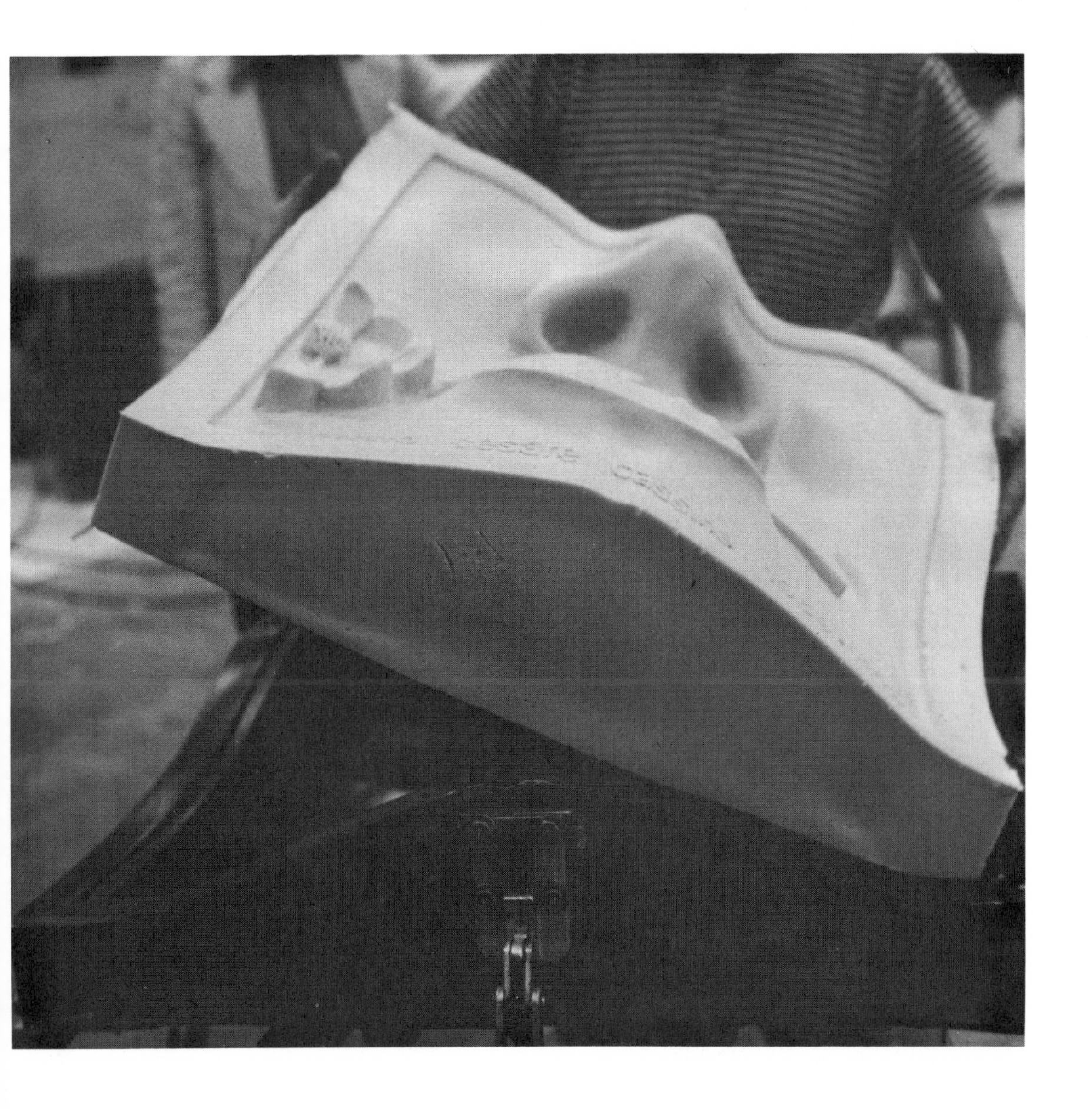

j. The form is pounded to release any trapped gasses.

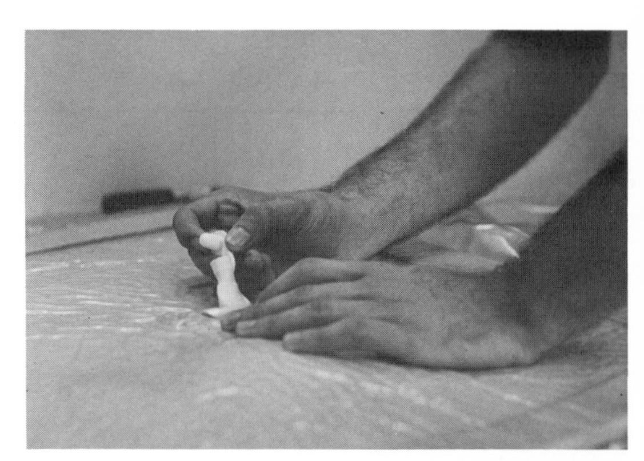

k. For packing and shipping, all air is pressed out of the foam and it is vacuum-sealed in a heavy polyethylene envelope.

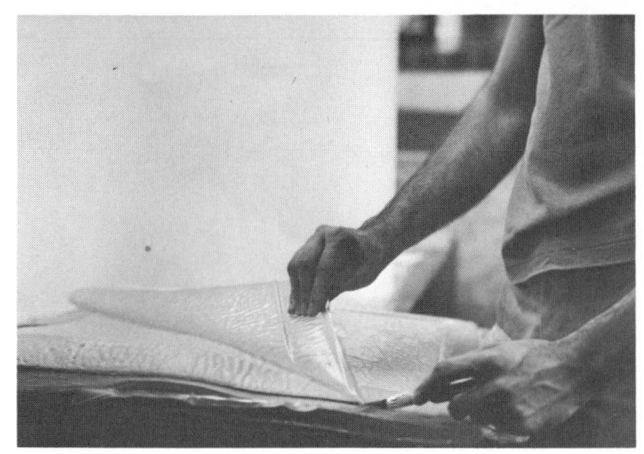

l,m. Upon arrival at its destination, the air valve is opened to allow air into the envelope and the polyethylene is cut loose.

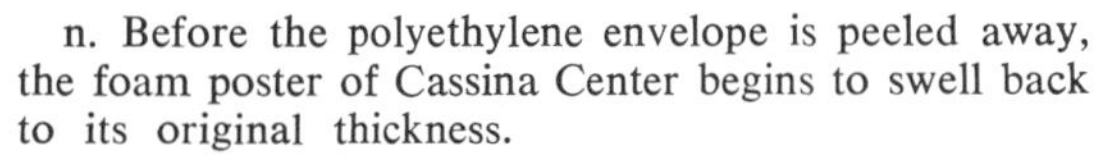

n. Before the polyethylene envelope is peeled away, the foam poster of Cassina Center begins to swell back to its original thickness.

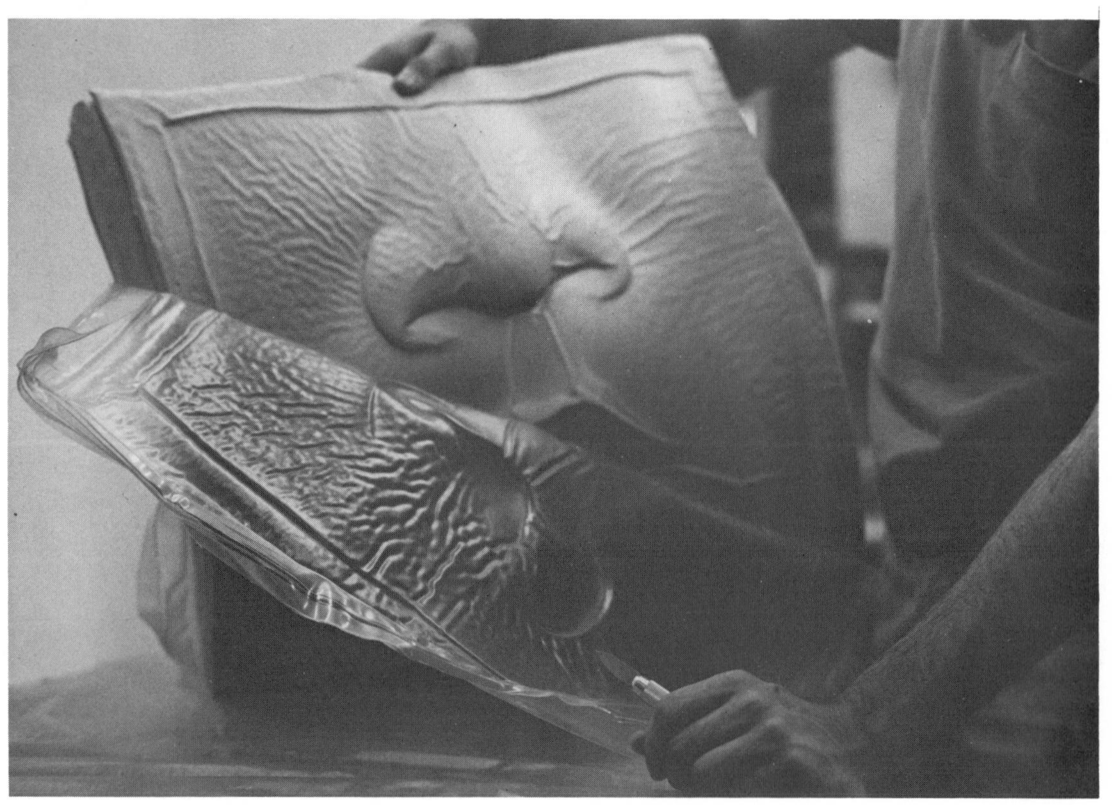

a. To demonstrate plastic as plastic possibility in furniture made with the Duromer cold-cure urethane processes, Mobay Chemical Company commissioned Peter Muller-Munk Associates to design these forms.

The free-form pieces include two couches—one with a back and arms and the other without—which utilize the same urethane base. The one-piece molded cold-cure cushioning weighs 50 pounds and is produced in a single molding. There is also a molded modular table consisting of two sections that could easily be converted to include many sections in its configuration, or each section could be used alone. The exposed rigid sections are painted white. Every piece of furniture in this room is made of urethane foam—flexible and rigid.

Production of free-form design made by Duromer structural and cold-cure flexible urethane.

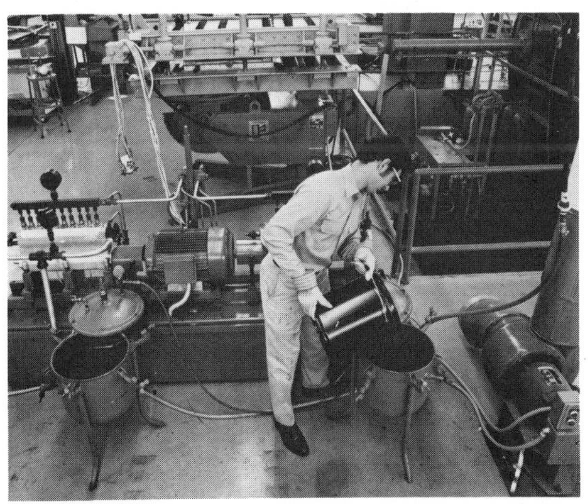

b. The metering and pumping machine is filled with liquid chemicals. In large production operations the machines could be fed from large storage tanks.

c. The metering and pumping machine is adjusted to take into account the urethane formulation being used and the flow rate desired.

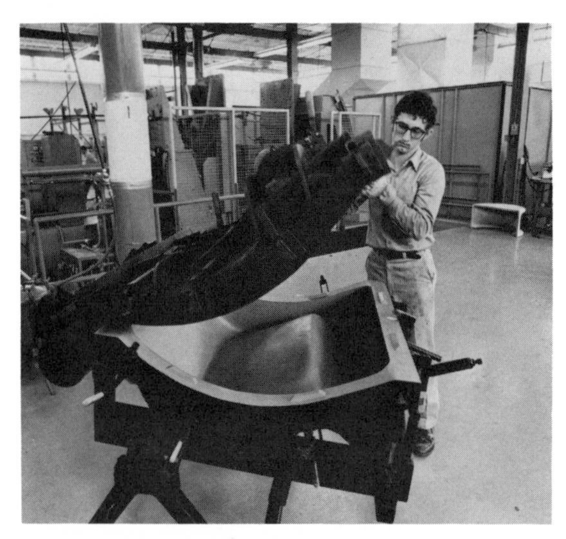

d. The easy-to-handle, hand-clamped mold is shown open.

e. The mold is sprayed with a release agent.

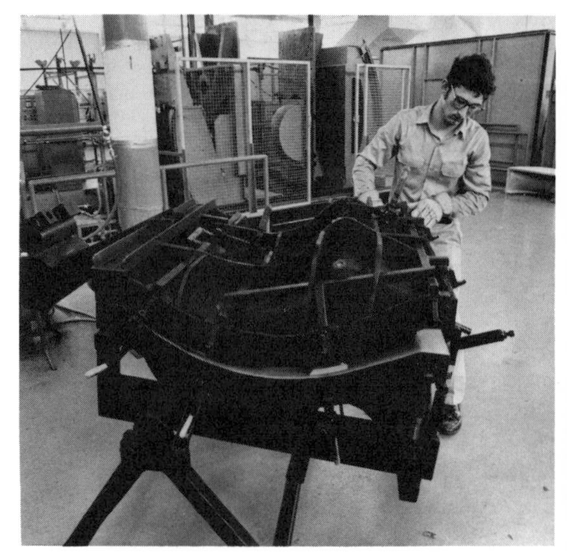

f. The mold is closed.

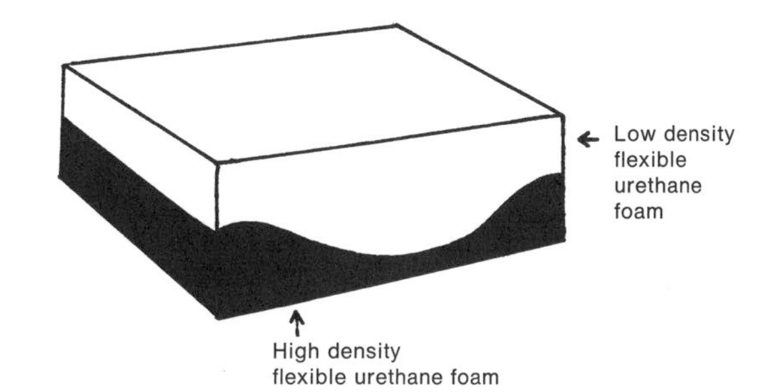

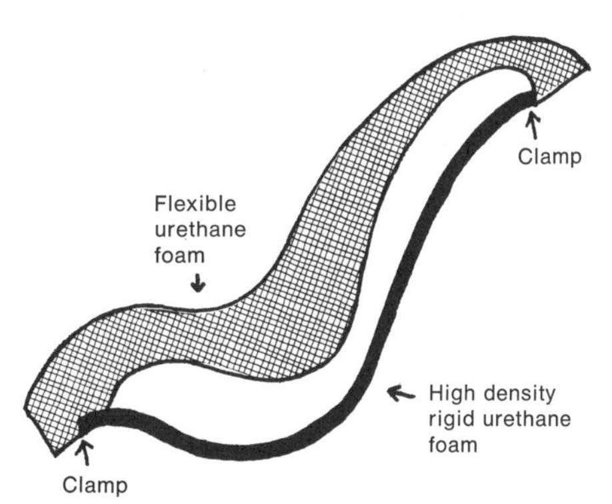

Urethane foams—both flexible and rigid—make possible new structures for seating that eliminate springs, straps, and other supports.

g. The machine operator introduces into the mold by the use of a hand-held mixing and dispensing head a metered amount of chemicals pumped from the metering machine. The mold is now moved to another area while the foam cures for demolding.

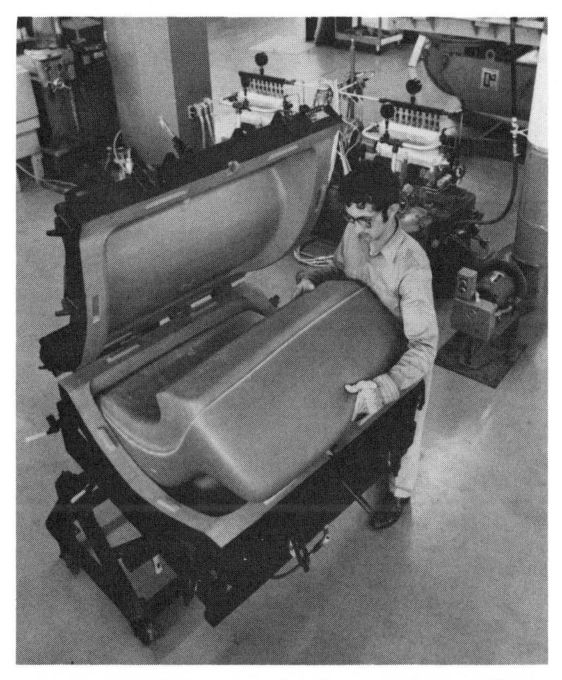

h. The mold is opened. The complete cold-cure flexible urethane piece, cushioning for a free-form love seat, is removed and is ready for upholstery. The highly resilient cold-cure cushioning weighs 21 pounds.

Depending on the size and design of the piece being produced, the formulations and equipment being used, and the production facilities, a complete cycle for producing a cold-cure flexible urethane foam piece may range from 10-15 minutes. (*Courtesy, Mobay Chemical Co.*)

a. A wood prototype for a flexible urethane chair at the Oppenheimer factory in Italy.

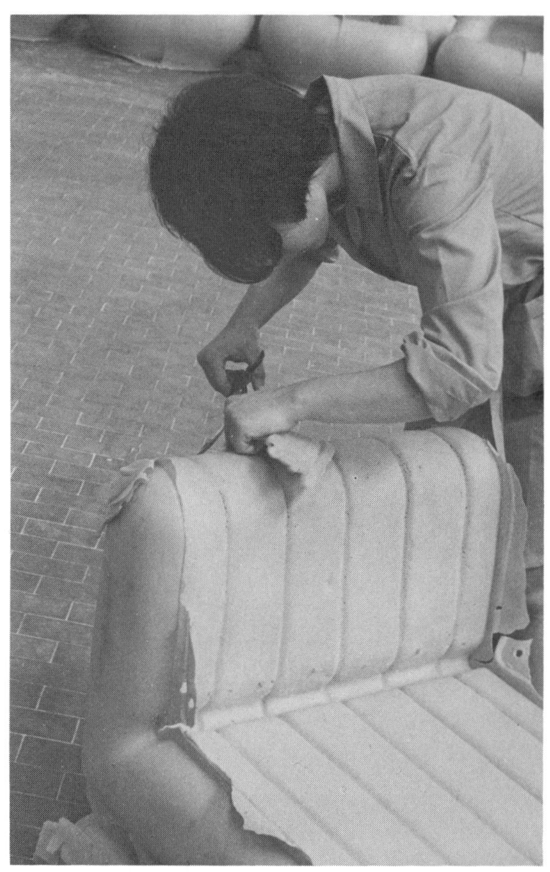

b,c. Trimming flash from the chair.

d. The foam form is ready for upholstering.
(*Courtesy, Oppenheimer of Italy*)

a. An interior foam part is being filled at the Oppenheimer factory in Italy, before shipping to the upholsterer's.

b. The chair designed by DePas, D'Urbino, and Lomazzi is covered in vinyl patent. (*Courtesy, Oppenheimer of Italy and DePas, D'Urbino, and Lomazzi*)

Sometimes supports are molded into the foam.

c. Closed version of the iron frame for "Palla."

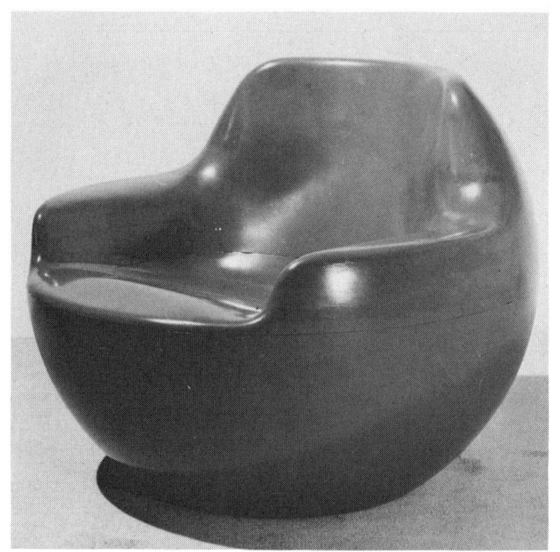

a,b. A wooden prototype of "Palla," a chair designed by Architect Claudio Salocchi for Sormanis spa.

d. Open version of the iron frame for "Palla."

e. The completed chair and ottoman have a rigid urethane core and a flexible overlay. The upholstery is a knitted stretch fabric. (*Courtesy, Claudio Salocchi*)

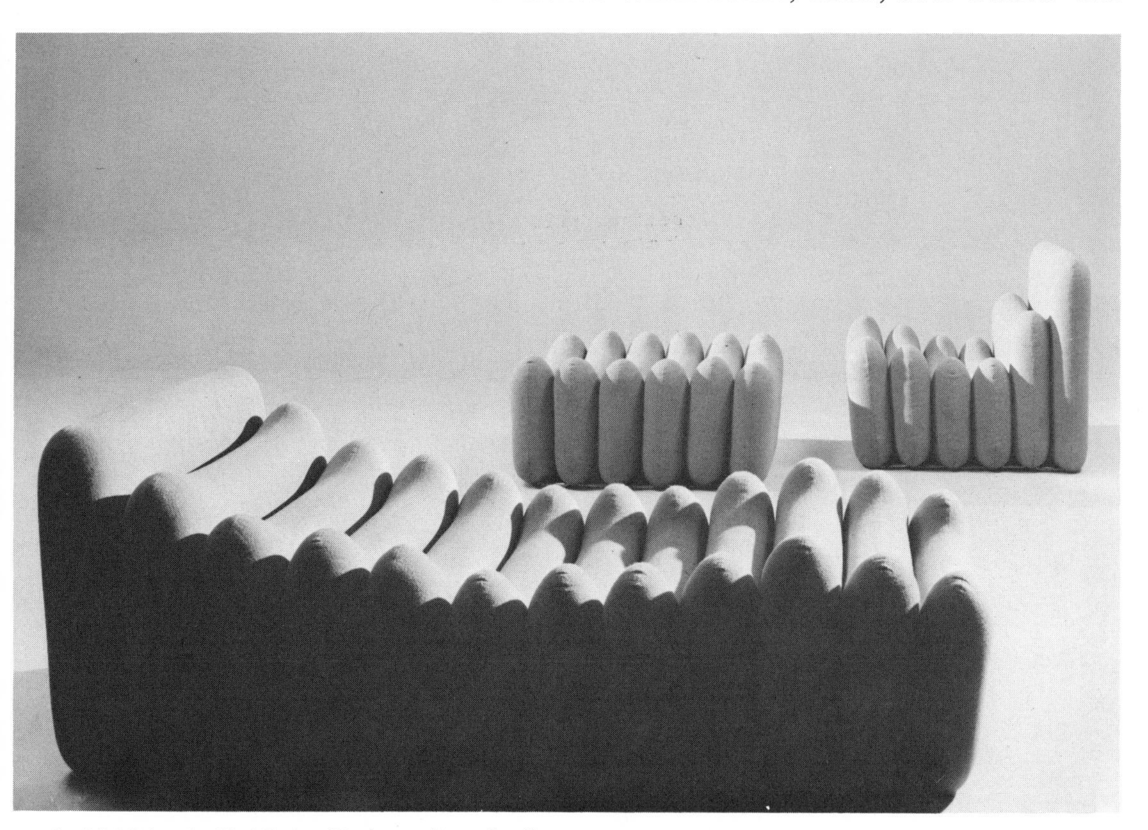

a,b. "Additional Chair" by Designer Joe C. Colombo can be uniquely combined into different versions depending on the length and how the components that have various heights are combined.

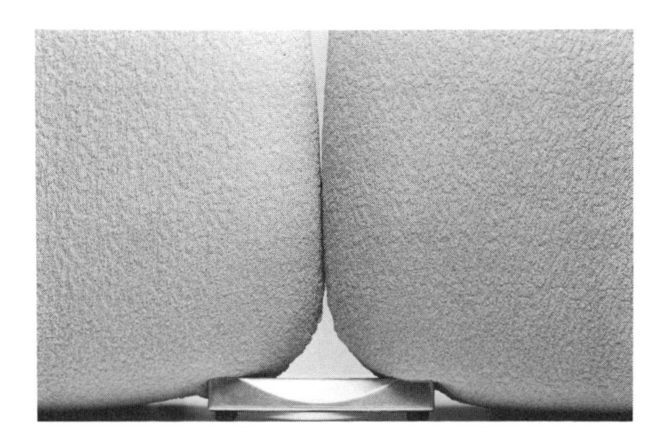

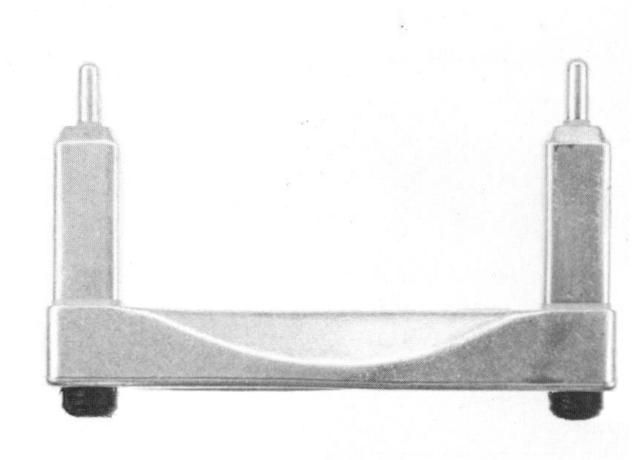

c,d. Details showing how the components are assembled into a secure unit and the "piece of hardware" that makes it possible. Sormani spa is the manufacturer. (*Courtesy, Joe C. Colombo*)

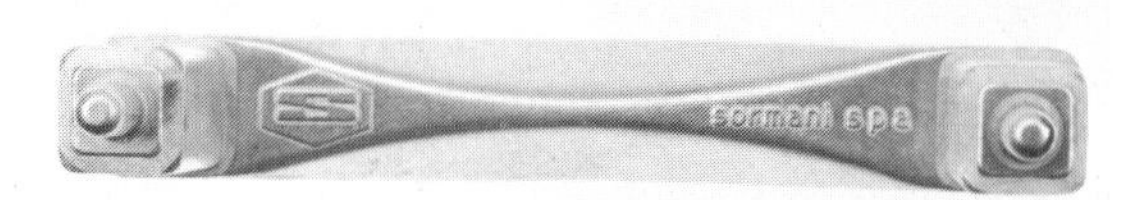

a. Wooden prototype of Antonio Rossin's "Colourstones" for urethane production.

b. A conversation group with "Colourstones." (*Courtesy, Antonio Rossin*)

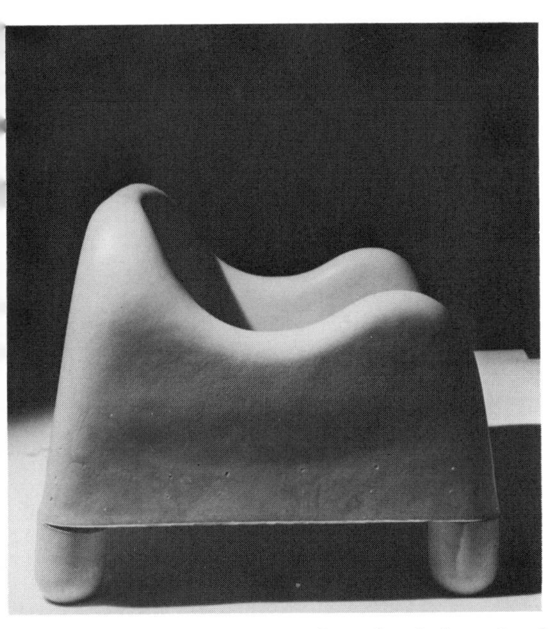

a. The raw view, before covering, of a chair made of part rigid and part flexible urethane foam.

b. The completed chair by Lomazzi, D'Urbino, DePas, and DeCursu. (*Courtesy, Lomazzi, D'Urbino, DePas, DeCursu*)

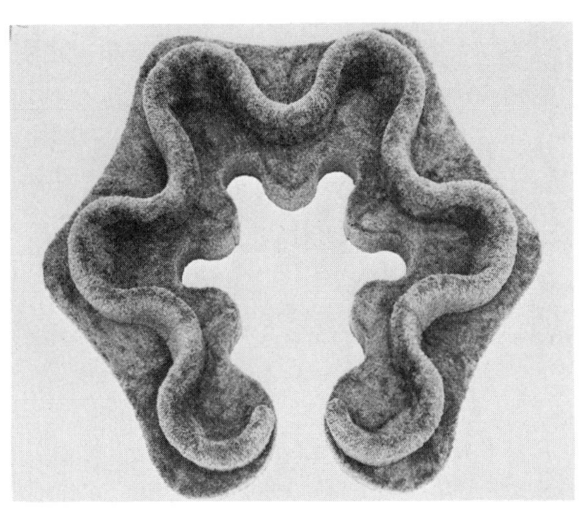

Verner Panton's conversation unit is of urethane. Its unique form makes the kind of seating one of "where you select to sit." The flexible and rigid foam unit is upholstered in a shaggy ruglike covering. (*Courtesy, Bayer-Leverkusen*)

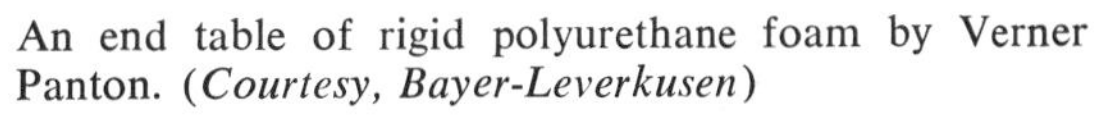

An end table of rigid polyurethane foam by Verner Panton. (*Courtesy, Bayer-Leverkusen*)

Three versions of many possibilities with Architect Claudio Salocchi's modular units of varying density flexible urethane foam. Sormani spa is the manufacturer. (*Courtesy, Claudio Salocchi*)

The "expo 11 sound chair" contains an audio system that can be enjoyed with full fidelity in private. It can be connected to any available sound source—AM-FM, stereo, radio, tape player, T.V., et cetera. The shell is one-piece molded rigid polystyrene with a fiberglass bolster headrest and polyurethane foam bonded to the polystyrene shell. The head bolster contains two four-inch speakers. Controls on a common shaft are on the right side of the chair. Dimensions: width, 29″; depth, 33″; seat height, 16″; base diameter, 19″. Weight, 40 lbs. It is manufactured in Australia. (*Courtesy, Eklektix, Inc.*)

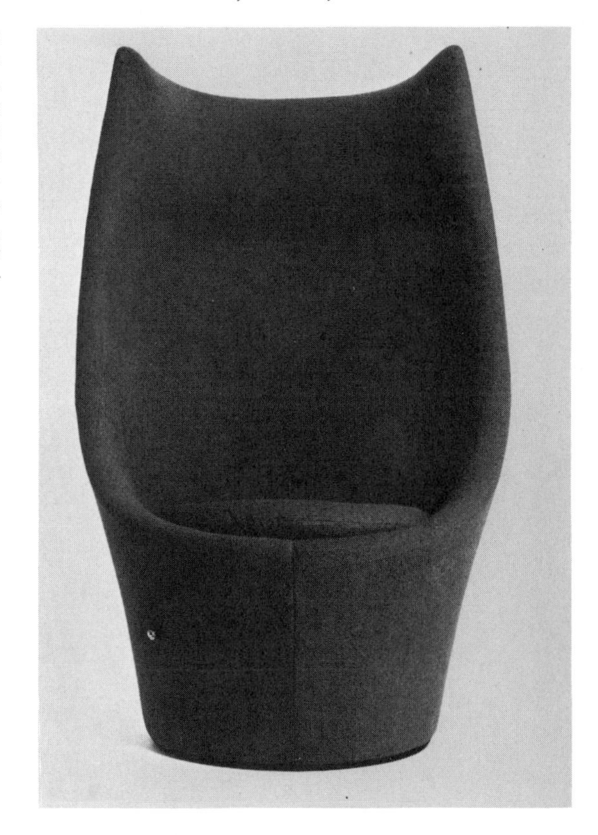

This Schjeldome 200 house has roof and wall sections made of ⅜″ rigid polyurethane foamboard. Factory-made, it sets up in a few minutes without special tools and folds down to a compact package that fits into a station wagon. Dimensions: 12′, diameter; 6′6″, walls; 8′6″, center height. Weight, 100 lbs. (*Courtesy, G. T. Schjeldahl Co.*)

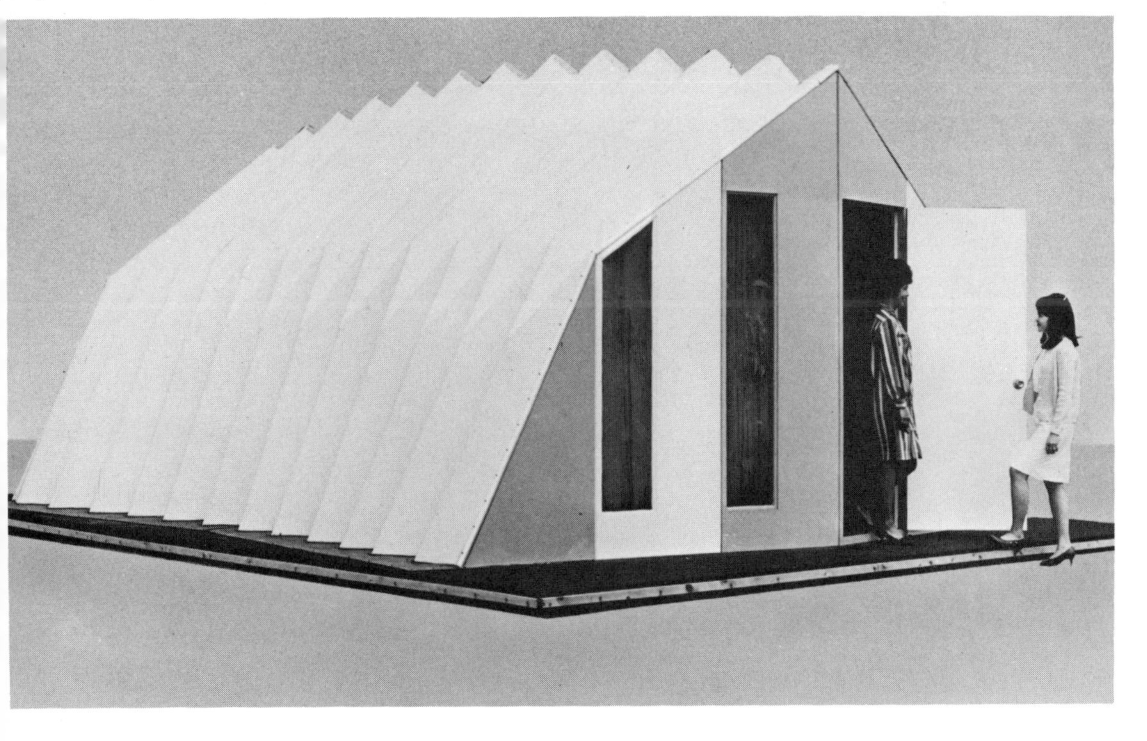

A blow-up tunnel made of P.V.C. Designed by Architect Lomazzi. Manufactured by Plasteco. (*Courtesy, Paolo Lomazzi*)

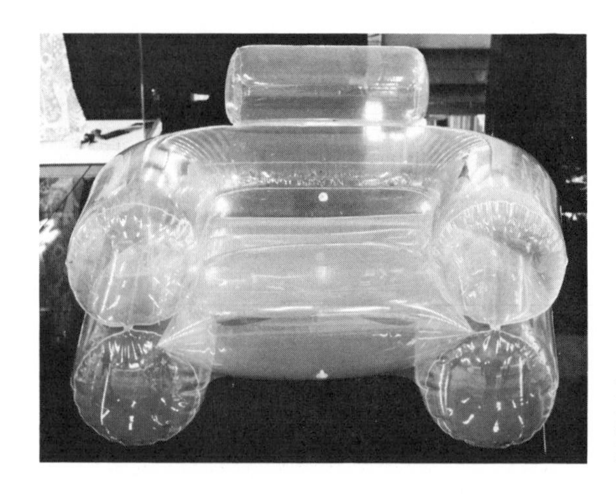

An original blow-up chair where shaft, springs, and padding are substituted for a simple air cushion. Made of P.V.C., sealing is done by radio-frequency waves. Designer, Paolo Lomazzi. (*Courtesy, Paolo Lomazzi*)

A temporary blow-up showroom by Lomazzi. Plasteco is the manufacturer. (*Courtesy, Paolo Lomazzi*)

A cable-air-roof structure housing the U.S. Pavilion at the World's Fair Expo '70, Osaka, Japan. The roof is Cordo—vinyl-coated fiberglass. (*Courtesy, Ferro Corp.*)

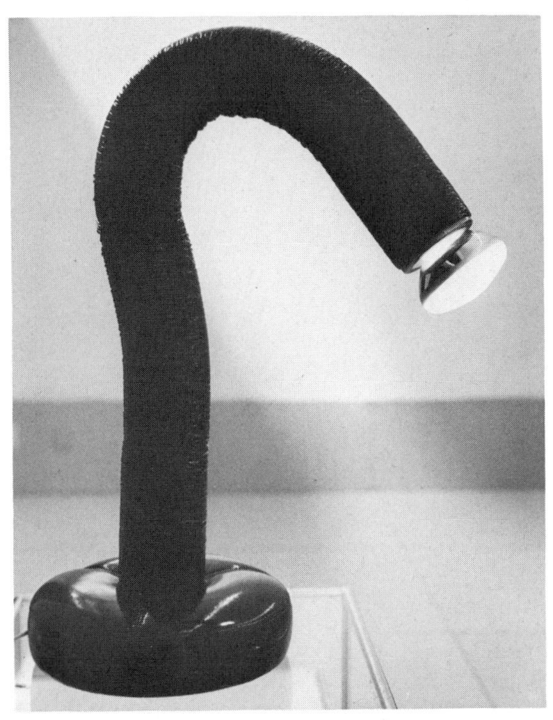

A soft lamp made of padded vinyl with a gooseneck stem; it has the capacity to be changed in shape and still maintain a balanced base.

Castiglioni and Frattini designed this polyethylene lamp for Artemide. It can be stretched snakelike or be coiled as seen here. (*Courtesy, Artemide*)

The jacket has feathery fringes of yellow Celanese acetate sheeting. The sheath underneath is made of Celanese nylon fiber. Designer, Elisa Stone. (*Courtesy, Celanese Plastics Co.*)

The fish are swimming through clusters of polypropylene monofilament seaweed in a test tank. It is the same material used to test ability to decrease erosion of land in New Jersey and England, by submerging the monofilament clusters on the sea floor. (*Courtesy, Avisun*)

a. In 1952 George Nelson created lamps made in a similar process. The name "Cocoon" has stuck both here and abroad. The technician is spraying a styrene webbing over a metal armature.

b. Here he is spraying the styrene webbing with coatings of vinyl until the vinyl builds up into a skin.

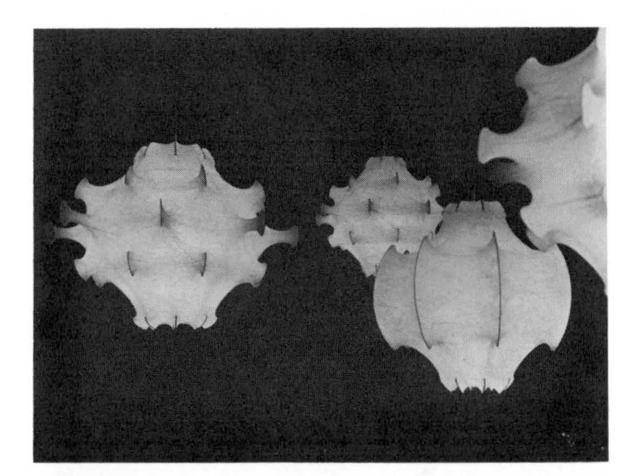

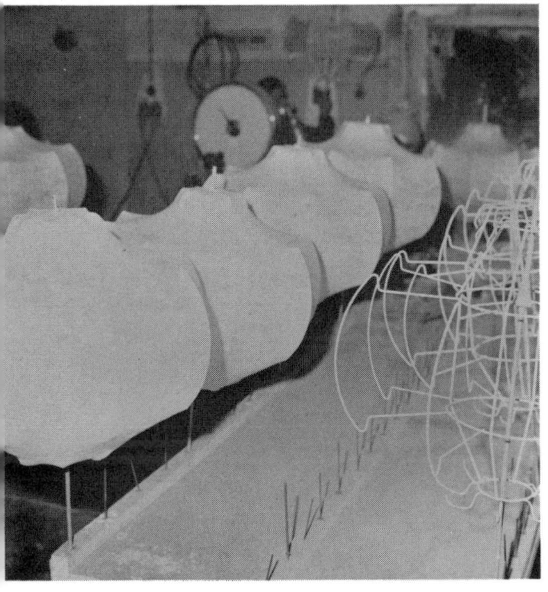

c. The lamps are "drying" before other coatings are applied. The material used in these lamps is made by R.M. Hollingshead Corp.

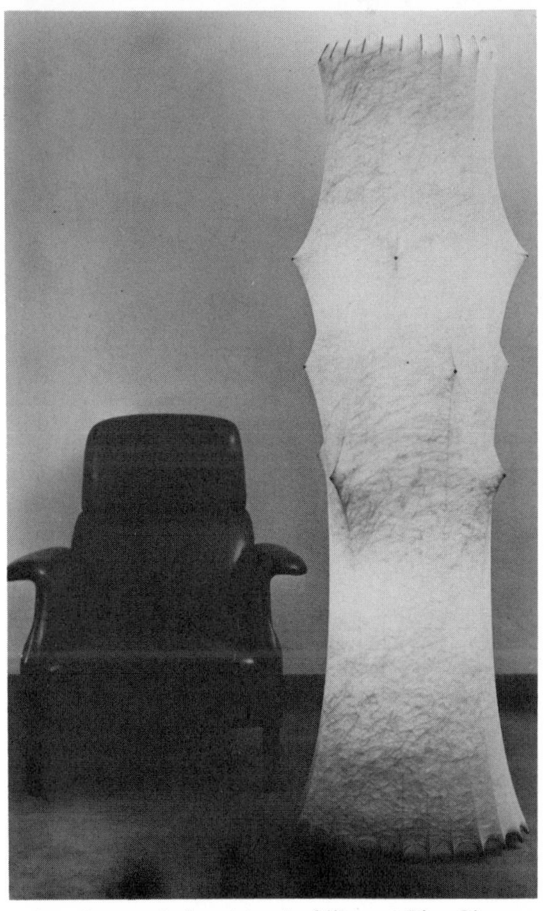

d,e. Lamps designed by Archille and Pier Giacomo Castiglioni and distributed by Atelier International, Ltd. (*Courtesy, Atelier International, Ltd.*)

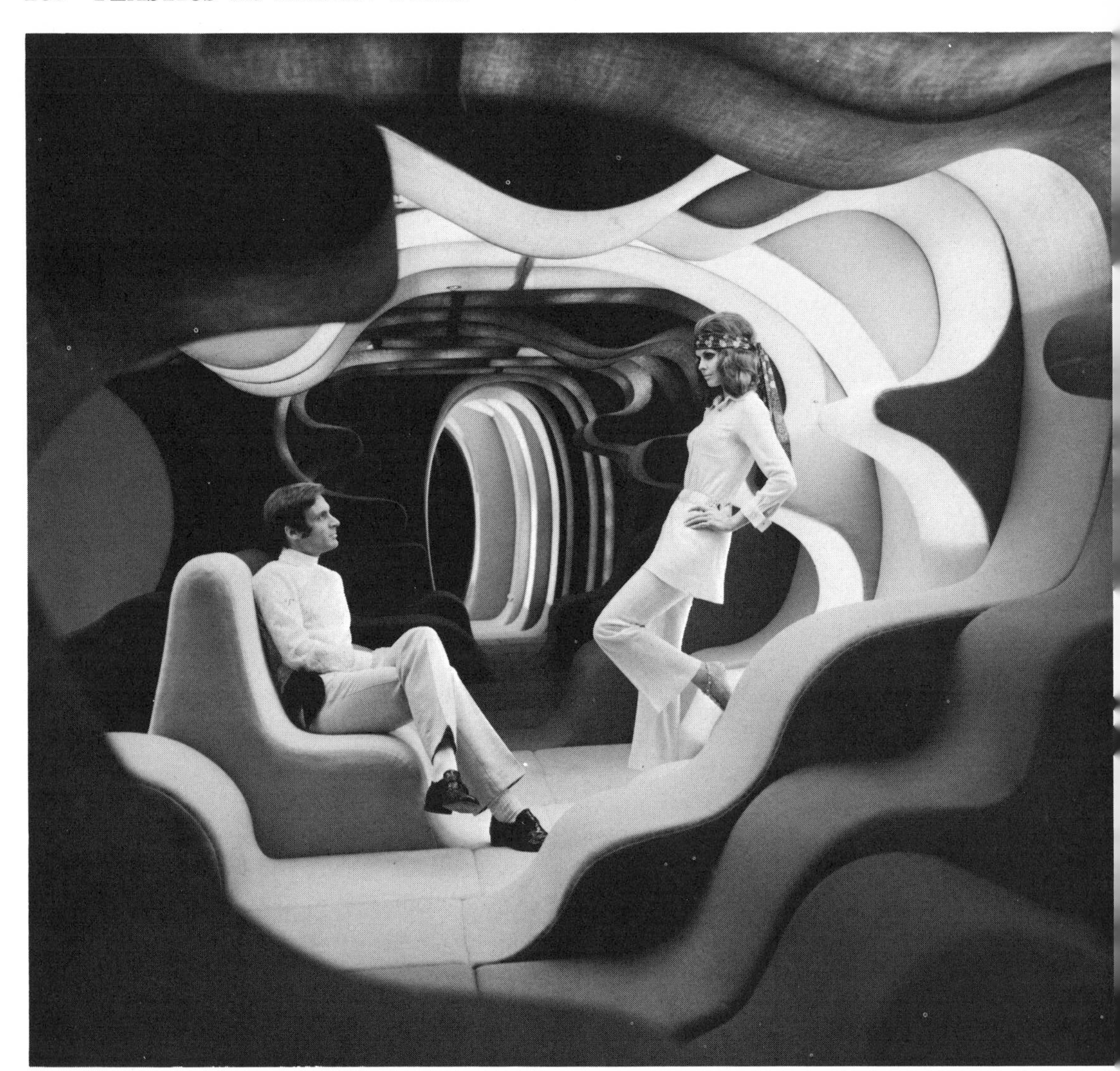

Verner Panton's "Visiona," an environment of tomorrow. Lighting, seating, and lounging forms evolve from the same structure. The ceiling is made of Dralon, translucent and colored in multihues of orange, red, violet, and blue. Armatures to contain the overhead lighting extend to urethane cushioning in continuous undulating patterns to create a soft environment from head to toe. (*Courtesy, Bayer-Leverkusen*)

Verner Panton's couch in a soft environment. Traditional units of chairs, tables, and couches are replaced by a totally integrated environment with each part functioning in more than one way. Fibers used are synthetic. Urethane is the structural base. (*Courtesy, Bayer-Leverkusen*)

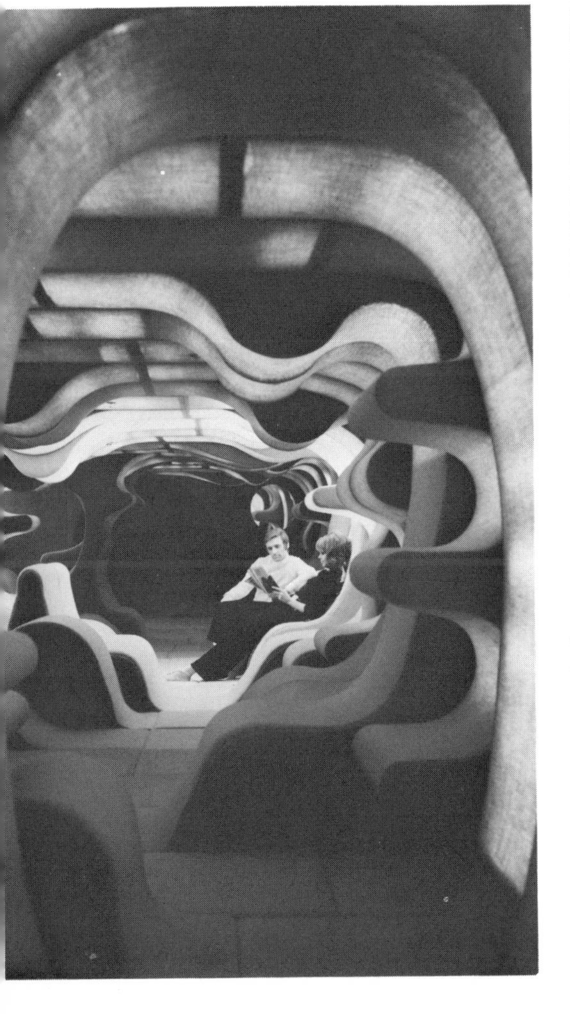

Ted Hallman's "Contemplation Environment" (from the exhibit of the same name) was designed to create a total atmosphere conducive to meditation. The use of Acrilan yarns (by Monsanto) in the space makes it warm, quiet, and comfortable. The floor, walls, and ceilings are composed entirely of fringed yarn in grays and beiges. The fringes are knotted to a yarn grid which is supported by a steel frame in the form of a six-foot cube. Very subtle concentric circles radiate outward from a point in front of the meditator to engulf the entire space. Individual fringes are between 1½′ and 4′ in length and the total weight of yarns is approximately 125 pounds. (*Courtesy, Museum of Contemporary Crafts*)

Michael Lax's concept is the use of synthetic fibers such as Acrilan, knitted into a fabric that can be stretched over lightweight armatures and then coated with resins to make them permanent. (*Courtesy, Chemstrand Co.*)

Cadon by Monsanto is used for this rug by Nell Znamierowski in neutral shades of silver gray, pewter, dulled white, and luminous beige. Dyeing is by a "total system," which allows all colors to maintain their inherent clarity in combination with one another. (*Courtesy, Monsanto Co.*)

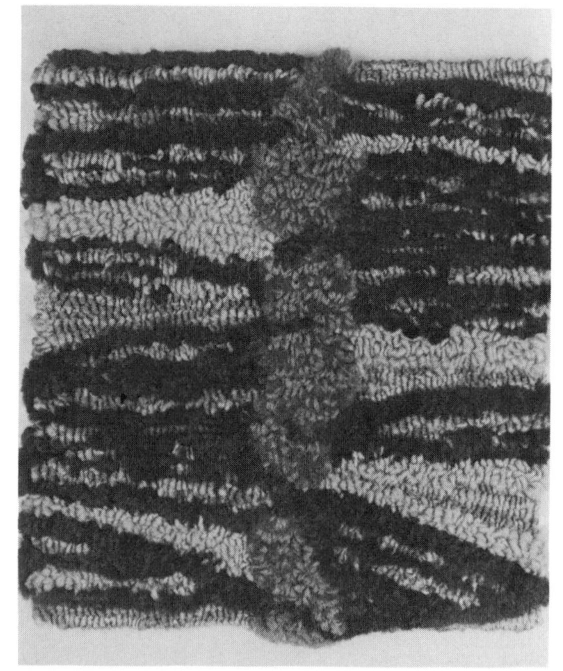

Colors ranging from luminous gray to bronzed earth to rich black are found in this experimental carpet design by Nell Znamierowski. Using Monsanto's new antisoil, antistatic Cadon nylon fiber, Miss Znamierowski proved that a piece of broadloom can have as many clear colors as a hand-knotted Persian rug, but at a cost no greater than traditional, solid colored carpet. (*Courtesy, Monsanto Company*)

"Pylos" by Nell Znamierowski for Regal Rugs looks like a sculptured rug that is hand-knotted. In fact, it is machine-made, with the synthetic fibers reacting differently when washed—some puffing out, some curling tighter. (*Courtesy, Regal Rugs*)

'Sunrise," a wall hanging by Nell Znamierowski, has ı plum-colored background that is made of Zefran. It s woven and dyed first, then returned to the stitching oom for the rest of the design. The hanging is vashed upon completion, which causes some fibers to :url, others to puff out. (*Courtesy, Nell Znamierowski*)

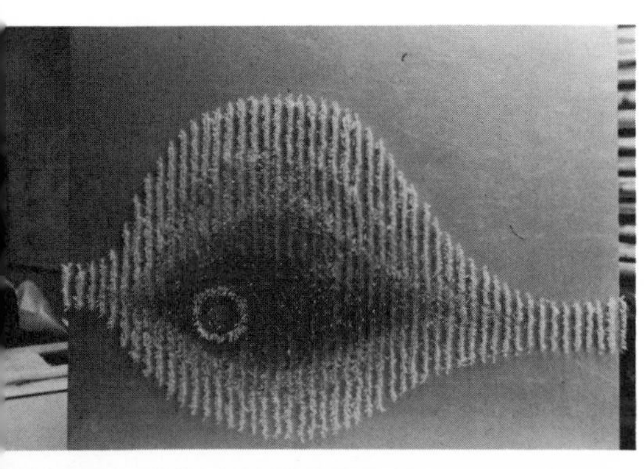

"Moonbird" by Nell Znamierowski is another example of her machine-made wall hangings. (A machine does the tufting through a backing by power equipment, not too dissimilar from hand hooking.) The hangings are of limited editions of about 100 each, made of Zefran acrylic. It is different from other acrylics, inasmuch as it is very soft, fluffy, pleasant to touch, and easy to dye. (*Courtesy, Nell Znamierowski*)

a,b. The upholstery of this not too exciting chair is made by a very exciting process at Cassina Center. A thermoplastic woven fabric is used for the covering.

c. The molding press for heat forming thermoplastic fabric at Cassina Center, Italy.

d. The mold is in place.

e. Fabric, cut to size, is attached to a frame.

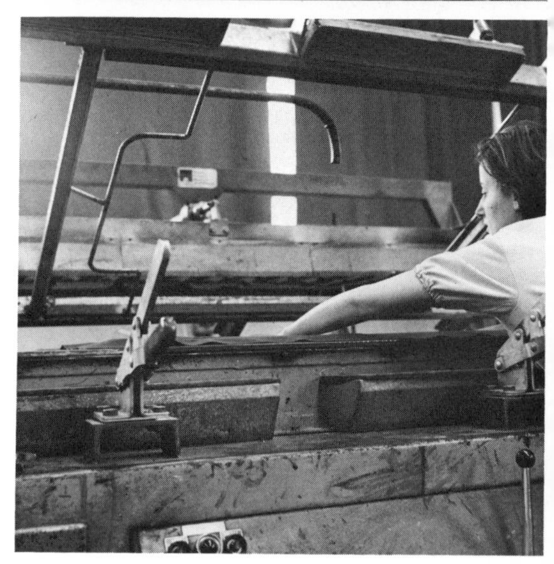

f,g. The mounted piece is placed into the molding press.

lungs. These fluorocarbons can develop pressure at temperatures above 75°F. Considerable care should be exercised when the shipping container is opened. Store the unused portion in a tightly sealed container in a cool place.

FILM

Films are available in a variety of thicknesses (up to 10 mils), metallized, textured, pressure sensitive, in all varieties from polyvinyl fluoride,[7] a very tough material that will weather well, to tough polyester films often called Mylar (also by Du Pont). The Velcro Corporation is manufacturing a unique mirror that is made of vacuum-coated aluminum on a polyester film .001 of an inch thick, stretched on a rigid frame with a urethane board backing. It is lightweight, shatterproof (but not punctureproof), nonmisting, highly reflective, and distortion-free. Syncel [8] is a new material basically a cotton-Dacron fiber based on a copolymer substrate, not unlike polyethylene. The top surface feels like cotton; the undersurface is heat-sealable with the pointed tip of a soldering iron and can be printed with acrylic-based paints.

Thicker than films, vinyl-coated beta fiberglass fabrics have been used to form air-supported and frame-supported structures. The United States Pavilion at the Japan World Exposition in Osaka had the world's largest cable air-supported roof of vinyl-coated fiberglass.[9] Some polyester vinyl-coated fabrics are high-frequency (for thermoset) welded, which pro-

[7] Tedlar, a Du Pont product.

[8] A product of Allied Synthetics Corp. See "Sources of Supply."

[9] Cordoglas, made by Ferro Composites Corp.

h. The fabric is heated.

i. The mold is pressed into place. After cooling, the formed fabric is peeled away and ready for attachment to the chair.

duces heat-set seams and eliminates older methods of cementing or sewing. Single skin materials such as these are strong enough to withstand high winds (150 mph), weather, can span large areas (as long as a city block) without the benefit of interior columns, and can be portable and demountable.

Sealing Films

Solvents can be used to cement films together but the simplest method is by heat sealing. Through thermal sealing, films are melted together by direct heat. Hand-held equipment is often used.

Commercially there are two variations of thermal sealing. Films to be sealed are placed between a pair of hot dies, or a hot and a cold die; pressure is applied to the dies for the time it takes the films to soften and fuse; then the dies are separated. Continuous seams can be made by passing the films between hot rollers under pressure, or over a hot platen and then applying pressure with cold rollers.

Thick and thin films also are sealed by radio frequency (dielectric). In dielectric sealing the flexible films to be joined act as the dielectric between the plates of a capacitor. These capacitor plates consist of a die face and a press platen and are connected to the output of a radio frequency (RF) generator. Molecular vibrations are induced by the RF energy and result in molecular friction—thus, heat.

FIBER

Fibers have been adapted readily to consumer needs. Much of our clothing is based on synthetic fibers of acrylic, polyester, et cetera. But there are nontraditional ways of utilizing synthetic fibers. They are basically plastic and can be engineered to make possible certain designs. Ted Hallman used acrylic yarn for a texture environment. Nell Znamierowski has been designing prototypes using various types of synthetic yarns for rugs and wall hangings. Synthetics react differently than wools and nature's other fibers. When subjected to a heat bath, for instance, synthetics will vary immensely in the way in which they will react. Some will curl down; others will bloom out—the tips of fibers open up into a "frothy look," to quote Nell Znamierowski. Cadon, to name one of many excellent synthetic fibers, is a new Monsanto fiber that is a second-generation nylon—antistatic, antisoiling, and recently nonflammable.

Some woven fabrics can be thermoformed to match the shape of the base they are to cover, so that upholstering requires simple tacking or stapling.

As with other plastics, synthetic fibers can be fabricated—e.g., woven—in traditional modes, but their unique qualities make possible a new technology in working with fibers to produce fabrics.

8

designing with casting and laminating liquids

One of the most versatile approaches to forming shapes is through pouring of a liquid that will harden and take the shape of the positive, or by layering liquid-saturated fabrics (fiberglass) over a mold form that finally, upon hardening, will produce a replication of the positive. All of this can be accomplished at room temperature with a minimum of equipment, except for casting of polymethyl methacrylate.

MOLD MAKING

Molds can be made of almost any material—glass, metal, wood, gypsum, and, of course, plastic. Two of the easiest materials to use for forming molds are those rubberlike plastics classified as room-temperature vulcanizing (RTV) and room-temperature curing (RTC). These are flexible (elastomeric) materials that allow for a certain amount of undercutting in the positive. Rigid molds are usually made of epoxy, fiberglass reinforced polyester or fiberglass reinforced epoxy, and vacuum-formed sheet such as polyvinyl chloride or polyethylene. Every mold is made from a master or pattern.

Preparation of the Master or Pattern

Since every detail of the master is reproduced in the mold and then transferred to the positive, meticulous preparation of the pattern is very important. The master may be constructed of almost any material—wood, plaster, wax, soap, metal, glass, or plastic. Light sanding with a 300–400 grit sandpaper is advisable, followed by blowing or brushing off sand and specks. Wax is usually used to fill in cracks and to lessen sharp undercuts. If silicone RTV is to be used, no other preparation is necessary; but if urethane RTC is to be used, then it is a good idea to coat the entire pattern with a thin layer of wax to seal the surface completely.

Pattern Mounting

The pattern, in finished form, is mounted on a rigid, flat base, usually made of plywood or

Worker's inflatable gloves and blowup vinyl shapes reach out of David Garber's exhibit. (*Courtesy, David Garber*)

A large inflatable by David Garber. (*Courtesy, David Garber*)

Roof of a subway station entrance in Milan, Italy, made of acrylic and FRP spanning trusses of steel and concrete.

Outside view of the subway station in Milan, Italy.

This "Obelisk" is part of a group of massive forms constructed with a polyester foam core that punctuates a square. Sculptor, Al Vrana.

Professor Felix Drury's rigid urethane foam environment for Pittsburgh Gateway Center.

"Chiquita Banana" couch by Ian Walker is made of flexible urethane painted with vinyl. (*Courtesy, Ian Walker*)

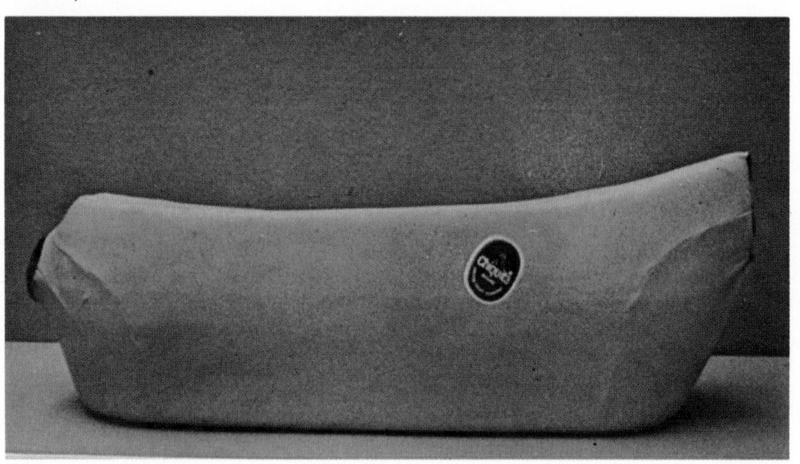

MAKING AN RTV MOLD

a. The master or pattern is prepared. Cracks are filled with wax and in this case covered with an acrylic gesso.

b. Roughness is sanded away because a smooth texture was desired.

c. A release is sprayed over the master.

d. Depth of mold is indicated on the mold wall.

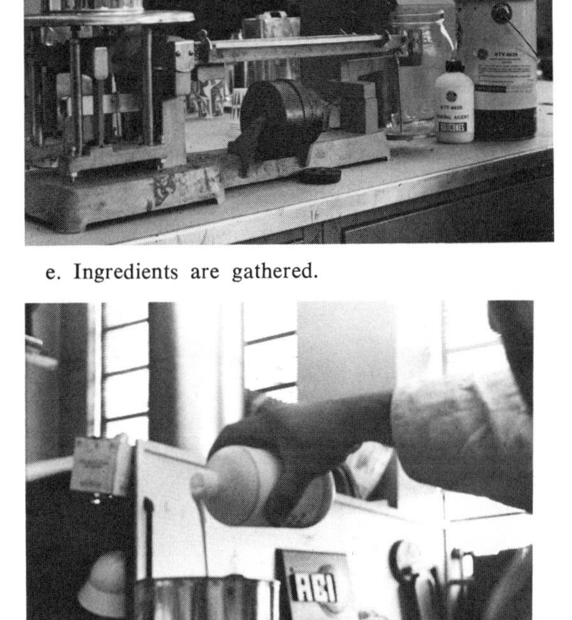

e. Ingredients are gathered.

f. The RTV is weighed and then catalyst is added and weighed.

g. The two components are mixed thoroughly.

h. After mixing, they are poured into a jar for vacuuming out the air.

Masonite. A detachable frame is fitted around the pattern for open molds, allowing ½″–¾″ spacing between the pattern and the frame, including ½″–1½″ above the pattern. Both the pattern and frame should be attached and then filled with wax or masking tape if necessary to prevent seeping of the mold material. All joints and corners have to be sealed as well.

Mold Release

If epoxy, fiberglass-polyester, or urethane is to be used as a mold material, a release coating is necessary to facilitate removal of the master. The

i, j. Air is vacuumed from the mixture. The process continues until the mixture has stopped rising to the top of the jar.

k. RTV is being poured from high onto the mold.

l. The next day, upon curing, the mold is cut away from the side of the mold box.

m. And the mold is pulled away from the master.

mold release must chemically or mechanically block the highly adhesive mold material from adhering to the pattern. Also, barrier coats as mold releases keep surface moisture from affecting the mold material. Silicone release agents are good, but create finishing difficulties if an object is to be painted later. Waxes have been used widely, but they act as a basic cause of age failure in paint films. Wax is difficult to remove from surfaces. Recently nonwax, nonsilicone releases have been developed.

Making of RTV Molds [1]

Although silicone rubber is the most expensive of the commonly used mold materials, it has good durability (count on about 1,000 polyester castings, 250 polyurethane castings), high tear strength, and good mold duplication of the original part, and it requires no mold release for polyester castings. (Mold costs can be reduced by pulverizing old molds and blending no more than 50% of the pulverized material with virgin silicone.) There are different classes of RTV silicones—a thin variety, good for narrow crevices; a paste variety for brushing on molds; more heat-stable varieties; and medium-viscosity systems. Catalysts are also interchangeable. There are those for normal cure time, for slightly accelerated room temperature curing time, and for fast curing (10 minutes–2 hours).

Select and estimate the amount of RTV silicone you will need. Weigh out the necessary amount in the proper ratio. Some small adjustment can be made in cure time by changing the catalyst ratio. Blend vigorously with a spatula by hand for one to two minutes until it is thoroughly mixed. Try not to fold in too much air. If possible, deaerate under a vacuum for 3–10 minutes. (Make certain your container will accommodate expansion.) Paint the surface of your pattern with the RTV silicone and then slowly pour the material, starting in one corner of the mold and allowing it to flow up the pattern to push air out. Hold the container high so that any remaining air bubbles will elongate and burst. If the mold is not completely filled, immediately mix more, following the same procedures. Do not wait for the first batch to cure.

[1] Some RTV silicone rubbers are: General Electric RTV Silicones (a full line) (GERTV 630 will handle intricate detail in molding); Silastic RTV by Dow-Corning Silastic E does a good job. (Do not use Silastic G; the color bleeds through to the positive.)

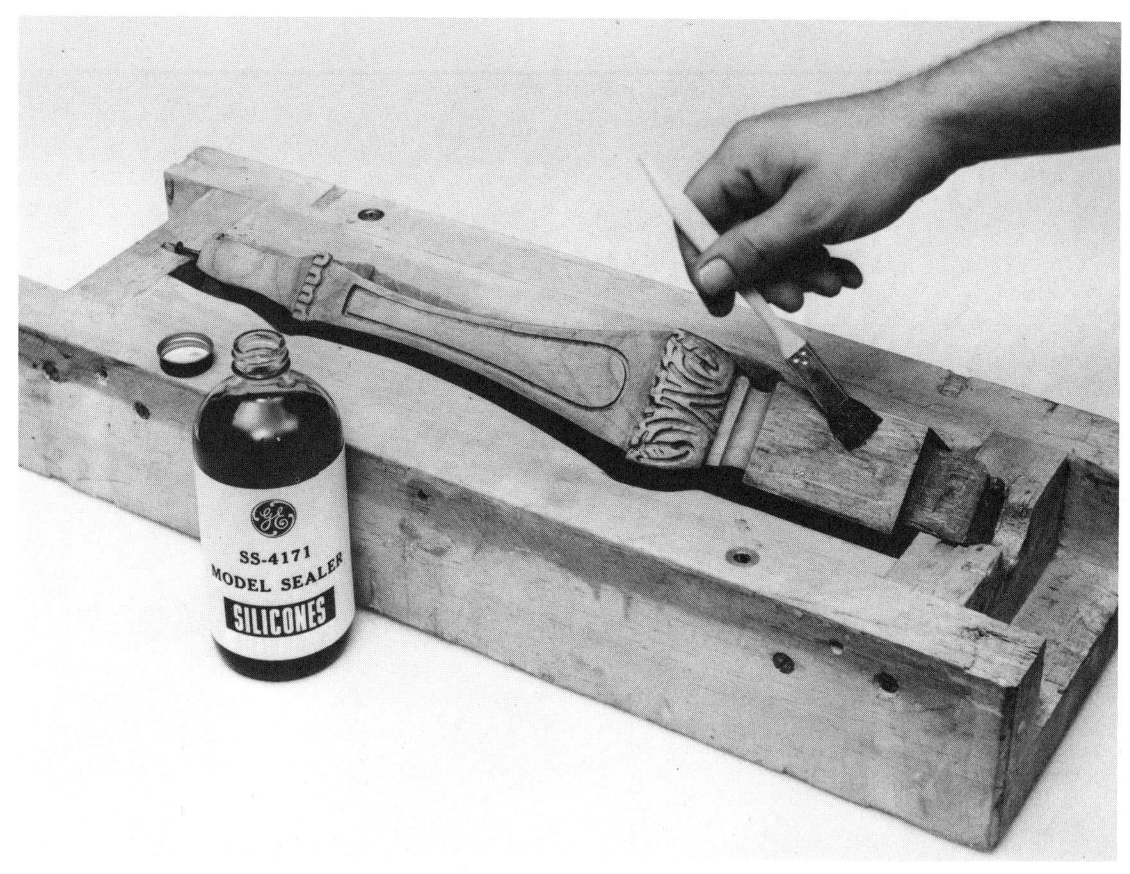

MAKING A TWO-PART MOLD

a. This is the way to make a two-part mold. Imitation wood furniture is made this way, as you can see by the carved wood leg. Pattern materials or finishes might inhibit the cure of the RTV (in this case GE RTV-662) and should be seal or barrier coated. The leg is braced in the mold so that RTV silicone can pour under the leg.

b. RTV silicone is poured into a container (cardboard) having four times the volume and weighed on a 10:1 ratio.

c. The catalyst is added.

d. After the curing agent is added, the combination is thoroughly mixed with a spatula to assure uniform cure. The sides of the container should be scraped clean several times during mixing.

e. If possible, entrapped air should be deaerated for best results. Vacuum should be maintained until the frothing action ceases and bubbles collapse.

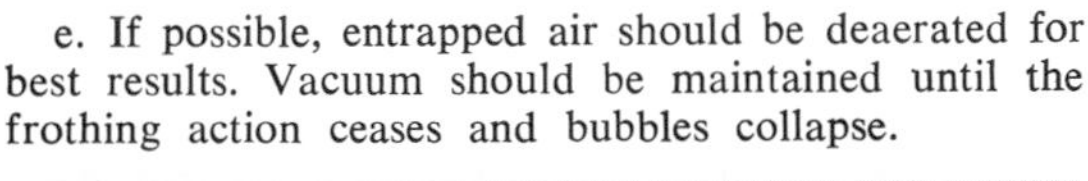

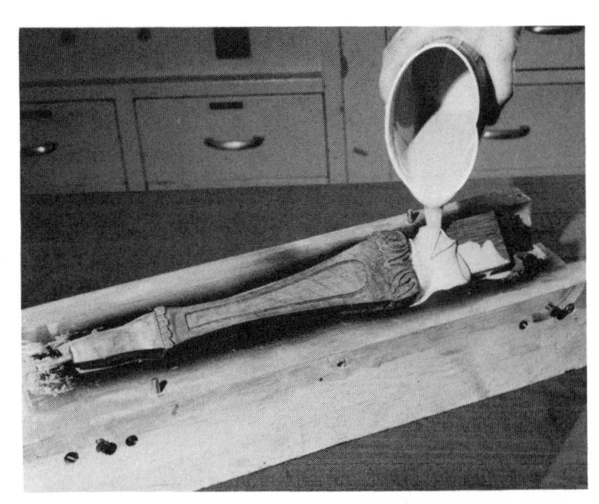

f. With the mold frame and mold in a level position the mixture is poured slowly and continuously from one end until the bottom half of the frame is completely filled. RTV cures in a minimum of 18 hours regardless of the thickness.

g. The cured first half of the mold and pattern is brushed with model sealer (SS-4171). This prevents bonding of the second half of the silicone mold to the completed section. The frame is closed and another batch of RTV silicone is pumped through a hole, drilled previously through the top of the box.

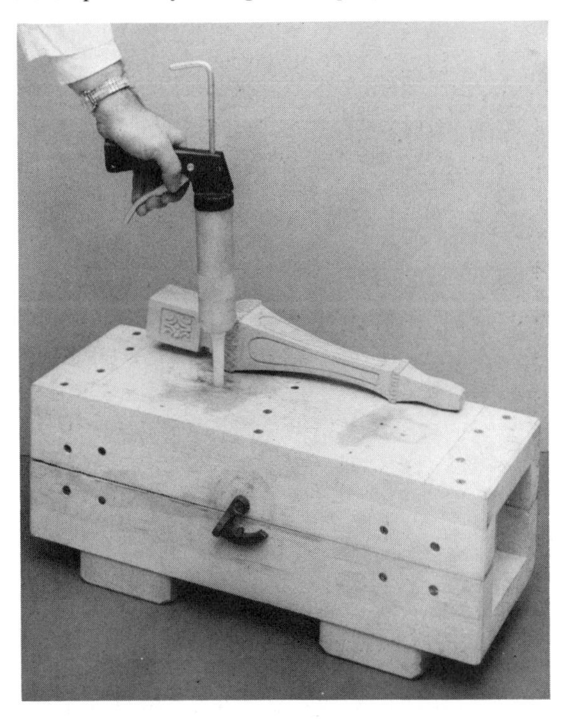

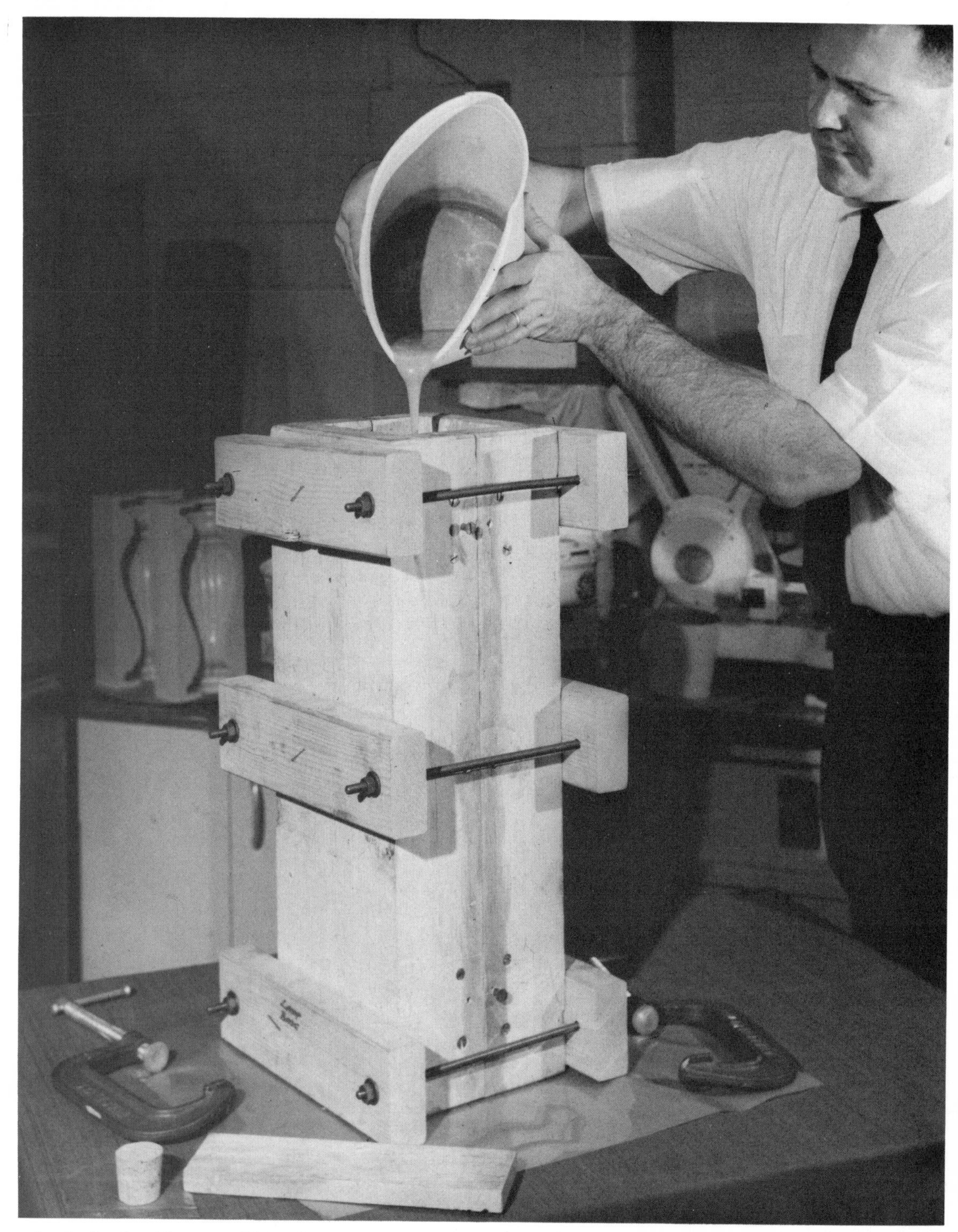

h. Again the silicone mold is allowed to cure overnight. The pattern is removed. The volume of the cavity is determined. Barrier release spray may be applied. The required amount of resin-urethane, polyester, epoxy is mixed. The mold is closed and clamped and the resin is poured into the cavity. (*Courtesy, General Electric, Silicone Products Dept.*)

Allow the mold to cure for 24 hours at room temperature, 75°-80°F., before stripping the mold from the pattern very carefully. Torn molds may be repaired with special RTV materials; or, cut away the damage, clean the mold surface with acetone, reseat the master in the mold, and pour in more catalyzed RTV. If there are difficult undercuts, the mold should be reinforced in those areas. Pour part of the rubber and put pieces of Dacron cloth mesh on that area; then pour the remaining rubber into the mold.

For "in-the-round" castings, molds have to be made in two parts. In this method one half of the model or pattern must be masked. The unmasked portion produces one part of the full mold. A simple method is to embed one half of the pattern in paraffin. Make grooves in the wax to form notches for mold matching and aligning. Metering, mixing, and pouring are accomplished as before. Upon curing, the silicone half of the mold is thoroughly coated with a paraffin wax–xylene mixture or a 5% vaseline and 95% methylene chloride solution. It is allowed to dry and the second half of the mold is poured. After the recommended cure time the master is removed and the two matching halves are now complete. Cut a sprue hole (for your pouring) and another one for an air vent. Tape the two halves together and you are ready for pouring. Before pouring your positive, if the pattern is very intricate, a parting agent of 3-5 parts household detergent soap (e.g., Joy, All, or Tide) in 100 parts of water will aid in easy release and help to protect the mold. After use, clean out the mold with acetone and dry before your next pouring.

Making of a Polyurethane [2] Elastomer Mold

All successfully used polyurethane elastomers are two-component systems which are processable and curable at room temperature, 70°-85° F. The cost is one fifth that of silicone rubbers, though generally they are not so durable as silicones. Mold definition is excellent and tear strength is moderate. In handling polyurethane elastomeric systems you must protect the two components against moisture. A moisture level above 0.06% may produce undesirable bubbles throughout the cured elastomer.

Preparation of the Pattern

Apply a barrier coat and if the barrier is not a release agent, also spray, brush, or wipe a release agent on the mounted pattern, the plate, and all the walls of the frame. Place the mold box on a level, clean, and dry surface.

[2] Some polyurethane elastomers are: Flexipol 9027/6104, Flexible Products Co. and Hardman Kalex, Hardman, Inc.

Mixing the Polyurethane Elastomer

Mixing of the polyurethane elastomer can be done by hand or machine. The hand method will be detailed here:

Stir each component separately by hand in a back-and-forth method to prevent air entrapment, until the consistency is uniform. Determine the amounts of both components needed to fill the topless mold box. Adjust the components to the proper temperature, if the manufacturer suggests this. Weigh the resin components accurately in a clean, dry container (tin cans, polyethylene); avoid the use of paper, unless it is "bone" dry. Mix components together with a spatula, using a figure-eight motion for 60 seconds; scrape the walls and mix again. If you use an electric drill, totally immerse the mixing blade and mix at high speed, but try to avoid an air-forming vortex around the blade. Make certain that the material on the container sides are also mixed. If possible, deaerate the mixture in a container having twice the volume of the mix. One manufacturer recommends painting the mixed urethane elastomer over the pattern and allowing it to cure for 10-15 minutes and then mixing another batch and pouring that until the mold is filled. The liquid level must rise ½″ above the pattern. Wash tools with methylene chloride. Allow the liquid to cure, undisturbed, until it resists permanent deformation when jabbed with a tongue depressor. It should take about one hour to cure at 70°-85° F. If a postcure is necessary, follow the manufacturer's directions. Then place the mold in a support box with a removable top. Brush or spray a suitable mold release agent, and the mold is ready to receive its contents.

Making of Fiberglass Reinforced Polyester Molds (FRP)

Clean and dry the master. Attach it firmly to a base and fill the joint around the master and base so that there will be no seepage. Wax the master with Butcher's wax and fill in any cracks and undercuts. FRP molds are rigid and cannot be pulled away from a form that has undercuts. Paint the master and base with two coats of polyvinyl alcohol and allow to dry. (PVA is water-soluble and washes from the brush easily.)

Apply catalyzed polyester gel coat to the master. Brush or spray over the master until all areas are covered; continue to work until the gel coat is almost gelled. It should build up to 1⁄16″-1⁄8″ thickness. The gel coat should cure in about one hour. As soon as heat dissipates, work may continue. Use acetone to clean tools.

For the fiberglass layup use strips of fiberglass matting or fabric that has been saturated with polyester resin. Lay piece over piece, four or five layers at a time, and then work it in by brush or roller, squeezing out all air bubbles. Use extra layers around the edges of the base to produce a thick flange for clamping purposes.

If there are undercuts then use 1⁄8″-1⁄4″ RTV silicone over the whole pattern, filling in undercut areas; then, after it has cured, apply the fiberglass-polyester matting.

Polyester shrinks, so adjustment has to be made for this in designing. Epoxy, which is more expensive than polyester but does not shrink so much, can be used instead following the same procedures but using epoxy instead of polyester resin. An epoxy mold will also yield more pieces (1,000-2,000). Solid epoxy, with fiberglass laminated between layers is also used for some mold applications and is very durable.

A fiberglass mold for a chair designed by Wendell Castle.

WORKING WITH "UNFILLED" POLYESTER RESIN

Polyester resin is a most versatile system. Its potential ranges from transparent, water-white castings close to the refractive index of acrylic to the density and opacity of wood and marble, depending on fillers that are used. Castings and laminations can be flexible or rigid. Highly reactive types can cure in minutes or polyesters can be slow curing.

Viscosity varies somewhat, but most often the consistency of polyester resin is like that of Karo sirup. Specifications are almost tailor-made to use. For example, polyesters may be chemically resistant; may have the capacity to cure with low exotherm and minimum warpage; may be fast curing and highly adhesive as for body patch; or may be designed for particular climatic conditions. There are light stabilized varieties, flexible resins and moisture-resistant gel coats. In fact, the technology has become highly sophisticated over the past few years. New demands stimulate the development of new resins.

GUIDELINES TO MOLD MATERIAL SELECTION

MOLD MATERIAL	ADVANTAGES	LIMITATIONS
Plaster	Good for prototypes, one part, inexpensive for small complicated shapes	Difficult to transport, can make only one part, generally not practical for large complicated shapes
Wood [a]	Easily machined and handled, good for few parts	Limited life, some dimensional instability
RP	Strong, moldable, durable, easily altered, light weight, doesn't deteriorate, can be stored, easy maintenance	
Steel	For high wear, complex shapes, and inserts in an RP mold	Expensive because machined
Aluminum	Good if mold requires machined sections	Relatively expensive (cheaper than steel)

[a] Life of wooden mold can be increased by using a 1-2 ply RP skin.

Polyester begins as a liquid. With the addition of a catalyst, the resin soon gels, takes its form, and then generates heat and hardness. The speed of curing (hardening) varies considerably; it can be built into the resin formulation by the manufacturer or can be controlled by the user by varying the amount of catalyst, the size of the casting, and air-temperature factors. Using a great deal of resin will increase the heat and if it is a large casting or a sphere casting,

FRP is used as molds for concrete. These cast concrete panels for the Guggenheim Museum Extension were made from FRP molds. (*Courtesy, Owens-Corning Fiberglas Corp.*)

CASTING A POLYESTER RESIN CHESSBOARD

a. The volume of the mold cavity is determined and polyester resin (Reichhold's Polylite 32-037) is measured. To this the proper amount of MEK peroxide catalyst is added.

b. The mixture is blended thoroughly, with care so as not to stir in unnecessary bubbles.

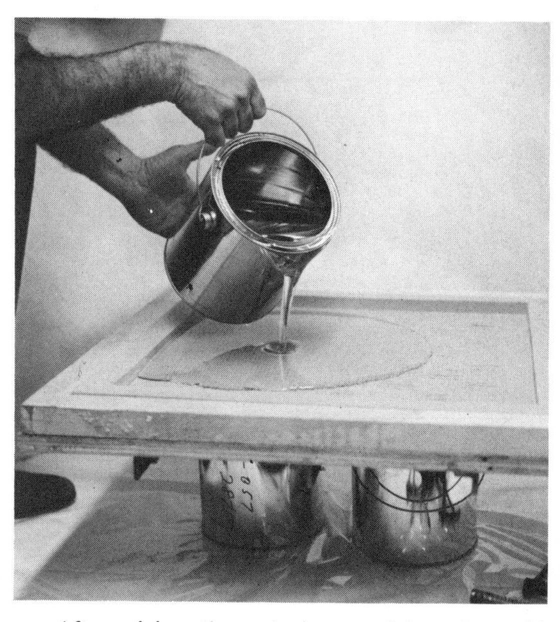

c. After mixing, the resin is poured into the mold (RTV silicone mold) to almost overfill—so that it is filled slightly above the top.

d. A Mylar release paper is carefully laid over the piece.

e. The cover plate is added.

f. And clamps are attached. Note the leaking of resin into a pan from the overfill.

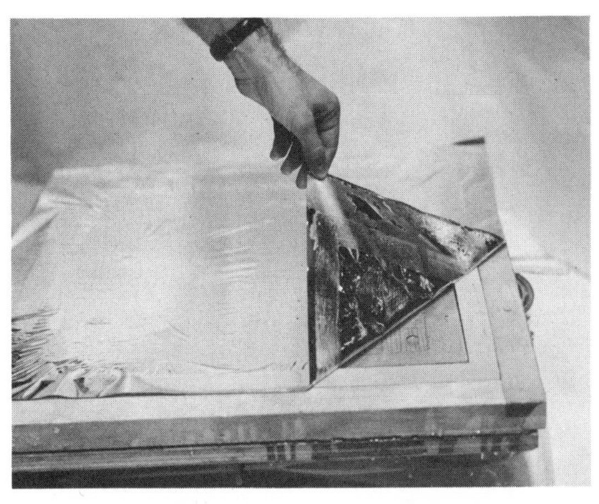

g. After curing has begun (the mold gets warm), remove cover plate and peel away Mylar.

h. Do not do this—warping will result. Instead, turn the piece—mold and form—upside down and remove mold from the positive. Weights can be added to the top to prevent warping as curing continues.

i. The completed transparent chessboard with carved acrylic chess pieces. Note how the pattern of chessboard refracts in the acrylic cylinders. Designer, Thelma R. Newman.

Pouring Polyester on an Acrylic Base

a. The acrylic surface is wet-sanded to give tooth to the surface.

b. Dirt is wiped off with methanol alcohol and a soft cloth.

c. To keep resin from overflowing, a dam is made by using putty strips.

d. Resin, color, and catalyst are added in that order and then mixed thoroughly.

e. A quantity is poured on the acrylic.

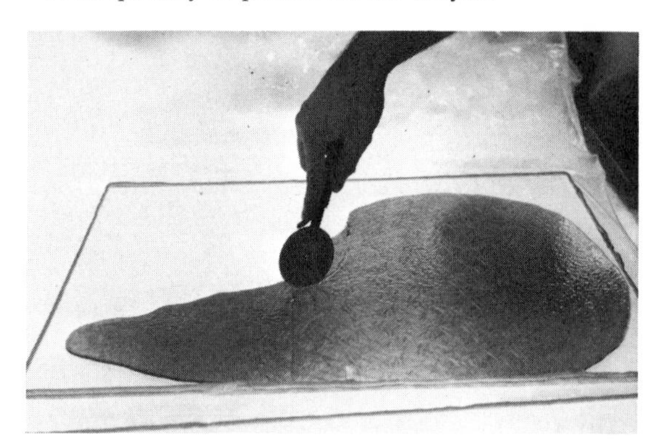

f. After gelling, the shape is defined with a pizza cutter, and excess resin is lifted away.

g. At this stage the resin-gel can be scrambled or lifted into patterns and textures.

h. A different color of catalyzed resin can be poured into the spaces.

i. Another sheet is made by pouring patterns of color.

j. Before the resin cures, while it is still a gel, the putty strip is pulled away from the acrylic base. Both panels will be framed in an aluminum frame after spacers of acrylic rod are added between the panels so that an air foil exists between the sheets, which allows light to enhance the textures.

k. This table was made in the same way. But after spacers were added, the sides were taped and sealed with fiberglass reinforced Mylar tape and 1½″ of polyester resin was poured into each side, after waiting each time for the resin to cure. Upon gelling, the tape was pulled away and the table mounted on a stainless steel base.

l. Detail of polyester-acrylic table. Note square spacers and pattern made by filling partially with polyester resin. Designer, Thelma R. Newman.

the form will split and crack. These castings cure faster as well. Therefore, when external heat is applied, such as casting in a warm room, the cure time will be accelerated.

Catalyst-Resin Proportions

The resin has a limited shelf life. It can self-cure without the addition of catalyst within six months to a year (depending upon the container and room temperature). Catalyzed resin should be used immediately. If there are ingredients to be added, such as fillers, colorants, or other additives, these should be mixed into the resin before the catalyst. The amount of catalyst used —and this is vital information—depends then upon the kind of resin it is in the first place, the thickness of your casting, the warmth of the room, and how much catalyst you are mixing into the polyester. Sara Reid's beautiful water-clear, crystal-like resin SR6912 [3] will gel in 14-16 minutes; it builds up to 250°F in 35-40 minutes and then becomes hard if the following formulation is used with MEK (methyl ethyl ketone) peroxide catalyst:

THICKNESS OF CASTING	MEK PEROXIDE CATALYST PER OUNCE OF RESIN
⅛″ or less	15 drops
⅛″ to ¼″	12 drops
¼″ to ½″	8 drops
½″ to 1″	6 drops
1″ or more	4 drops

If less catalyst is used, the color will not bleach out to crystal-clear.

The Catalyst

Most polyesters have promoters already incorporated into their formulation. Unpromoted polyester will not cure with the addition of MEK peroxide catalyst. Usually cobalt naphthanate is used as a promoter. If it is not, then it has to be added *before* the catalyst. *Mixing a promoter with MEK peroxide catalyst may result in a violent explosion!*

Typical concentrations of MEK peroxide used for room-temperature-curing polyesters is in the range of 1.0% based on the weight of the resin. Maximum concentration is generally 2.0% and the minimum is 0.25%. MEK should be stored in a cool area or a refrigerator. Only glass, polyethylene, or ceramic funnels or measuring devices should be used.

MEK is a severe irritant and can cause damage to eye tissues. Wear goggles when handling the material.

Casting Procedure

A typical casting would require preparation of the surface [4] or mold (release agents needed sometimes); measuring the proper amount of resin into a paper cup or clean can; adding color, filler, or texturing material; mixing these additives; and then finally metering the catalyst and stirring the catalyst into the resin thoroughly (30-60 seconds). The mixture is now ready for pouring.

When embedding objects, the best procedure is through multiple pours. Pour a layer; when it gels, position the object. (Dip the object into catalyzed resin first to eliminate air bubbles.) Then pour another layer; when that gels, add another object. Repeat the process until your objects are embedded and the mold is filled.

Selectron 50111 (PPG Industries) was designed as a crack-resistant polyester that does not stick to glass and can be poured into spheres or other thick castings. Keep the catalyst level low.

Colorants and Fillers

Colorants may be transparent and brilliant to translucent and opaque. Fillers such as Cab-O-Sil make the resin thixotropic so that it can be applied on a vertical surface to chalky (calcium carbonate) or wood-like (pecan flour) materials.

Molds

Glass breakaway molds that require no release agent can be used. They impart a glasslike gloss to the final surface. Reusable glass that requires a release agent (polyvinyl alcohol, wax, or silicone) can also be used. Others are metal, such as aluminum or stainless steel, that require release agents (as for glass); ceramic, which also requires a release agent and gives a good surface translation; polyethylene, no release agent necessary, but has limited life; RTV, no release agent, and polyurethane elastomer, which requires a release agent (both of the latter give excellent reproduction); and polyester-fiberglass that also requires a release agent and will hold up for castings that have no undercuts.

Surface Tackiness

Quite often castings will be tacky on the back. This is because air inhibits the cure. Eventually the resin will harden. You can hasten the hardening of the surface in various ways. One way is to add a bit of wax to the resin, but this dulls the surface. Brushing a very heavily catalyzed resin-catalyst mixture on the surface will speed the surface cure. Carefully rolling a sheet of Mylar or cellophane on the casting so that it covers the surface, will keep air away (try not to trap air bubbles) and give you a hard, glossy surface. Cellophane may pucker into a wavy texture. Mylar also acts as an excellent surface upon which to pour polyester and to

[3] SR6912-Sara Reid Designs. Reichhold's Polylite 32-037 is very close to SR6912. See "Sources of Supply."
[4] There is a detailed development of how to use polyesters in *Plastics as an Art Form* by the author. See "Bibliography."

Ben Shahn and a Menorah he designed for Temple Beth Zion, Buffalo, New York. Christopher Ray for Willet Studios translated his design into wrought iron and then it was coated with 23-karat Hastings gold leaf. Jan Ozog cast the clear polyester chunks and added them to the wrought iron. The base alone required 20 gallons of resin. Because of the thickness, half a gallon at a time was poured and cured. The completed work is over 7′ tall. (*Courtesy, Willet Studios*)

The Willet Studios have brought the development of faceted glass to a fine art. Brilliantly colored glass, usually an inch or more in thickness, is cut to the desired size by hammer blows. The inner surface of some pieces may be faceted and chipped further to catch light. A matrix of epoxy (concrete is used sometimes) is poured around the glass and holds the pieces together with great strength. The thickness of the glass, which makes it unnecessary to control the light by use of texture painting, allows light to gleam through while being further defined by the matrix of epoxy. This is a window in the Hillel Foundation Building at Case Western Reserve University, Cleveland, Ohio. In this case the epoxy matrix is sand-textured in light tan. (*Courtesy, Willet Studios*)

Faceted glass with an epoxy matrix, this window is in the Congregation Ezra Bessaroth, Seattle, Washington. Another variation of this approach known as "gemmaux" combines scraps and chunks of colored glass of varying thickness and texture, together with crushed glass, all embedded in a clear epoxy. (*Courtesy, Willet Studios*)

laminate. Polyester mirror-images anything it touches, if it can be released.

Solvents

Acetone is a volatile solvent that easily cleans tools. *Do not use it on your hands.* Methylene chloride is also a solvent for polyester but is not a fire hazard, as is acetone. Since polyester makes hands sticky, it is a good idea to wear gloves, or to use barrier creams that will permit the resin to wash off your hands with hot soapy water. Kerocleanse [5] is a cream that releases resin from your hands without water.

A word of caution: adequately ventilate your working area. Acetone and MEK peroxide are not good for you to inhale and although there is no incriminating evidence against polyester resin, it could cause trouble. Do not wear resin-saturated clothes. Immediately flush away catalyst if it gets on your skin. And do not spray resins without wearing a mask for organic vapors. Take adequate precautions and polyesters will be safe materials to use.

Fastening

Polyester can be adhered to polyester with more heavily catalyzed polyester resin or with epoxy resin. Since solid polyester can be cut, drilled, routed, et cetera like wood, fasteners such as those used for acrylic can mount polyester to itself or other materials.

EPOXIES

Epoxies usually are amber to opaque although there are some almost water-white [6] varieties available. They will do whatever polyesters will do, except they do not shrink so much, and they are more expensive.

Working with Putty-Type Epoxies

Epoxies usually are mixed on a 1-1 ratio. To keep putty epoxy from sagging on a vertical surface, let it harden *a bit* after it is mixed, before applying.

To obtain a smooth finish cover the uncured putty epoxy with a piece of wax paper or polyethylene. When the cure is complete, remove the covering.

[5] Kerocleanse. Ayerst Laboratories. See "Sources of Supply."

[6] Stycast 1264 & 1266—water-white room-temperature coating epoxy resin. Emerson & Cumming Inc. See "Sources of Supply."

VARIATION IN GEL TIME WITH TEMPERATURE AND CATALYST CONCENTRATION FOR A TYPICAL RAPID ROOM-TEMPERATURE CURING POLYESTER RESIN

Temperature ° F.	Gel Time in Minutes for the Indicated Concentrations of MEK Peroxide Catalyst		
	0.5%	0.75%	1.0%
60	89	51	35
78	48	30	18
90	25	18	10

Courtesy Allied Chemical Corp.

To shape putty-type epoxies use a small bit of water on your trowel or tool.

Working with Liquid Epoxies

In general, larger masses cure faster and smaller amounts take longer to cure. The higher the temperature, the faster it cures and the lower the temperature, the more slowly it cures. Typical curing time for a pound of epoxy at 70°F will be 1½-2 hours. Pot life is 45 minutes.

To get maximum adhesion roughen the surface to be joined; make certain it is clean, free of dirt, oil, and moisture. To keep epoxy from sticking use a release agent designed for epoxy.

To overcome trouble with bubbles brush or coat the first layer of epoxy. Allow it to set for five minutes before casting the remainder. Mix the epoxy and hardener slowly to keep air from being trapped in the mixture. Pour the epoxy slowly, from high, starting in one corner. A few drops of acetone on the top surface of an epoxy casting before it completely cures will help to eliminate any bubbles that might have risen. To remove tackiness from a cured piece of epoxy wash it with soap and water.

The slower the heat cure, the less shrinkage there will be.

To reduce the cost of large castings use fillers inside the casting, such as wooden blocks or foam.

To avoid warpage of large castings use a span to the depth ratio of 8-1. This means that for every eight inches of length there should be an inch thickness of epoxy. If thinner sections are desirable, then use reinforcing materials such as rods, wire mesh, or fiberglass cloth. If a casting is greater than four inches thick, pour it in layers of two inches and allow it to cure for half an hour before pouring the second layer. Another way to reduce excessive heat and the resulting shrinkage is to use thick walled containers (one inch) so that heat will be carried off.

Curing Agents

One of the difficulties in working with epoxies has been the irritating and corrosive characteristics of the polyamine hardeners. New hardeners have been introduced, such as the polyamides, which exhibit reduced skin irritation tendencies but cure at slower rates. A very good curing agent for epoxy is diacetone acrylamide. When reacting with polyamines, it exhibits better weathering ability, greater transparency, and is less irritating to the skin.

SILICONE CRYSTAL-CLEAR CASTING

Clear silicone potting compounds [7] can be used to make thick castings, and is popular for embedding objects or specimens. The end product is elastic and somewhat friable at the edges; therefore casting in clear boxes or forms of polystyrene or acrylic is recommended. This is a two-part system—a resin and catalyst. Application is much the same as for polyester resin. Castings cure in 24 hours at room temperature without exotherm. Optimum properties develop in from 4-6 days. And the pot life is 2-3 hours at room temperature. Air bubbles that become entrapped need to be vacuumed out; otherwise air produces cloudiness.

WATER-EXTENDED POLYESTER RESIN (WEP)

Unsaturated polyesters are often compounded with a filler or extended in order to reduce cost, moderate heat during the curing process, and relieve shrinkage stresses. WEP [8] has been developed to do all of these things. Water used as an extender is inexpensive, cuts down curing temperature, and has a high bulking efficiency. WEP is easy to use and can be mixed with as much as 90% water.

WEP composites are cellular and white or beige and resemble plaster of Paris. (It is also used as a superior substitute for Hydrocal.) A wide variation of characteristics is available—flexibility, compressive strength, mechanical strength, and nonflammability. Specially formulated MEK peroxide catalysts that are compatible with water are also used.

Because water is retained in the material, it is not recommended for outdoor use (it might freeze and crack) or for attaching and combining with other pieces. Shrinkage of WEP during curing varies from 0.4-3.9%, and, in aging, a WEP piece will lose additional water—¾-1%. Dimensional stability is not very good.

Finishing or any surface coating must be a water compatible, breathable, porous type such as polyvinyl acetate-type paints.

Working with WEP

Typically WEP is brownish. Water is added to the resin slowly while it is mixed with an electric mixer *at fast speed.* After it has been mixed, coloring and then a special MEK is added. Reaction occurs within seconds and the mixture must be poured into a mold. No release agent is needed for silicone RTV molds. The working life of the resin is 2-3 minutes and the object can be demolded in 8-10 minutes. Reinforcement with wire can be added before curing. Finishing should be delayed until the majority of water drives off. This can be accelerated with postoven curing.

REINFORCED POLYESTER (RP)

Polyester resin can be brightly colored; when layers are laminated with reinforcing materials, the product becomes very strong and durable. Superimposed layers of reinforcing materials that have been impregnated with polyester resin (or epoxy) can be molded in room-temperature operations.

Reinforcing materials are of many types, each suited for a separate application. Nylon, cotton, asbestos, paper, and fiberglass are most often used. Fiberglass use will be detailed here.

Some resins used to impregnate reinforcing materials may require high pressure; Formica is made that way with melamine.

The basic principles of reinforced plastics that the designer should keep in mind relate to strength, characteristics, economics, and production process. Strength of an RP product, for instance, depends upon the percentage of reinforcement in a composite and the way the fibers are arranged. Parallel arrangements afford the highest reinforcement loadings; right-angle

[7] General Electric LTV-602 and SRC-05 catalyst; Emerson & Cumming, Inc. Eccosil 2 CN; Dow Corning 93-500 Space Grade Encapsulant.

[8] Reichhold Polyite 32-180; Ashland Chemical-Aropol WEP 41, 42. See "Sources of Supply."

CASTING A CHESSBOARD OF WEP AT REICHHOLD CHEMICAL COMPANY LABORATORIES

a. Resin is weighed and then water is weighed.

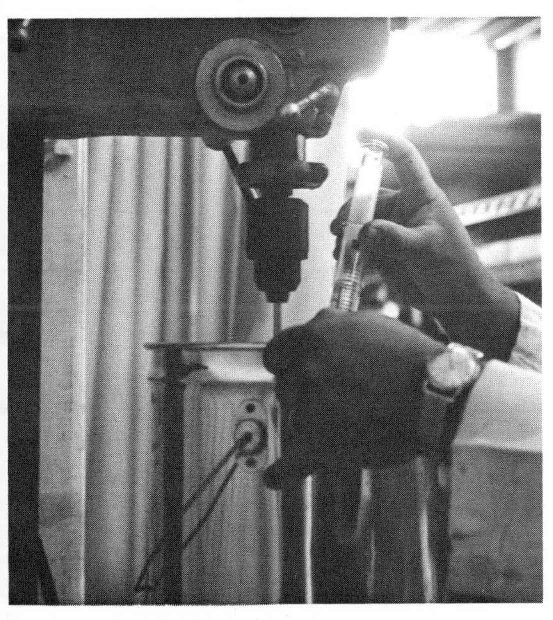

b. The water is slowly added to the resin while a mixer is going at full speed. When all the water has been "absorbed," catalyst is then metered into the mixture and mixed.

c. Immediately the water extended (or filled) polyester is poured into the RTV silicone mold.

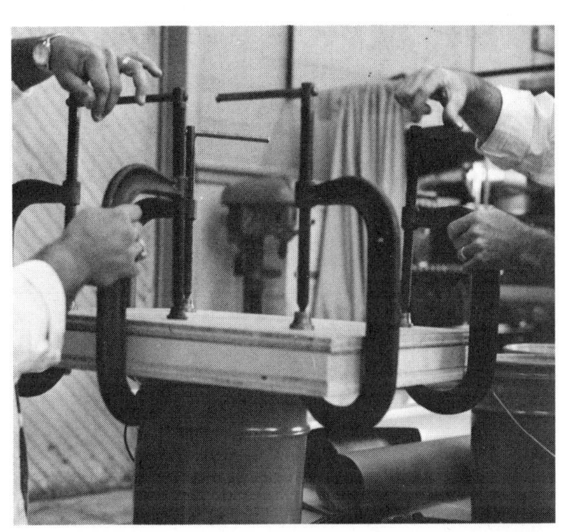

d. The mold cover is added and the mold is sealed.

e. Within minutes the mold is pulled away from the chessboard. Weights are placed on top of the board until most of the curing is completed.

patterns are next; and random arrangements have the least physical strength. The resin selected also determines performance, weatherability, and the like, as well as the kind of production feasible. This whole picture accrues to determine economic factors. Polyester resin is used 80% of the time for RP. Epoxies, phenolics, silicones, melamines, and acrylics are also used. Nylon, polystyrene, polycarbonate polyethylene, and fluorocarbons are also combined with fiberglass in extrusion and injection molding as well.

VISUAL DEFECTS IN REINFORCED PLASTICS

DEFECT	NATURE	REMARKS
Air bubble, void	Air entrapment within and between plies of reinforcement; noninterconnected; spherical in shape	Pierce and fill with resin; roll or daub after applying the resin
Blister	Rounded elevation of surface of laminate whose boundaries may be more or less sharply defined, somewhat resembling in shape blister on human skin	Pierce and fill with resin; roll or daub after applying the resin
Burned spot	Showing evidence of thermal decomposition through some discoloration, distortion, or destruction of surface of laminate	Too highly catalyzed
Chip	Small piece broken off edge or anywhere on surface	Saturate woven roving with resin and bind the edge. Stipple out excess air and resin with a brush
Crack	Actual separation of molded material, visible on opposite surfaces of part and extending through thickness	Poor wetting of fibers. Fill with body putty; sand and paint it
Crack (surface)	Crack existing only on surface of part	Fill with body putty; sand and paint it
Crazing	Fine cracks at or under surface of part	Coat with viscous paint that will saturate and fill the cracks
Delamination	Separation of layers of material in laminate	If fully cured, remove and start again
Dry spot	Area of incomplete surface film on laminated plastics; area over which interlayer and glass have not become bonded	Inject resin and apply some pressure; daub out air and excess resin
Fish eye	Small globular mass not blended completely into surrounding material; particularly evident in transparent or translucent material	Sand away and fill with body putty
Foreign object	Particles included in a plastic that are foreign to its composition	If its contraction and expansion ratio will not affect the FRP, leave as is
Fracture	Rupture of surface without complete separation of laminate	Fill with body putty; sand and paint it
Lack of fill-out	Area of reinforcement not wetted with resin; mostly seen at edge of laminate	Saturate with resin; bind with woven roving that has been saturated in resin
Orange peel	Surface roughness somewhat resembling surface of an orange	Too much heat applied to gel coat; sand and paint with a filled, viscous paint
Pimple	Small, sharp, or conical elevation (usually resin-rich) on surface of a plastic; form resembles pimple	Sand and refinish area

DEFECT	NATURE	REMARKS
Pit (pinhole)	Small regular or irregular crater in surface of a plastic, usually with width approximately of same order of magnitude as depth	Spatula body putty into craters; sand and paint
Porosity	Presence of numerous visible pits (pinholes)	Sand and paint with a filled, viscous paint
Pre-gel	Extra layer of cured resin on part of surface of laminate (not including gel coats)	Leave as is, unless it needs refinishing for cosmetic reasons
Resin pocket	Apparent accumulation of excess resin in a small, localized area	If cured, leave as is, unless it needs refinishing for cosmetic reasons. If moist, squeegee off excess resin
Resin-rich edge	Insufficient reinforcing material at edge of molded laminate	Bind with resin-saturated fiberglass
Scratch	Shallow mark, groove, furrow, or channel normally caused by improper handling or storage	Fill with body putty; sand and paint
Shrink mark or sink	Dimplelike depression in surface of part where it has retracted from mold; has well-rounded edges and no absence of surface film	Leave as is
Wash	Area where reinforcement has moved during mold closure, resulting in resin richness	Leave as is
Wormhole	Elongated air entrapment either in surface of laminate or covered by thin film of cured resin	Try to inject resin into hole. If this cannot be done, slice into hole and fill with body putty, sand, and paint
Wrinkle	Surface imperfection that has appearance of crease or wrinkle in one or more plies of fabric or other molded-in reinforcement	If air is entrapped, apply the above solution

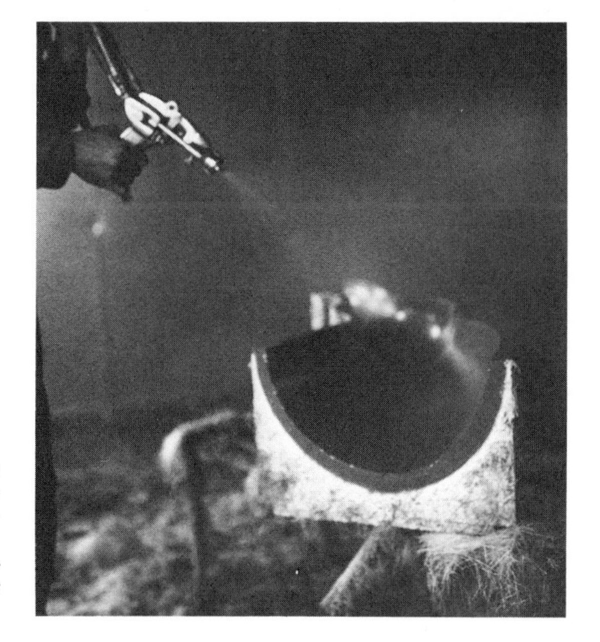

FRP Layup

a. Over a pre-carved urethane foam form (master) a mold is made of FRP. The positive which is described here is made in the same way as the mold except that a harder tooling resin is used.

This mold, by Wendell Castle, is sprayed with release agent, possibly polyvinyl alcohol. Then, after drying, a gel coat is sprayed on.

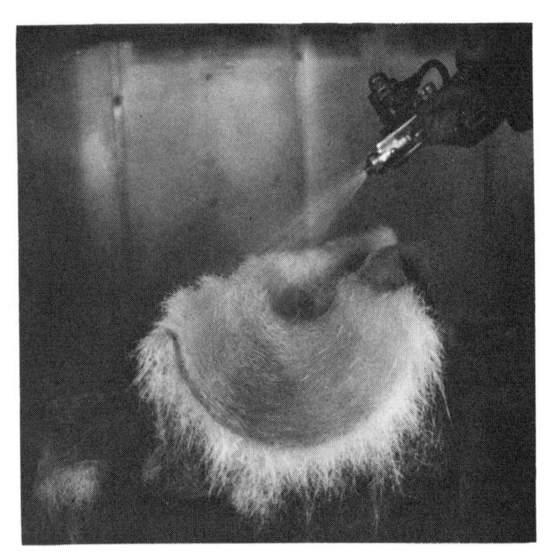

b. Through a dispensing machine fiberglass roving is chopped and polyester resin and catalyst are mixed all in the same action of the trigger.

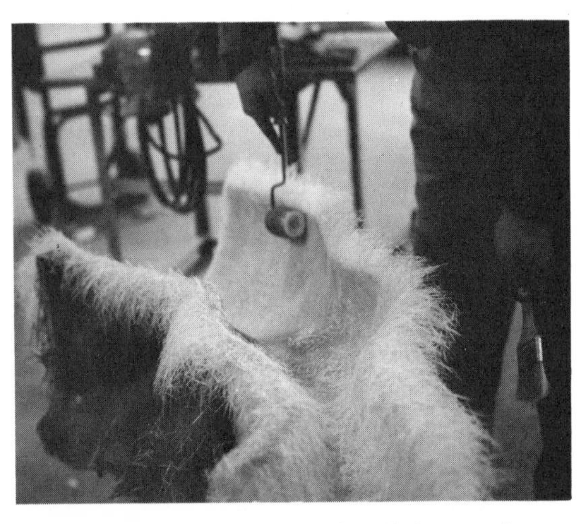

c. The fiberglass-resin layup is rolled, squeezing out air bubbles and distributing the resin evenly.

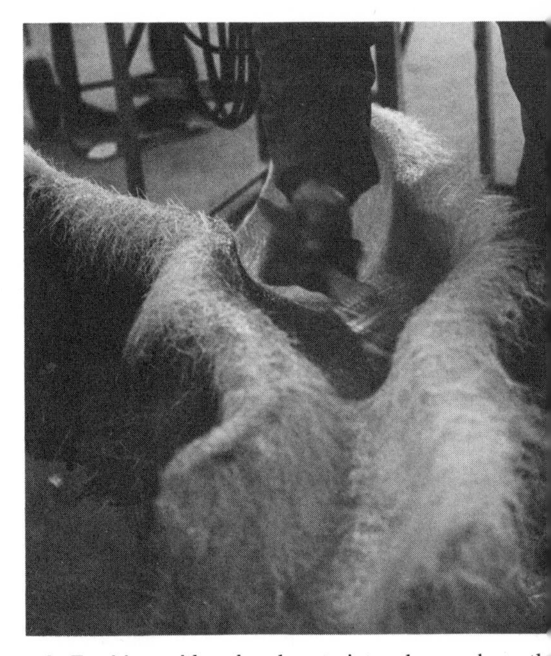

d. Daubing with a brush gets into places where th roller cannot go.

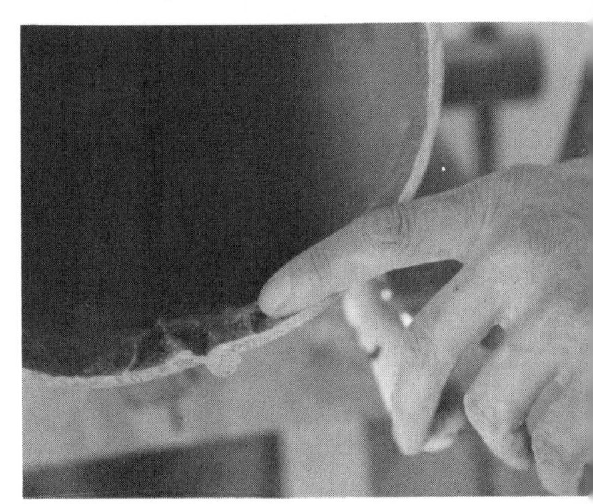

e. Upon curing, the fiberglass form is pried from th mold; if it is half of another part, it is joined wi body putty (on the inside).

f. Wendell Castle mixes body putty.

g. He applies the putty with a scraper.

h. Upon curing, he sands away imperfections and, vhere necessary, fills the area with more body putty; ands it again until the form is smooth.

i. After spray painting the form, he polishes the entire form to reach a high gloss.

j,k. Attachments are added to hold neon or bulbs. (*Courtesy, Wendell Castle*)

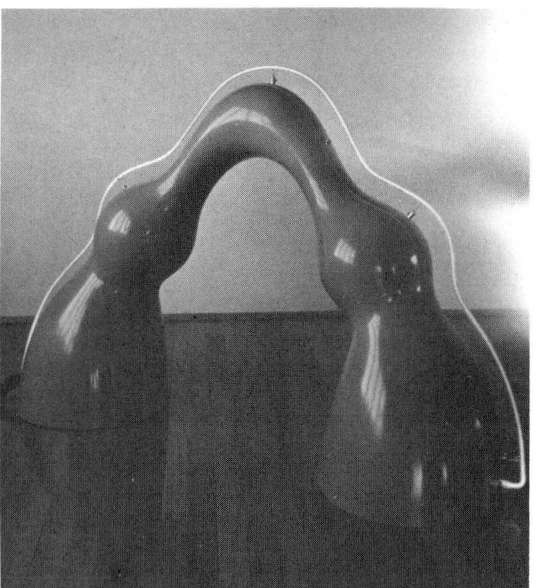

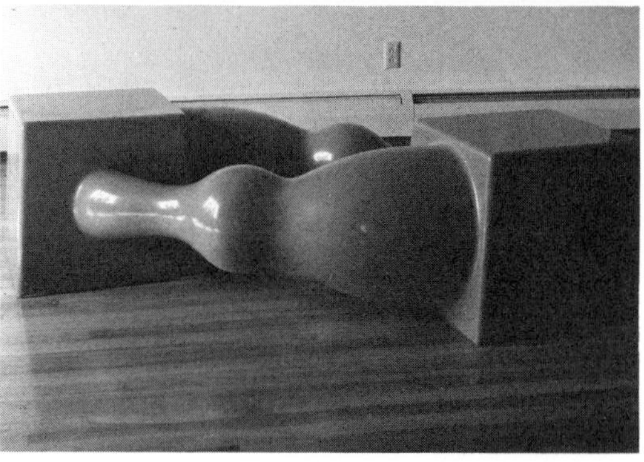

Wendell Castle, known for his furniture fantasies in wood, is working with FRP in making "one-of-a-kind floor lamps," as he calls them. They are brightly colored and glow with either edge-lit Plexiglas, neon or incandescent bulbs. A basic mold provided parts for all these variations; they were merely combined in different ways. (*Courtesy, Wendell Castle*)

Table or bench of FRP by Wendell Castle. (*Courtesy, Wendell Castle*)

FRP chairs by Wendell Castle.

"Floor lamp" by Wendell Castle is FRP flocked with an electric blue.

An acrylic sandwich filled with polyester forms. The table top by Thelma R. Newman.

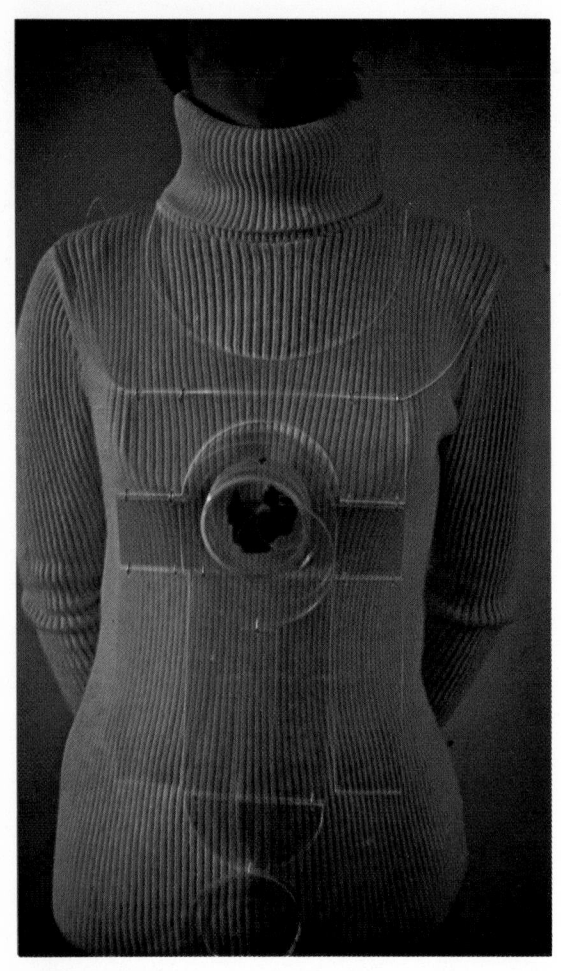

Carolyn Kriegman, "Body Ornament" in warm acrylic colors.

Table lamp of melamine for Artemide S.p.a. was designed by Architect Vico Magistretti. (*Courtesy, Artemide S.p.a*)

Mario Marenco designed this compression molded FRP chair for Artemide S.p.a. (*Courtesy, Artemide S.p.a*)

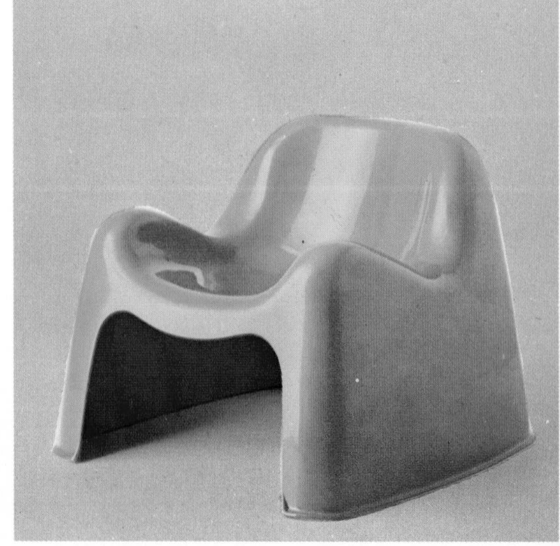

Joe C. Colombo's "Visiona" environment of the future for Bayer. (*Courtesy, Joe C. Colombo*)

Plastico created this translucent showroom of blo up fiberglass impregnated with polyvinyl chloric (*Courtesy, Plastico*)

The Gomma Gomma factory outside Milan, Italy, developed a technique for combining rigid and flexible polyurethane foam in the same forming operation. Designer Joe C. Colombo's chair "Astrea" is a combination of both types of foam. (*Courtesy, Gomma Gomma S.p.a*)

Two densities of flexible polyurethane foam are con bined in the "Cuscini" chair by Lomazzi, DePas, D Cursu, D'Urbino. Four identical cushions are joined create the chair. (*Courtesy, Gomma Gomma S.p.a.*)

Chair by Wendell Castle of FRP painted in hot and bright colors. (*Courtesy, Wendell Castle*)

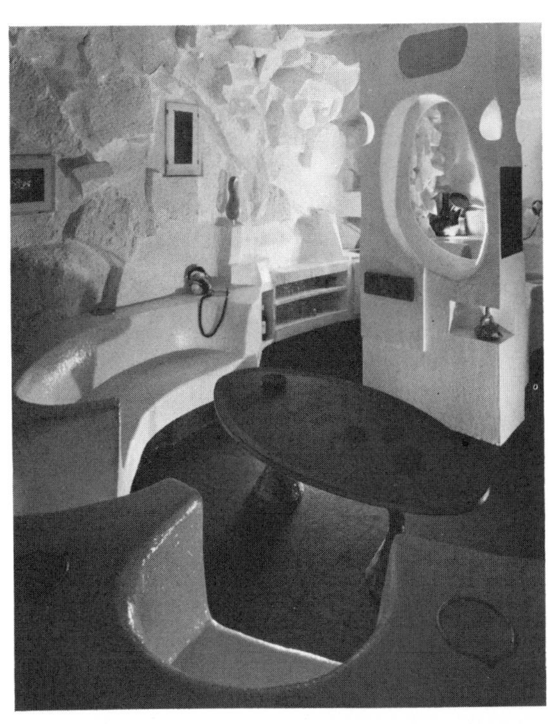

Free-form walls of plaster co-ordinate with cast polyester and fiberglass furniture. A homogenous environment created by Daniel Grataloup of Switzerland. (*Courtesy, Daniel Grataloup*)

A mosaic by Jeanne Martin in Birmingham, England, which uses a background of fiberglass reinforced polyester (Scott Bader's Crystic polyester resin). (*Courtesy, Scott Bader Co. Ltd.*)

A relief panel, part of a wall in Forte's autogrill in London is made of molded FRP. Balclutha Ltd. made the panels of Scott Bader's Crystic polyester resin. (*Courtesy, Scott Bader Co. Ltd.*)

A fiberglass reinforced polyester sculpture over an epoxy matrix embedding faceted glass. First Presbyterian Church, Youngstown, Ohio. Designed by Willet Studios. (*Courtesy, Willet Studios*)

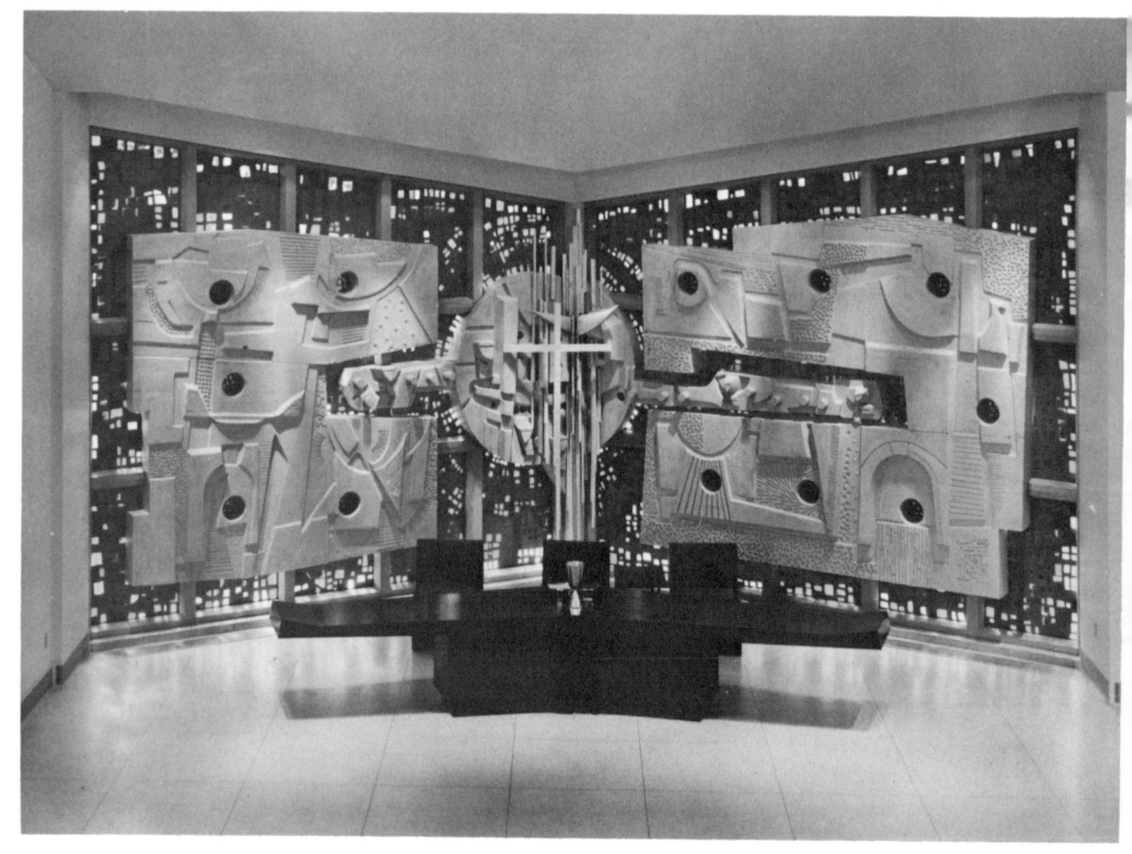

The Rondo House was designed by Casoni and Casoni of Basle, Switzerland. The entire house is molded of FRP rigid urethane foam by a sandwich construction method using FRP as the "bread" and urethane as the "filler." It is completely finished in the factory and ready to live in. (*Courtesy, Scott Bader Co. Ltd.*)

The body of this two-seater coupe (molded by Glassport Motor Co., South Africa) is of FRP. (*Courtesy, Scott Bader Co. Ltd.*)

A prize-winning bus by Alberto Rosselli and Isao Hosoe made of FRP. (*Courtesy, I.D.A., Milan, Italy*)

Compression molding of premix FRP for this table by designers Anna Castelli Ferrieri and Ignazio Gardella. Colors are bright and intense. Compression molding of premixed layers of fiberglass saturated polyester creates a uniform molding with an excellent surface texture. (*Courtesy, Kartell-Binasco*)

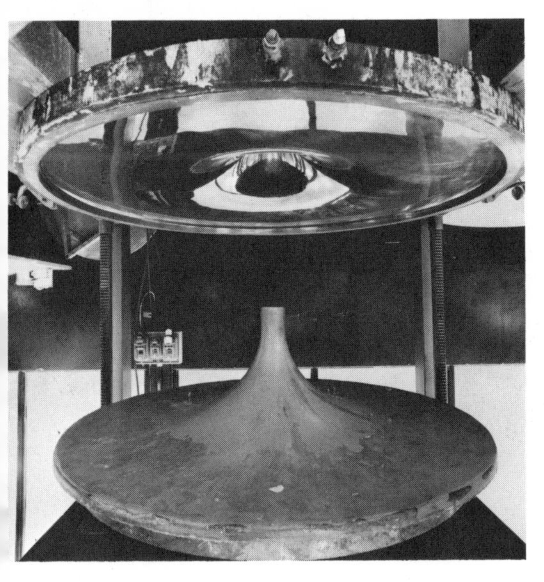

Strong FRP facings on thick lightweight cores form the prefabricated basis for this house being erected in Germany. Sandwiches are used for the floor and interior surfaces as well. The structural sandwich behaves in very much the same way as an I-beam. The facings act as the flanges and the core acts as a tie between the facings. (*Courtesy, Professor Albert G.H Dietz*)

FRP cantilevered lounge chair and footstool for John Stuart Inc. (*Courtesy, John Stuart Inc.*)

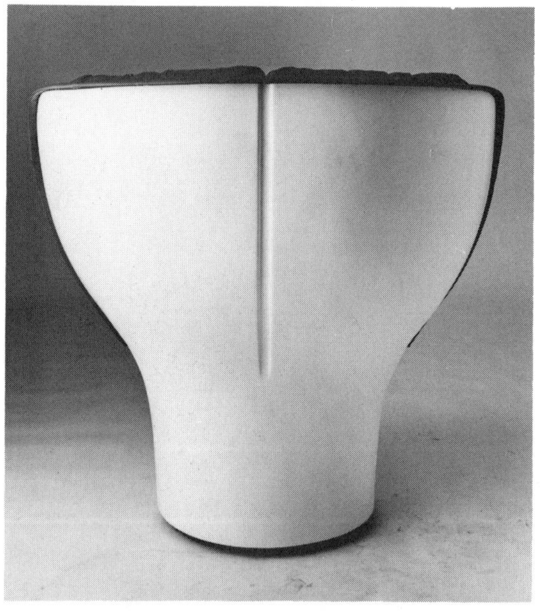

The "Elda" chair designed by Joe C. Colombo specifically for the female form. It is compression molded of FRP premix. The original mold was plaster of Paris. (*Courtesy, Joe C. Colombo*)

The "Bravo Chair" for the Charlton Co. has an FRP base. (*Courtesy, Charlton Co. Inc.*)

The shell of this chair is compression molded of FRP. The flat sides permit ganging into groups. Designed by Architect Mario Bellini for C & B Italia. (*Courtesy, Mario Bellini*)

Architect Alberto Rosselli designed this FRP shell for Saporiti. Skeleton construction honestly reveals the inside and outside at the same time. (*Courtesy, Alberto Rosselli*)

The same technique was used as for high-pressure laminates except that polyester was used over particle board and a bent substrate of plywood. Upholstery is hooked in place. Compound radii and intersections of planes made thermoforming a good choice for molding and kept costs down. The chair was designed by J. R. Strignano for Project One. Width, 24″; height, 35½″; depth, 31″. (*Courtesy, Project One*)

A stacking chair of FRP for indoor and outdoor use by Vico Magistretti. Note the convoluting of the legs for extra strength and the use of curves. (*Courtesy, Moreddi Inc.*)

A variation on the pedestal theme in FRP. (*Courtesy, Stendig, Inc.*)

Another variation on the pedestal theme by Daystrom. It is now becoming a staple design idea, as much as FRP is becoming a dependable and practical material. (*Courtesy, Daystrom*)

A table desk also a variation of the pedestal with adjustable heights made of FRP by Saporiti. Designer, Alberto Rosselli. (*Courtesy, Alberto Rosselli*)

A small living-room table in FRP by Architect Sergio Asti. Note that the convolutions in the legs become both the design's uniqueness and structural necessity. (*Courtesy, Sergio Asti*)

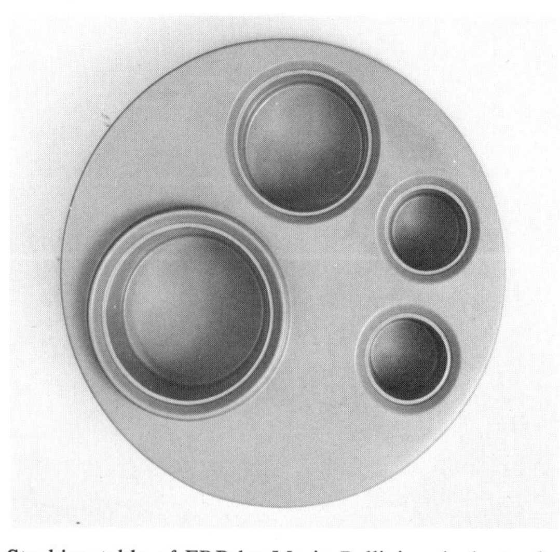

Stacking table of FRP by Mario Bellini and planter by Paolo Caliari for C & B. Both compression molded of FRP premix. (*Courtesy, C & B Italia*)

Basic Processes with RP

Open single-cavity molds either male or female are often used. They require little or no pressure. Results have one finished side. Complex and large shapes can be formed in this manner. Variations of this approach are: hand layup, spray-up, vacuum-bag and pressure-bag molding, autoclave, filament winding, centrifugal casting, and continuous pultrusion.

Closed molds are two-piece male and female molds, usually of metal. Both sides are finished with excellent reproduction of detail. Match-die molding, injection molding, and continuous laminating used closed molds.

Hand Layup

This is the oldest and simplest way of forming with RP. The mold is coated with a release agent. (It could be polyvinyl alcohol.) If a high-quality surface is desired, a gel coat (pigmented surfacing resin) is applied, usually by spraying, on the mold prior to layup. Then fiberglass mat or cloth in strips is saturated with resin and placed on the mold. The strength of the finished object is directly related to the amount of glass in the finished object. Strength increases directly in relation to the amount of glass. Eighty percent of glass and twenty percent by weight of resin makes for a piece four times stronger than a part containing opposite amounts of the two materials. Strength is also related to the arrangement of the glass fibers. Maximum strength is obtained in parallel arrangement of glass, because more glass can be placed in a given volume. Resin helps prevent abrasion of glass fibers by maintaining the position of the fibers and keeps them separate. In an isotropic arrangement, where fibers are distributed in random directions, strength is divided in all directions; the result is lower strength in all directions.

After about four layers have been applied, a brush or roller is used to daub or roll out air bubbles, to equalize the distribution of resin, and to make certain that all fibers are saturated with resin. More layers are applied until the desired thickness is reached. Thicknesses should be at least $3/16''$. The layup cures at room temperature but heat may be used to accelerate the cure. The exposed surface is generally covered with cellophane or Mylar for a smoother finish.

SPRAY-UP

A fiberglass or other type mold made of tooling resin (polyester that is harder and shrinks less when cured) is highly polished and waxed or coated with polyvinyl alcohol. Polyvinyl alcohol is water-soluble and can be sprayed on. It dries in 20-30 minutes.

A gel coat which is a thixotropic polyester resin with color that cures wrong side out (cures to the mold) is sprayed on the mold. It takes about an hour for the gel coat to cure; this process cannot be speeded up with heat because it will alligator (contract). Then chopped glass fibers and resin are simultaneously deposited in or on the mold from special spraying equipment. Roving is fed through a chopper and into a resin-catalyst stream in the same forward action that leaves a spray of this FRP on the mold. After about 3/16″ is deposited, the glass-resin mix is rolled with a hand roller to remove entrapped air, lay down fibers, and smooth the surface. A brush is also used to daub out air from difficult, hard-to-get-to spots. The layup is cured at room temperature or with an accelerated heat cure. After the FRP gels, fuzz is trimmed off with a mat knife. It takes several hours to cure and then, with some difficulty, it is carefully pried out of the mold. If the mold is in two halves, they are put together with body putty (on the inside). Bottom edges are reinforced with glass mat or cloth on the inside. Then body putty is put on the outside and sanded smooth. The painting process completes the job.

RECIPE FOR BODY PUTTY

Ingredients	*Parts by Weight*
Polyester Resin (unpromoted)	42
Cobalt Napthanate 6%	0.4% based on resin
Talc (46 microns max.)	30
Talc (6 microns max.)	10
Talc (98% through 325 mesh)	10
Talc (90% through 325 mesh)	7
Black iron oxide pigment or other color	1

Mixing: Use a spiral or Hobart mixer.

Time in Minutes	*Step*
0	Add resin to mixer
1	Add any additional promotor and color blend thoroughly
10–20	Slowly add dry filler to mixer and mix until fillers are thoroughly "melted" and dispersed
30–60	Mix should be completed to a smooth pastelike consistency, ready for packing

SHELF LIFE: (uncatalyzed) 20 days at 130°F. Room-temperature cure may be induced by adding MEK. Gel time should be 10 minutes.

VACUUM-BAG MOLDING

Vacuum-bag molding is a refinement of hand layup. It uses a vacuum to eliminate voids and force out entrapped air and excess resin. A suitable film such as cellophane, polyvinyl alcohol, or Mylar is placed over the layup and sealed at the edges with a special sealing compound. The vacuum is then drawn on the bag formed by the film and the laminate; cure is completed.

PRESSURE-BAG MOLDING

After application of FRP a tailored bag, usually of rubber, is placed over the layup and then, by using air or steam pressure to eliminate voids, air and excess resin are driven out.

AUTOCLAVE MOLDING

This is a modification of pressure-bag molding. After the layup, the entire bagged assembly is placed in a steam autoclave, usually at 50-100 p.s.i. The high pressure helps to achieve high glass loadings and excellent removal of air. Pressure can be controlled to give the desired degree of uniformity and quality.

FILAMENT WINDING

Continuous reinforcements of glass fiber roving or single strands are fed from a creel through a bath of resin and wound on a suitably designed mandrel. Special winding machines lay down the glass in a predetermined pattern. When the desired number of layers has been applied, the wound mandrel is cured at room temperature or in an oven.

CENTRIFUGAL CASTING

Chopped glass fibers and resin are positioned inside a hollow mandrel. The assembly is rotated forcing the glass and resin against the walls. Cure may be at room temperature, or cure time may be shortened by placing the assembly in an oven.

Continuous Pultrusion

Continuous strands in the form of roving or other reinforcement are impregnated in a resin bath and drawn through a die which sets the shape of the stock and controls the resin content. Final cure is effected in an oven through which the stock is drawn by a pulling device.

Matched Die Molding

This is a mass production method. Mat, fabric, or preform is combined with resin at the press prior to or just after placement in the mold. In the mold matched metal dies under pressure form and cure the part at a certain temperature. Cure cycles may range from less than one minute to as much as five minutes.

COMPARISON CHART

	POLYESTER REINFORCED FIBERGLASS	EPOXY REINFORCED FIBERGLASS
Specific gravity	1.21–1.23	1.16
Density	78 lbs. per cubic foot	69 lbs. per cubic foot
Volumetric shrinkage	6%	2%
Linear shrinkage per foot	2%	Less than 1%
Warpage	Some	Little
Removal from mold	Relatively easy	Very difficult
Color stability	Good	Not good
Weatherability	Acrylic modified varieties last over 10 years	Not as good performance
Gel coats	Good	Good
Cost factors	Relatively inexpensive	More expensive

Durability of FRP

There is no simple answer to the weatherability of FRP. This is a multicomponent process and decisions all along the way can affect the long-range life of a product. There are also extreme variables in weather—location, season of initial exposure, duration of exposure, et cetera. Generally the gel coat is a very important key to outdoor weatherability of FRP. Small flaws and isolated cracks can create problems later. Moisture, freezing, and thawing will aggravate defects. Once glass reinforcement is exposed, weather will take its toll. Moisture, spreading by capillary action, will multiply the damage. If unattended, ten years will see the breakdown of that part. Distortion of the form while it is still curing can affect the shape. If cut edges are not sealed well enough to prevent exposed glass from being attacked by water, delamination can start later. Good maintenance will undoubtedly increase the life of an FRP form.

Premix

This is a ready-to-go system consisting of fiberglass in some form, mixed with resin, pigment, filler, and thickening agent. The premixed material is formed into accurately weighed charges and placed in the mold cavity under heat and pressure. A plastic film covering separates the layers to permit stacking without contamination. The new premix compounds produce high strength and are easily moldable into forms. All surfaces parallel to the mold movement should have at least 1° of draft to facilitate removal. Walls should be kept thin and constant in all parts. Ribs and bosses are advisable for strengthening areas, but ribs and bosses should be integrated into the design of the piece, e.g., textured area, raised line, or the like, so that they are not so noticeable.

Finishing Materials

There are various kinds of finishes for FRP. Some are plain colored gels, exotic metallic finishes, flocking, and just plain paint.

A new marine gel-coat [9] finish can be applied aerosol-can fashion to finished FRP pieces. This gel coat resists salt water, extreme temperature changes, and UV radiation. To use it, the FRP surface has to be prepared by sanding and filling with a polyester-type body putty, cured and sanded smooth, dust removed; then the Spray-Gel can be applied.

Metallic flakes [10] are epoxy- or vinyl-coated aluminum foil specks. They have a high order of reflectance and come in brilliant colors. Application is with normal spray equipment fitted with

[9] Spray-Gel-Seymour of Sycamore, Inc. See "Sources of Supply."

[10] Metallic Flake—Meadowbrook Inventions; Fireflake—Mt. Vernon Mills. See "Sources of Supply."

A coating of aluminum filled epoxy was placed over the WEP chessboard. Lustre is brought out by first rubbing with steel wool, followed by burnishing with the back of a stainless-steel spoon.

CASTING WITH ACRYLIC RESINS

This is a difficult process needing exacting controls and specialized equipment. The process of working with methyl methacrylate is included here for the few intrepid souls who want to try it and for those who just want to know about it. The lure is great, because of all plastics the acrylics are a premium material. Cast methyl methacrylate has good tensile and impact strengths, excellent weathering properties. They can be machined easily and the material can be formed into almost any shape. Best of all, acrylics have high optical clarity, luster, sparkle, and are pleasing and warm to the touch.

Casting Using a Sirup

As received from the manufacturer, methyl methacrylate monomer contains 200-600 p.p.m. of hydroquinone inhibitor. In order to obtain maximum clarity, the inhibitor has to be removed. This is usually accomplished by vacuum distillation under reduced pressure or by washing. When the inhibitor is not removed, higher temperatures, longer times may be needed and greater pressure to reduce a tendency to bubble in thick castings. (Du Pont, in "Embedment and Casting Techniques," describes the distillation and washing process in detail.) Uninhibited methyl methacrylate monomer can be stored in a refrigerator for a while.

The next step is to thicken the uninhibited monomer to a sirup. The monomer is partially polymerized by heating in a water bath in a

a lacquer air cap (e.g., DeVilbiss #36) which reduces overspray and thus yields a glossier finish (pressure cup 10 lbs., spray gun 30-35 lbs. pressure). An underbase of Du Pont acrylic lacquer, close to the metallic flake color, should be applied before the metallic flake. Then to 1½ quarts of clear metal lacquer add one pound of metallic flake and reduce it to 2¼ quarts with lacquer thinner; then spray. (Add one or two ½ inch nuts to the spray cups so that they will help to keep the flakes in suspension as you spray.) Six to eight top coats of lacquer are recommended after spraying the metallic flakes.

Urethane, acrylic polyester, or epoxy based paints will provide excellent coverage and a wide range of color possibility if applied over a dry, clean, greaseless surface.

Jeff Low's table was at first a plywood cube form coated with acrylic polymer gesso. For achieving texture he varied his tools from trowels to sticks of wood. Jeffrey Low added sand and gravel to the putty-like mixture for additional surface qualities. Color is in earth tones of acrylic paint; finished by coatings of acrylic matte medium and varnish. (*Courtesy, Jeffrey B. Low*)

The bedroom setting designed by Milo Baughman for Thayer Coggin is made through a proprietary process by Monsanto that permits *casting* of nylon plus a silica-type reinforcement. Low pressure is used in this process. The reinforcing material and nylon are metered directly into the mold, where a *chemical* bond takes place (fiberglass and polyester form a mechanical bond), which makes the resultant material as strong as cast zinc or aluminum. Size and time are not design deterrents. Since heat is involved in this process, aluminum or steel molds generally are used. Monsanto does not sell the technology of this process or the materials, but rather makes the parts for companies. The white cases of the high chest and bedside commodes are made through this process. The case fronts and headboard are olive ash burl veneer. (*Courtesy, Monsanto Co.*)

steam-jacketed kettle under a reflux condenser with mechanical agitation at a temperature of 194°-212°F. until polymerization has progressed to the point of turning the monomer into a viscous sirup, like molasses. Cooling is accomplished by circulating water through the steam jacket of the kettle. The sirup is then run off into containers and stored in a cool area.

Casting of the sirup is done in closed containers (to avoid losses by evaporation). Glass, metal, clay coated with gelatine, cellulose acetate, tin foil, or plaster of Paris coated with polyvinyl alcohol are proper mold materials. Rubber or cork stoppers must be covered with tin foil to avoid extraction of inhibitors.

Anything embedded must be treated. One must also take into account 20% shrinkage and the fact that exothermic heat of the reaction occurs suddenly; special care must be taken to draw off the heat.

Polymerization can be started by heat, light, or catalyst. Once started, it will continue until completion. The risk of bubbling is very great. Shrinkage is evidenced by internal voids in the mass or by the casting pulling away from the sides of the container.

Polymerization carried out in a water bath minimizes the possibility of bubbling or runaway because water surrounding the mold serves to absorb the heat. One of the simplest methods is to place the mold containing the sirup in an oven or water bath held at a temperature of 105°F. for 4-7 days. Polymerization may be hastened by increasing the temperature or by

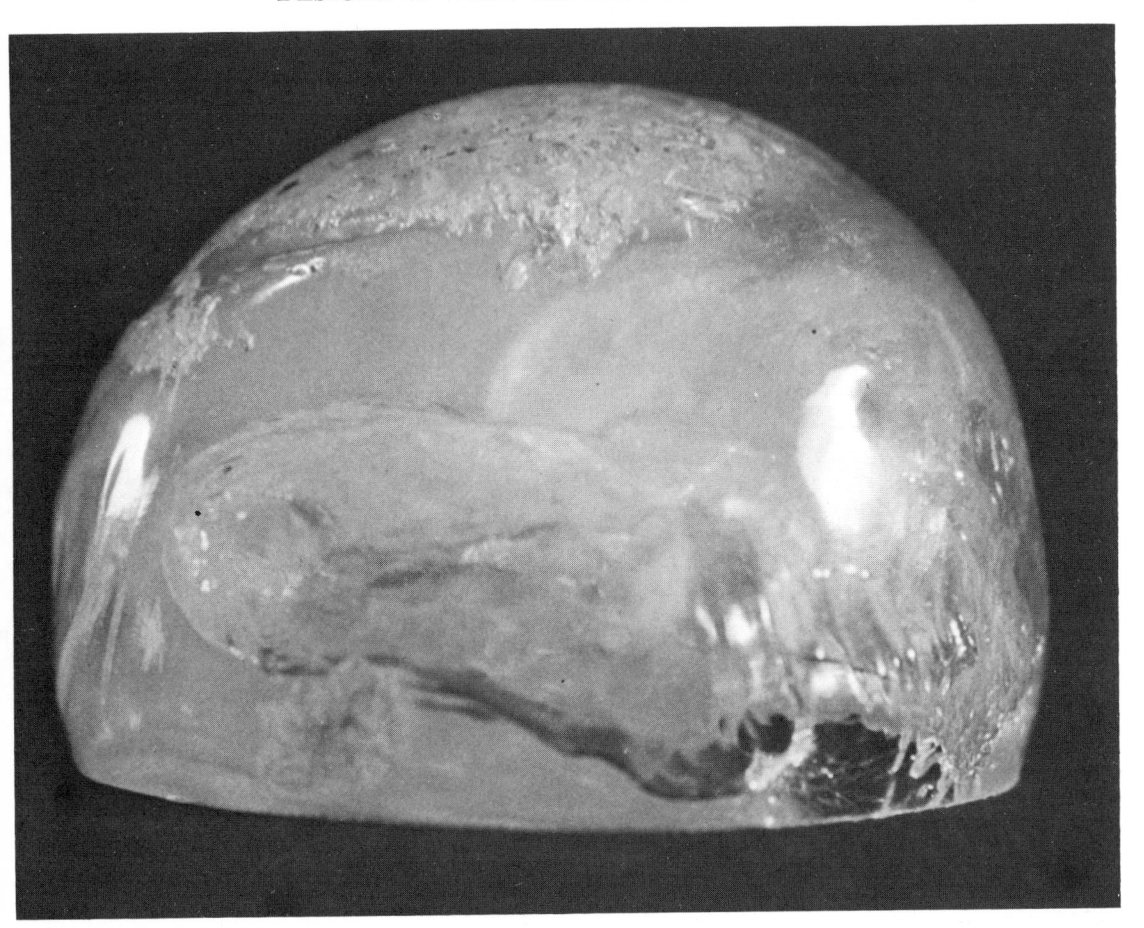

Acrylic monomer runaway. Note cloudiness caused by bubbles.

adding catalyst, but runaway may be likely to occur. Another method involves pouring of 0.5″ thick layers held at 160°F. every 24 hours. This method shows lines of demarcation in a cross-section.

Polymerization also may be carried out in an oven at controlled elevated temperatures and normal atmospheric pressure with periodic vacuuming of the sirup to eliminate gas; or by autoclave polymerization at controlled elevated temperatures and pressures up to 150 p.s.i.

HEALTH HAZARDS

It is said that methyl methacrylate is "only moderately toxic when absorbed into the body by any route, but for some few individuals it is a powerful sensitizer of the skin." [11] In animal experiments Du Pont reports that absorption into the body of methyl methacrylate vapor is not cumulative. Evidences, though, have not been written up and properly disseminated. The usual cautions of covering the skin, keeping skin and clothing clean, and adequately ventilating the working area should be followed.

[11] "Methyl Methacrylate Monomer, Properties and Procedures for Handling and Use," Du Pont, Wilmington, Delaware 19898.

9

plastic prospective

What does the future hold for plastic technology in the service of man? Pause for a moment and think back just one or two decades about what has happened. Speed of change is accelerating at a phenomenal pace. Man has unlocked the atom, harnessed the electron, traveled to outer space, stood on the moon. He has created mechanical servants for all sectors of life to free man from repetitive and dreary labors. But ironically the quality of life has regressed, in part just because of these machines and their wastes and because of commercial contiguities—forced obsolescence, polluting disposables. The means that were meant to improve life actually are destroying life. And the plastics industry cannot be absolved of guilt.

PLASTICS AND ECOLOGY

Just a decade or two ago people thought of plastics as temporary materials. Artists and designers hesitated to use plastics because they thought of them as temporary, fragile, and degradable. Actually modern plastics are resistant to heat, cold, corrosion, and, unfortunately, to decomposition. The reason is that they are made of giant synthetic molecules (macromolecules) whose atoms are linked in intertwined and folded chains up to 10,000 times the length of those found in nonsynthetic molecules. Microorganisms just cannot degrade these man-made materials so easily as other organic matter having shorter molecular chains. Some scientists postulate that it will take millions of years to evolve microorganisms that could eventually decompose macromolecules.

Meanwhile, scientists are wrestling with the problem of how to decompose plastics and reduce our waste piles. A group of Canadian researchers are developing a sensitized group of plastics that will degrade through ultraviolet radiation and are concentrating on plastics used for disposables. Other scientists are searching for ways to reconstitute plastics by breaking down the finished piece to other molecular forms. Designer Spiro Zakas sees forced obsolescence as a boon when an "old" piece of thermoplastic furniture can be melted down and remade into another shape. Members of the Society of Plastics Engineers have been conferring about plastics and ecology and have been addressing themselves to problems in toxicology, pollution, waste disposal, safety, flammability and smoke factors, and biodegradability. Trade journals have been covering progress made in this area. All in all, there seems to be a strong sense of responsibility about pollution et al. on the part of the plastics industry. Designers, too, are trying to redesign the city and control pollution. One solution is to contain the industrial sector under giant, transparent, blow-up domes. We can look forward to a different set of values about purchasing for the sake of acquisitiveness rather than need and forced obsolescence. Solutions to pollution problems will take a little longer.

A VISUAL LANGUAGE FOR PLASTICS

Looking back two decades at design form, we see that, on one hand, designers have been "hooked" on past centuries and in their forms have been trying to capture yesterday's glories; on the other hand, a relatively small group of designers and museums have been working to establish a visual language that addresses itself to the special needs of today's condition and have been making amazing progress. There are lags, though; we have copied the designs of other ages for so long that we are finding it very difficult to discover a true visual language of our day that the public understands. The dishonesty of imitation is making itself felt. Relatively few people understand and can use the vocabulary and grammar of contemporary visual expression. Look at the homes we are building and the products we are buying to furnish our homes and ask yourself whether the contemporary world is being embraced, lived in and understood, or shunned. There is no reason why values of the past cannot exist today, but there is reason to exclude yesterday's values when they no longer are relevant.

There are efforts on the part of important designers who work for giants in industry, small independent designers, and hand craftsmen to raise the level of design and to free man from the limits of tradition. They are trying to place in proper perspective bits and pieces from three thousand years of influence. And these designers are effectively matching man with today's environment by considering his needs in modern living. Their view is broad and they recognize diversity by trying to offer a wide range of choice within which man can exercise his own creative spirit and individuality.

Independent designers and small producers of plastics as plastics need to join forces and to speak in a loud public voice about the significance of what they are doing. Their leadership will benefit all designers and help the public to acquire a contemporary design language. Giant industries, with courage, ought to lead strongly toward contemporary design forms by sponsoring research and producing valid designs of today as pace setters. Industry also should provide a showcase for experimental design projections that do not have immediate commercial connotations.

On another front, designers and industry should transmit to a clearinghouse their discoveries about how designs behave with particular plastics, processes, and environments. With the help of computers industries could sponsor research into the complexities of standard allowable design values. These designs parameters should become a fund for all designers to draw upon from the clearinghouse. Time and money are now being wasted in duplicating procedures that lead only to failure; this is a costly waste that could be avoided.

If more creative research is done and shared, we may not have the rampant imitation that we have today, particularly in the furniture field. There is one recent case of an original and a copy coming onto the market on nearly the same day (by two different producers).

SOME DESIGN CONCEPTS FOR FUTURITY

With population compounding and space shrinking, forms need to be more adaptable to multipurposes, lightweight, compact, stackable, foldable. Space needs to be utilized all the time rather than part of the time. Furniture, for example, can become storage units or suitcases for moving; or it can become seating modules that function as seats and as tables, beds, and storage areas. Funiture can also grow, as a child grows, by adding parts or changing its purpose; for example, a bed could become a couch or a desk, and a chair could convert to an end table.

Factory-made housing, stamped on site, delivered, and stacked, is within the realm of our potential. Fiberglass and urethane could be com-

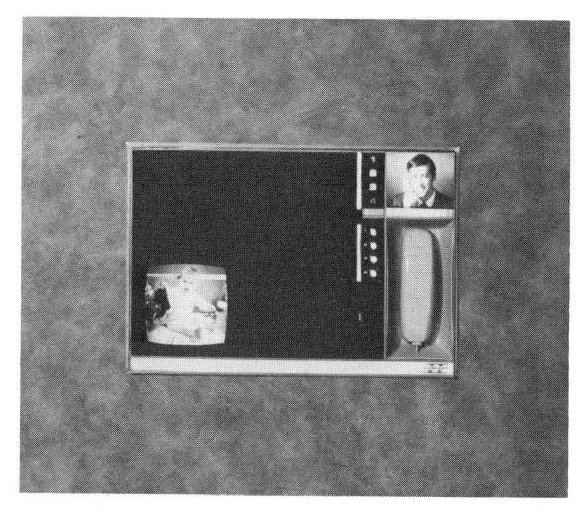

An exploratory concept sponsored by Rohm & Haas for future possibilities. This is a home communications control center including a video telephone, a home intercom unit, color TV and radio, plus remote TV cameras for monitoring various rooms. (*Courtesy, Rohm & Haas Company*)

used to create heat and dispose of waste. Water can be brought in by tank truck. Formable plastics in sheet or fabrics can be used to mold rooms on site over electrical and plumbing conduits.

Since we eat some plastics, it is possible for packages containing foods to be heated through microwave energy along with the food content and then be edible, which would thus eliminate waste. Cans, glass and paper cartons can be replaced by preirradiated and sterilized plastic containers for food and drinking that will degrade in garbage pails equipped with an ultraviolet lamp. Kitchens can become a string of small containers holding miniaturized, transistorized power packages that perform preparation, cooking, serving, and cleanup tasks. A device

bined to create integrated interiors free of woodwork, hand fitting, and free of the need to paint and repaint. Plastic windows can react to light by lightening or darkening in color, depending on the time of day. Other housing can be sprayed on site, no matter how remote and inaccessible to water and electricity. Chemical units can be

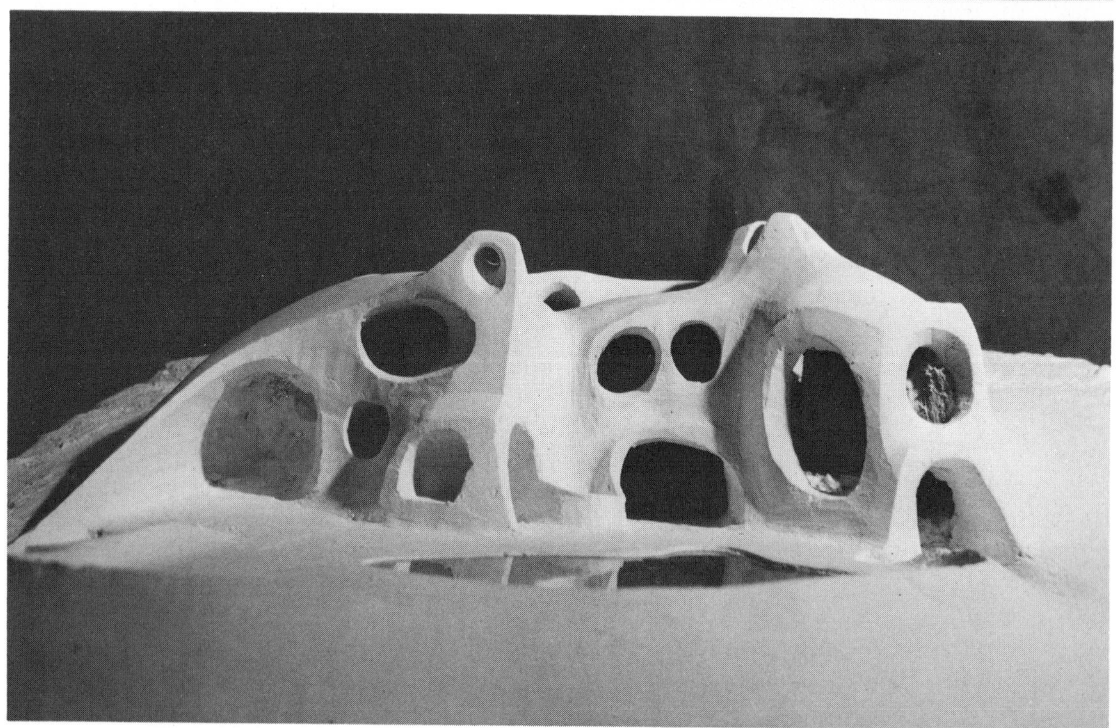

Future houses, single and multidwellings. Possibility: urethane foam and FRP. Designer, Daniel Grataloup of Switzerland. (*Courtesy, Daniel Grataloup*)

could stamp out needed dishes and utensils on the spot as necessary.

The Brookhaven National Laboratories have developed a plastic-irradiated concrete said to be four times as strong as concrete, with double the life span. This material is impermeable to water and weighs 7% more than concrete. Whole cities built underwater at shore lines would increase living space and at the same time reduce erosion of coast lines.

Transportation in our cities can be carried underground through filament-wound tunnels. Mail and messages can be transmitted through an underground network of these tunnels just as electricity and telephone service is distributed today. High-voltage power can be transmitted the same way and thus free our landscape of power lines.

Releasing the imagination is not a fantasy trip but an excursion into what is practical and possible with plastics.

TECHNOLOGY AND POTENTIAL

Think back a decade or two to plastics technology when plastics was still an infant. Processing machines were small and produced small numbers of products. Processes were very limited and quality was not so uniform and excellent as it is today. In these few years the growth of plastics technology has been phenomenal. Our

Designer-architect Daniel Grataloup and his conception of multiunit prefabricated housing. (*Courtesy, Daniel Grataloup*)

The Grataloup multiunit prefabricated housing. (*Courtesy, Daniel Grataloup*)

machinery has become very sophisticated relative to that which we had a few years ago, but no doubt immature in relation to what we will be using two decades hence. We are now capable of producing very large forms, tank-car size, and the speed of production of smaller shapes has reduced prices considerably. The computer will be combined with additional production machinery and will speed production even more. Larger shapes will be feasible. And, in combination with new plastics that will be created to take over all the favorable qualities of metal, new processing techniques will follow.

At the present time we can make gears and bearings for heavy machinery via lamination techniques. We can build car bodies and ship hulls; we can support large buildings on plastic beams or pontoons. Project into the future by magnifying size and one can see that there will be seagoing craft made of plastic, airplanes and space ships traveling at supersonic speeds. All the chemist needs to accomplish this is to design a new macromolecular combination, and the engineer will devise new techniques for processing these new plastics.

We have called this age the Atomic Age, but future generations may likely name it the *Chemical Age*. Consider the wonders that have been brought about through the use of chemicals. Besides creating plastics, scientists are discovering the chemical signaling systems of the brain, the creation of life in a test tube, mood-and intelligence-bearing chemicals, the use of new forms of chemical energy. If we discover the secrets of photosynthesis we will be able to harness new forms of energy to create food. Fibers, fabrics, dyes, insecticides, hormones, vitamins, rubbers, et cetera—all are synthetically produced today. If we can liberate ourselves from our bondage to petroleum as an energy fuel, we will be able to release for use a wide storehouse of chemical building blocks to reconstitute new plastics; and therefore we will save other natural materials that are growing scarce. Urethanes and polyesters, for instance, will have made wood obsolete in about ten years. Chemistry will no doubt find a way to stop the need to rape our earth and at the same time improve the quality of living for more people. A continuing relationship between man's need and research

Entire industrial complexes can be housed under blow-up structures. (*Courtesy, DeCursu, D'Urbino, DePas, Lomazzi, and Lapietra*)

CONTINUOUS GENERATION OF LARGE ENCLOSURES
A shell form can be created with fast-rising and hardening plastic foam. The foam can be pumped into a mixer and transferred to a small molding form at the end of an arm. As the arm rotates, the foam is deposited in a continuous layer, spiraling up the form. A fixed arm generates a sphere; an articulated arm determines various shapes. For instance, a hyperbolic paraboloid can be formed by a moving slit running on inclined edge ribs. (*Courtesy, Albert G. H. Dietz*)

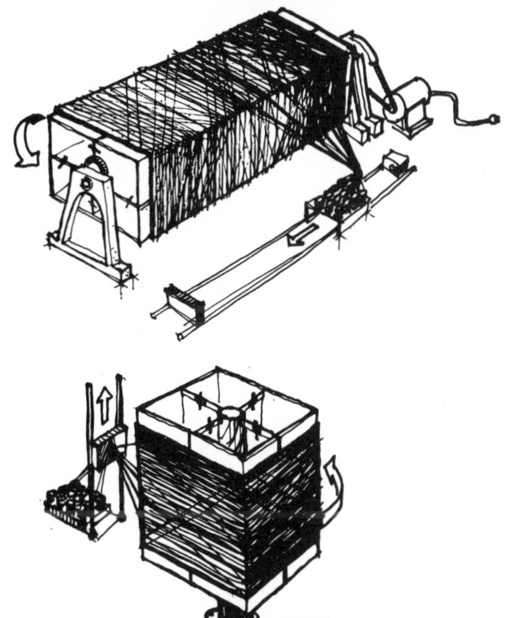

FILAMENT WINDING HUGE FORMS
By winding continuous strands of resin-coated glass filaments on a collapsible mandrel, high strength and lightweight structures can be formed. It is possible, in the process, to incorporate strengthening ribs and supports. Size is of no import. As illustrated in the diagrams, the mandrel can rotate vertically or horizontally. Concept and illustrations of the following materials owe their origin to Professor Albert G. H. Dietz of M.I.T. School of Architecture and Planning. (*Courtesy, Professor Albert G. H. Dietz*)

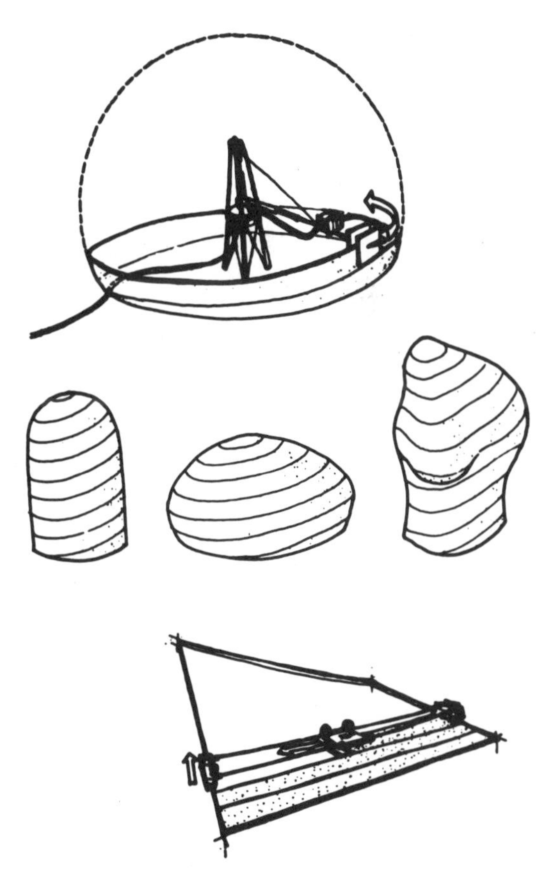

Planks of polystyrene foam laid down at the end of a pivoting arm are heated and welded to the previous plank of foam. An articulating arm can form various shapes, as seen in these diagrams. (*Courtesy, Albert G. H. Dietz*)

is the means. We are on the threshold of even more remarkable development. Plastics has yet to reach maturity.

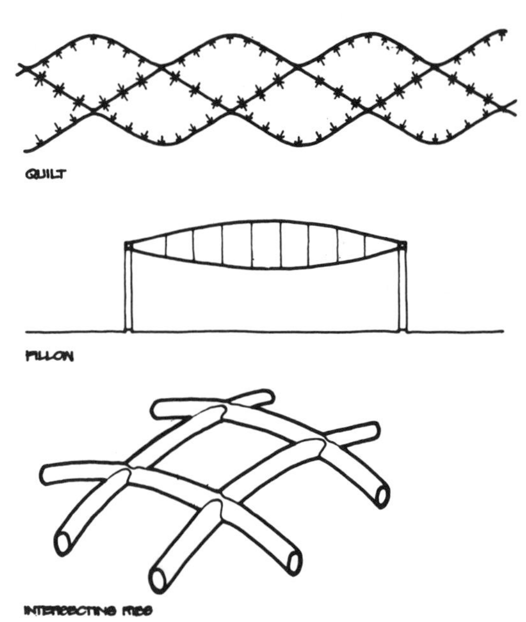

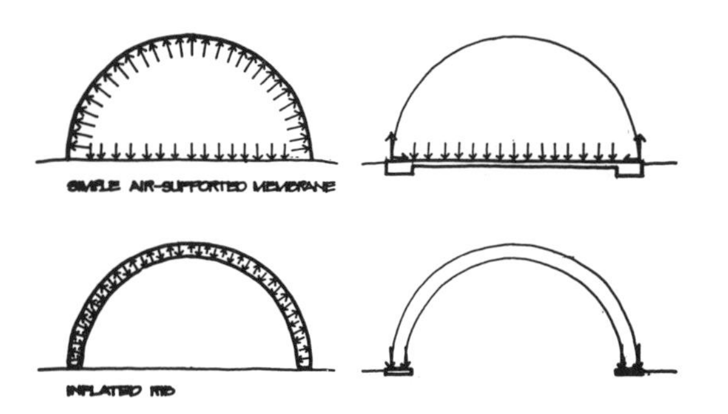

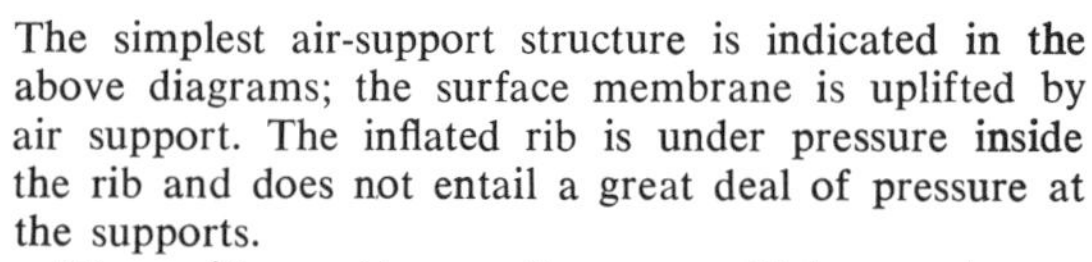

The simplest air-support structure is indicated in the above diagrams; the surface membrane is uplifted by air support. The inflated rib is under pressure inside the rib and does not entail a great deal of pressure at the supports.

The quilt provides continuous multiple membranes and, in the pillow type, two membranes are held separated in internal ties. Intersecting ribs provide a two-way support system with membranes between the ribs. These concepts also were developed by Professor Dietz. (*Courtesy, Albert G. H. Dietz*)

appendix a

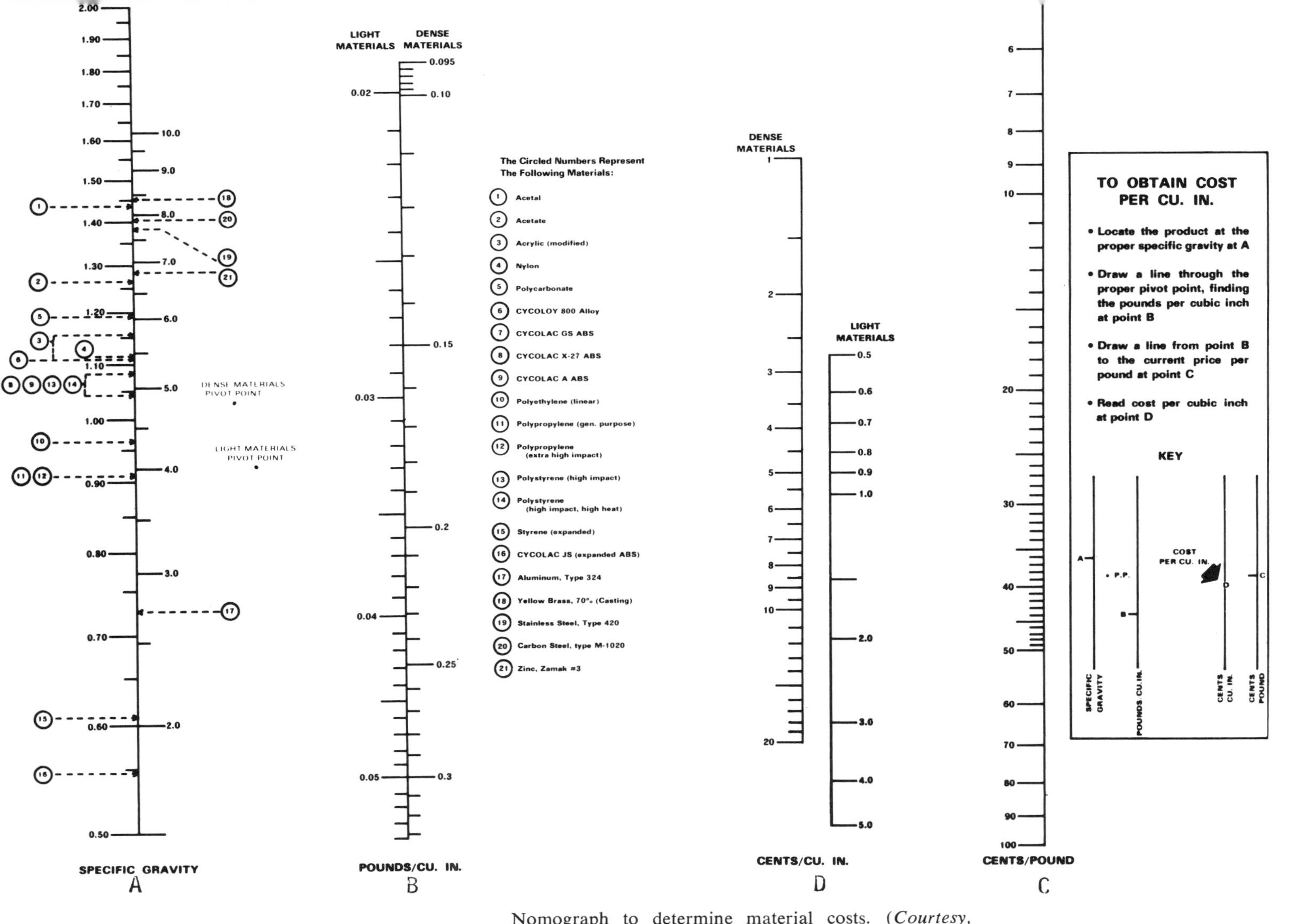

Nomograph to determine material costs. (*Courtesy, Marbon Division, Borg-Warner Corp.*)

appendix b

SOURCES OF SUPPLY

Sources listed here are a random sampling; other places of supply can be found listed in the classified section of your telephone directory. Most manufacturers prefer to sell only large quantities. In that case they will recommend a jobber for smaller orders.

ADDITIVES

Fire Retardants:

Hooker Chemical Co.
North Tonawanda, N.Y. 14120

Ultraviolet Light Stabilizer:

Geigy Chemical Corp.
Saw Mill River Road
Ardsley, N.Y. 10502
Tinuvin P.

General Aniline and Film Corp.
435 Hudson Street
New York, N.Y. 10014
Univul.

ADHESIVES

Adhesive Products Corp.
1660 Boone Avenue
Bronx, N.Y. 10460
A variety of adhesives.

Columbia Cement Co.
159 Hanse
Freeport, N.Y.
Quik Stik, rubber cement for urethane foam.

Eastman Chemical Products, Inc.
Chemicals Division
Kingsport, Tenn. 37662
Eastman #910 and activator, glues almost anything to anything, water white, very strong.

Guard Coating and Chemical Corp.
58 John Jay Avenue
Kearney, N.J. 07032
Adhesive cements for: acrylics, butyrates, Mylars, nylons, cellophane, acetates, styrenes, PVA's, PVC's, epoxies, polyesters, cellulose.

Rezolin Division of
Hexcel Corporation
20701 Nordhoff Street
Chatsworth, Calif. 91311
Uralite Elastomeric Adhesives, for polycarbonates, ABS, polysulfones, reinforced fiberglass, polyester, epoxy, acrylic, aluminum, and steel.

Schwartz Chemical Co., Inc.
50–01 Second Street
Long Island City, N.Y. 11101
Solvent bonding agents.

CATALYST

Argus DB VIII:

Argus Chemical Corp.
633 Court Street,
Brooklyn, N.Y. 11231

Dowanol EP:

Dow Chemical Co.
Midland, Mich. 48640

MEK Peroxide, etc.:

Lucidol Division
Wallace and Tiernan Inc.
1740 Military Road
Buffalo, N.Y. 14217

Reichhold Chemicals Inc.
RCI Building
White Plains, N.Y. 10602

Sara Reid Designs
RR 1, Long Valley, N.J. 07853

COLOR

Carbon Black:

Columbia Carbon Co.
380 Madison Avenue
New York, N.Y. 10017

Dry Color, Polyester Paste, and Liquid Dispersions:

American Hoechst Corp.
Carbic Color Division
270 Sheffield Street
Mountainside, N.J. 07092

Chempico Pigments and Dispersions Co.
P.O. Box 203
South Orange, N.J. 07079
Polyester color pastes, transparent to opaque.

Ferro Corp.
Color Division
Cleveland, Ohio 44105

Patent Chemical, Inc.
335 McLean Boulevard
Paterson, N.J. 07504

Pfizer Co., Inc.
235 E. 42nd Street,
New York, N.Y. 10017

Plastic Molders Supply Co., Inc.
75 South Avenue
Fanwood, N.J. 07023
Potent freeze-dried colorants for styrene, polyethylene, polypropylene, ABS.

Potters Bros., Inc.
Industrial Road
Carlstadt, N.J. 07070
Colored powdered glass filler.

Pylam Products Co., Inc.
95–10 218th Street
Queens Village, N.Y. 11429
Dyes (acetamine) for acrylics.

Schwartz Chemical Co., Inc.
50–01 Second Street
Long Island City, N.Y. 11101
Dyes.

FILLERS AND REINFORCEMENTS

Asbestos:

Carey-Canadian Mines, Ltd.
320 S. Wayne Avenue
Cincinnati, Ohio 45215

Raybestos Manhattan, Inc.
123 Steigle
Manheim, Pa. 17545

Calcium Carbonate:

Diamond Alkali Co.
300 Union Commerce Building
Cleveland, Ohio 44115

Clays:

Whittaker Clark & Daniels, Inc.
100 Church Street
New York, N.Y. 10007

Cotton Flock:

Plymouth Fibres Company, Inc.
Traffic and Palmetto Streets
Brooklyn, N.Y. 11227

Limestone:

National Gypsum Company
Buffalo, N.Y. 14225

Metal Granules:

Metals Disintegrating Company
P.O. Box 290
Elizabeth, N.J. 07207

Mica:

English Mica Co.
26 Sixth Street
Stamford, Conn. 06905

Pecan Shell Nut Flour:

Southeastern Reduction Co.
Valdosta, Ga.

Perlite:

Pennsylvania Perlite Co.
Allentown, Pa. 18103

Thixotropic Filler:

Cabot Corp.
125 High Street
Boston, Mass, 02110
Cab-O-Sil.

Henley & Company, Inc.
202 East 44th Street
New York, N.Y. 10017
Ultrasil VN-3.

Interchemical Corp.
Finishes Division
1255 Broad Street
Clifton, N.J. 07013
Thixogel.

FILMS

Acetate, Mylar, and Polyester:

Coating Products
326 South Street
New Britain, Conn. 06051
Adhesive stock silver sheets #002.

E.I. du Pont de Nemours & Co., Inc.
Wilmington, Del. 19898
Mylar polyester film and Tedlar vinyl fluoride film.

Mirro-Brite Coating Products, Inc.
580 Sylvan Avenue
Englewood Cliffs, N.J. 07632
Chrome and gold Mylar-vinyl with adhesive, and .0075 gauge acetate with adhesive, comes in silver, different patterns and thicknesses.

3M Co.
Reflective Products Division
15 Henderson Drive
West Caldwell, N.J.
Chrome polyester film, #530, with adhesive back, in roll or sheets.

Fabric-Films:

Applied Synthetics Corp.
59th and Industrial Avenue
East St. Louis, Ill. 62204
Syncel, cotton-Dacron flocked on a film not unlike polyethylene.

Ferro Composites
34 Smith Street
Norwalk, Conn. 06852
Cordoglas, vinyl coated fiberglass used in air supported and frame supported structures.

FINISHES

Hot-Stamping:

Admiral Coated Products
8 Empire Boulevard
Moonachie, N.J.
Manufactures foils and colors for hot-stamping surface treatments, also has machines for experimental work.

Metallic:

Art-Brite Chemical Co.
Jersey City, N.J. 07306
Modeling epoxy-aluminum compound called Model-Metal.

Meadowbrook Inventions, Inc.
Bernardsville, N.J. 07924
Metallic Jewels, pigmented, baked epoxy coated aluminum foil. Compatible with polyester, epoxy, acrylic lacquers, vinyl, or plastisol. Flakes used with polyurethane too.

Western Products Division
Mount Vernon Mills, Inc.
1015 Western Avenue
Haverhill, Mass. 01830
Metal flakes in color, Kingston Series, epoxy coated, can be sprayed, molded, extruded, cast, knife-coated on all or a part, solvent resistant, and light-fast.

Surface Color:

Ditzler Automotive Finishes
Detroit, Mich. 48204
Acrylic lacquer for spray coating finished forms.

H & H Paint and Lacquer Co., Inc.
64 Spencer Street
Rochester, N.Y. 14608
Wholesale and retail autobody lacquers and finishes.

National Solvent Corp.
3751 Jennings Road
Cleveland, Ohio 44109
TUF, transparent urethane finish.

PPG Industries, Inc.
1 Gateway Center
Pittsburgh, Pa. 15222
Durethane 600 Elastomeric lacquer and Durethane 100 Elastomeric thermosetting enamel, both for use over urethane foam.

Reliance Universal, Inc.
Progress and Prospect Streets
High Point, N.C.
Toners, stains, glazes, top coats.

Schwartz Chemical Co., Inc.
50–01 Second Street
Long Island City, N.Y. 11101
Transparent lacquers and dip dyes for acrylics.

FOAM SYSTEMS

Aircraft Specialities Co., Inc.
37 West John Street
Hicksville, N.Y. 11801
AscoFoam.

Foam-Form, Inc.
6701 Governor Prinz Blvd.
Wilmington, Del. 19809
Blocks for carving.

Flexible Products Co.
Marietta, Ga. 30060
Solid and flexible foam systems.

Molded Products Co.
Dept. of United Elastic Co.
Easthampton, Mass. 01027
Sheets, rods, tubes both flexible and rigid.

Mr. Plastics and Coatings, Inc.
11460 Dorsett Road
Maryland Heights, Mo. 63042
Mistafoam.

Stauffer Chemical Co.
Plastics Division
Franklin, N.J.
Flexible and solid foam systems: 2002 soft, 2011 harder, 2033 hardest, 2007 flame retardant.

MACHINES AND ACCESSORIES

Drill Bits:

AAA Saw & Tool Service and Supply Co.
1401–07 Washington Blvd.
Chicago, Ill. 60607
Router bits and wheel brushes.

ACE Drill Corp.
Adrian, Mich. 49221
Ace Type "M" drills with polished flutes, slow spiral drills.

Aetna Mfg. Co.
Bensenville, Ill. 60106
Mistic mist coolant system.

Americana Rotary Tool Company
44 Whitehall Street
New York, N.Y. 10004
Router bits.

Edstrom-Carlson & Company
1400 Railroad Avenue
Rockford, Ill. 61101
Router bits.

Henry L. Hanson Co.
25 Union Street
Worcester, Mass. 01608
High speed drills for acrylic, developed by Rohm & Haas Company.

Jiffy Mixer Co., Inc.
515 Market Street
San Francisco, Calif. 94105
All-purpose mixers for drills: models P and PS and larger, model HS for smaller quantities; does not suck in air or splash.

Lehigh Metal Products Corp.
134 Alewife Brook Parkway
Cambridge, Mass. 02240
Self-tapping screws: Parker-Kalon Type "F" tapping screws, and Shakeproof Type 23 and Type 25 thread-cutting

Procunier Safety Chuck Co.
5815 West Lake Street
Chicago, Ill. 60644
Tapping attachments for drill press.

Finishing Equipment:

Divine Brothers Co.
Hardware Products Division
200 Seward Avenue
Utica, N.Y. 13503
Complete line of buffing and polishing machinery, wheels, and supplies.

Kramer Industries, Inc.
1189 Sunrise Highway
Copiaque, N.Y. 11726
Polishing, burnishing, deflashing compounds and equipment for plastics.

Setco Industries
5880 Hillside Avenue
Cincinnati, Ohio 45233
Heavy industrial type machinery.

Foam Cutting Tools:

Arzdorf Machinery Corp.
P.O. Box 94
Towaco, N.J. 07082
Slicing, cutting, and trimming machines for foams.

Atlas Sandt Corp.
240 West 23rd Street
New York, N.Y. 10011
Slicing, cutting, and trimming machines for foams.

Dura-Tech Corp.
1555 N.W. First Avenue
Boca Raton, Fla. 33432
Hot-wire cutters for foams.

Fecken-Kirfel KG Engineering and Machine Co.
51 Aachen
West Germany, Goebbelgasse 1–15
Plastic foam cutting machines.

V. Gredzen's Tools and Machines
240 Pine Street
Biloxi, Miss. 39533
Slicing, cutting, and trimming machines for foams.

Le Brecht Machine Co., Inc.
12–90 Plaza Road
Fair Lawn, N.J. 07410
Vertical and angle cutting machines for foams.

Foaming Machines:

Admiral Equipment Corp.
305 West North Street
Akron, Ohio 44303

Automatic Process Control, Inc.
1123 Morris Avenue
Union, N.J. 07083

Binks Mfg. Co.
3136A Carroll Avenue
Chicago, Ill. 60612

The DeVilbiss Co.
300 Phillips Avenue
Toledo, Ohio 43601

Glas-Craft of Calif.
3225 North Verdugo Road
Glendale, Calif. 91208

Lake Erie Machine Co., Inc.
3156 Belleville Road
Toledo, Ohio 43606

Martin Sweets Co.
3131 Market Street
Louisville, Ky. 40201
Metering equipment for urethane foam dispensing systems.

Impregnating Rollers for Fiberglass:

Venus Products
1862 Ives Avenue
Kent, Wash. 98031
Rollers for rolling out air bubbles in fiberglass impregnations; gun and hydraulic systems.

Materials for Masters and Patterns:

American Cyanamid Co.
P.O. Box 425
Wallingford, Conn. 06492
Laminac polyester resin.

Devcon Corporation
Danvers, Mass. 01923
Devcon C.

Rezolin, Inc.
20701 Nordhoff Street
Chatsworth, Calif. 91311
Rezolin R-72S.

Thermoset Plastics, Inc.
5101 East 65th Street
Indianapolis, Ind. 46220

U.S. Gypsum Co.
101 So. Wacker Drive
Chicago, Ill. 60606
Hydrocal A-11 and B-11, and Ultracal 30.

Ovens:

Blue M Electric Co.
Corporate Headquarters
Blue Island, Ill. 60406

Electric Hotpack Co.
5083 Cottman Street
Philadelphia, Pa. 19135

The Grieve Corp.
1350 N. Elston Avenue
Chicago, Ill. 60622

Trent, Inc.
201 Leverington Avenue
Philadelphia, Pa. 19127

Safety Masks:

Acme Protection Equipment Corp.
1201 Kalamazoo Street
South Haven, Mich. 49090
A full line of protective masks for face, chest, and back.

Saw Blades:

Atkins-Nicholson
Greenville, Miss. 38701
Circular saw blades, high speed steel, hollow ground.

DoAll Co.
254 No. Laurel Avenue
Des Plaines, Ill. 60016
Band saw blades: Dart-Precision with raker set.

Forrest Mfg. Co.
P.O. Box 189
Rutherford, N.J. 07070
Circular saw blades, carbide tipped.

Lafayette Saw & Knife Co.
87 Guernsey
Brooklyn, N.Y. 11222
Circular saw blades, carbide tipped.

Lemmon & Snoap
2618 Thornwood S. W.
Grand Rapids, Mich. 48506
Circular saw blades, carbide tipped.

Radial Cutter Manufacturing Co.
831 Bond Street
Elizabeth, N.J. 07201
Circular saw blades, carbide tipped.

Simonds Saw Division
Fitchburg, Mass. 01420
Circular saw blades in high speed steel and carbide tipped.

Spring and Toggle Clamps:

Adjustable Clamp Co.
417 No. Ashland Avenue
Chicago, Ill. 60622

De-Sta-Co. Division
Dover Corp.
350 Midland Avenue
Detroit, Mich. 48203

Lapeer Mfg. Co.
1144 W. Baltimore
Detroit, Mich. 48202

Strip Heaters:

Electric Hotpack Company, Inc.
5083 Cottman Street
Philadelphia, Pa. 19135

E. L. Wiegand Division
Emerson Electric Co.
7500 Thomas Blvd.
Pittsburgh, Pa. 15208

General Electric Co.
1 Progress Road
Shelbyville, Ind. 46176
Calrod heaters.

Hydor Therme Corp.
7155 Airport Highway
Pennsauken, N.J. 08109
Plastiheater.

Industrial Engineering & Equipment Co.
425 Hanley Ind. Ct.
St. Louis, Mo. 63144

Reusable Syringes for Applying Adhesives:

The Biggs Company
1547 Fourteenth Street
Santa Monica, Calif. 90404
Reusable polyethylene syringes with removable tips.

Phillip Fishman
7 Cameron Street
Wellesley, Mass. 02181
Full line of syringes.

Unusual Sundry Materials and Supplies:

Brookstone Company
Peterborough, N.H. 03458
Hard-to-find sundry sandpapers, hardware items, materials, and supplies.

Vacuum Forming Machines:

Brown Machine Division
Koehring Co.
330 N. Ross Street
Beaverton, Mich. 48612

Comet Industries, Inc.
1320 North York Road
Bensenville, Ill. 60160

Dymo Products Co.
P.O. Box 1030
Berkeley, Calif. 94701
Dymo-form.

O'Neil-Irwin Mfg. Co.
Lake City, Minn. 55041
Di Arco Plastic Press

Plasti-Vac, Inc.
1901 North Davidson Street
Charlotte, N.C. 28205

Spencer-Lemaire Industries, Ltd.
Edmonton, Alberta
Canada

Welch Scientific Co.
Skokie, Ill.

Westflex Machine Co.
Norwalk, Conn. 06851

MISCELLANEOUS

Concrete Plastic Additives:

Dewey and Almy Chemical Division
W.R. Grace Company
Cambridge, Mass. 02140

Flocked Rubber Sheeting:

Archer Rubber Co.
Milford, Mass. 01757
Stretchable suede.

Gypsum Plasters:

U.S. Gypsum Co.
300 W. Adams Street
Chicago, Ill. 60606

Hand Cleaner:

Ayerst Laboratories
Department A
685 Third Avenue
New York, N.Y. 10017
Kerocleanse, hand cleaner for removing resin from hands. Kerodex #71 blocks skin against irritants, corrosives, or allergenic compounds does not wash off but wears off with skin should not be put on irritated tissue.

Lamp Parts:

Lamp Products
P.O. Box 34
Elma, N.Y.

Marking Pencils:

Blaisdell, Inc.
Huntington Valley, Pa. 19006

Masks, Sealants, and Tapes:

Borden Chemical Co.
360 Madison Avenue
New York, N.Y. 10017
Mystik, clear Mylar tape, #8063, tear resistant.

DAP Inc.
P.O. Box 999
Dayton, Ohio 45401
DAP Flexiseal Polysulfide.

Intercoastal Corp.
P.O. Box 4060
Baltimore, Md. 21222
Incolastic 7500 butyl tape.

PPG Industries, Inc.
225 Belleville Avenue
Bloomfield, N.J. 07003
Butyl tape.

J.L.N. Smythe Co.
1300 West Lehigh Avenue
Philadelphia, Pa. 19132
Clearmask.

Spraylat
1 Park Avenue
New York, N.Y. 10016
Spraylat is a latex that is water soluble and can be brushed on to protect acrylic while one is working on it. It takes 45 min. to dry after applying; when dry, it can be peeled off like paper.

3M Company
St. Paul, Minn. 55101
Masking paper, No. 343 and No. 344 "Scotch" protective tape.

Tremco Mfg. Co.
10701 Shaker Blvd.
Cleveland, Ohio 44101
Lasto-Meric polysulfide sealant.

Metal Frames:

Contract Products Corp.
636 Broadway
New York, N.Y. 10012
Extruded metal frame units, 8" to 40" long; purchased broken down, easily assembled. An excellent system, aluminum or gold finish.

Multilensed Plastic Sheeting:

Edmund Scientific Company
101 E. Gloucester Pike
Barrington, N.J. 08007

Neon:

N. Glantz and Sons
437 Central Avenue
Newark, N.J. 07107
All parts and materials for making neon signs. Transformers, paint, plastic sheeting, et cetera.

MOLD MATERIALS

RTV Silicone:

Dow Corning Corp.
Midland, Mich. 48640

General Electric Co.
1 River Road
Schenectady, N.Y. 12306

RTV Vinyls, Urethanes, Silicones:

Adhesive Products Corp.
1660 Boone Avenue
New York, N.Y. 10060

Calresin Co.
4543 Brazil Street
Los Angeles, Calif. 90039

Flexible Products Co.
1007 Industrial Park Drive
Marietta, Ga. 30061

General Fabricators
Van Nuys, Calif. 91408

Hardman Co.
Belleville, N.J. 07109

Smooth-On Manufacturing Co.
572 Communipaw Avenue
Jersey City, N.J. 07304

Silicone Rubber Granulator:

Entoleter, Inc.
P.O. Box 1919
New Haven, Conn. 06509

PAINT FOR PLASTIC

Glidden Co.
11001 Madison Avenue
Cleveland, Ohio 44102
Glidden acrylic sign finishes.

Keystone Refining Co., Inc.
4821–31 Garden Street
Philadelphia, Pa. 19137
Grip-Flex. Attacks surface of acrylic and sticks like cement. It can be built up and air-brushed; high acrylic content.

Wyandotte Paint Products Co.
P.O. Box 255
Norcross, Ga. 30071
Grip-Flex and Grip-Mask.

PLASTIC FINDINGS

Ace Plastic Co.
91–30 Van Wyck Expressway
Jamaica, N.Y. 11435
Acrylic balls, rods, and tubes.

Hastings Plastics, Inc.
1704 Colorado Avenue
Santa Monica, Calif. 90404
Full gamut of plastic supplies, from release agents to Mylar film to resins.

Holophane Co.
Edison, N.J.
Acrylic prismatic lenses.

Orange Products, Inc.
Passaic Avenue
Chatham, N.J. 07928
Solid precision-ground acrylic balls from 1/8" to 1½" in diameter.

Standard Plastics Co., Inc.
Attleboro, Mass. 02703
Acrylic half domes.

PLASTIC MOSAIC TILES

Poly-Dec Company, Inc.
P.O. Box 541
Bayonne, New Jersey 07002
Poly-Mosaic tiles, versatile heat-fusable, gluable, ¾" square tiles.

PLASTIC POLISHES, SANDPAPERS, AND COMPOUNDS

Antistatic Cleaners and Polishes:

Chemical Development Corp.
Danvers, Mass. 01923
Anstac-2M Cleaner.

Chemical Products Co.
3014 No. 24th Street
Omaha, Nebr. 68110
Kleenmaster Brillianize.

Mirror Bright Polish Co.
P.O. Box CT
Irvine, Calif. 92664
Mirror Glaze Cleaner.

Permatex Co., Inc.
P.O. Box 1350
West Palm Beach, Fla. 33402
Permatex Plastic Cleaner.

Surefire Products Co.
6445 Bandini Blvd.
Los Angeles, Calif. 90022
Surefire Plastic Cleaner and Surefire Scratch Remover.

Polishing Compounds and Wheels:

The Butcher Polish Co.
Boston, Mass. 02148
Butcher's white diamond wax.

Lea Mfg. Co.
E. Aurora Street
Waterbury, Conn. 06720
Muslin buffs. Coarse, fast cutting compound: Learok #765, #857. Medium cutting, medium finish: Learok #155, #884.

M & T Chemicals, Inc.
Rahway, N.J. 07065
Muslin buffs. Coarse, fast cutting compound. Bobbing compound. Medium cutting, medium finish: M & T #PC-52.

Matchless Metal Polish Co.
Glen Ridge, N.J. 07028
Domet flannel buffs, with one row of sewing Coarse, fast cutting compound: Matchless #327 and #962. Medium cutting, medium finish: Tripoli #114, Triple XXX Diamond.

Park Chemical Co.
8074 Military Avenue
Detroit, Mich. 48204
For fiberglass molds and chairs: fast cut hand compound #711 Gray (concentrate).

United Laboratories
E. Linden Avenue
Linden, N.J. 07036
Muslin buffs. Coarse, fast cutting compound Plascor #205, #215. Medium cutting, medium finish: Plascor #708, #726.

Sandpapers and Sanding Belts:

Carborundum Co.
Niagara Falls, N.Y. 14302
Fast cut, wet-or-dry, silicon-carbide cloth sanding belts.

Norton Company
Coated Abrasive Division
Troy, N.Y. 12181
Tufbak-Durite, wet-or-dry, abrasive paper.

Rockland Dental Co., Inc.
S.E. Cor. 21st and Clearfield Sts.
Philadelphia, Pa. 19132
Flex-i-Grit, aluminum oxide coated Mylar abrasive.

3M Company
St. Paul, Minn. 55101
Three-M-ite Resin-Bond cloth belts, aluminum oxide sanding belts, grit sizes 80 and 100, for dry sanding only. Wet-or-dry Tri-M-ite paper

PLASTIC POWDERS AND PELLETS

Polyethylene:

Koppers Co.
Plastics Division
Koppers Building
Pittsburgh, Pa. 15219

Spencer Chemical Co.
Dwight Building
Kansas City, Mo. 64105

U.S. Industrial Chemicals Co.
99 Park Avenue
New York, N.Y. 10016

Polystyrene Pellets:

Poly-Dec Company, Inc.
P.O. Box 541
Bayonne, N.J. 07002

Shell Oil Co.
Plastics Division
110 West 51st Street
New York, N.Y. 10020

PLASTIC PUTTIES

Atlas Minerals and Chemicals Division
ESB Inc.
Mertztown, Pa. 19539
Two-part claylike epoxy putty-adhesive in various colors.

Devcon Corp.
Danvers, Mass. 01923
Plastic steel and epoxy bond.

H&H Paint and Lacquer Co., Inc.
64 Spencer Street
Rochester, N.Y. 14608
Plastic auto body putty.

Sculpmetal Co.
701 Investment Building
Pittsburgh, Pa. 15222
Plastic steel and epoxy bond.

Woodhill Chemical Co.
18731 Cranwood Pky.
Cleveland, Ohio 44128
Duro-plastic, Aluminum, Liquid Steel, Gook, Celastic.

REINFORCEMENTS

Fiberglass:

Burlington Glass Fabrics
1450 Broadway
New York, N.Y. 10018

Ferro Corp.
Fiber Glass Division
200 Fiber Glass Road
Nashville, Tenn. 37211

Pittsburgh Plate Glass Co.
Fiberglass Division
One Gateway Center
Pittsburgh, Pa. 15222

Western Fibrous Glass Products
739 Bryant Street
San Francisco, Calif. 94107

Polyester Fabric:

Du Pont
Old Hickory Plant
Nashville, Tenn. 37211
Reemay.

Preimpregnated Fabric:

Ferro Corp.
Cordo Division
200 Fiber Glass Road
Nashville, Tenn. 37211
Pre-preg reinforcement material.

RELEASE AGENTS

General:

Brown Paper Co.
Kalamazoo, Mich. 49004
Wallhide 626, self-releasing paper (#40 base).

Dow Corning
Midland, Mich. 48640
Silicone release paper.

Martin C. Hutt Sales Inc.
105 Church Street
Berea, Ohio 44017
Clear parting film V-28.

Johnson Wax
Service Products Division
Racine, Wisc. 53403
Wax Plate, Bright-Plate, Solve-Cote, Permacote, wax coatings, and mold releases.

Miller-Stephenson Chemical Co., Inc.
Route 7
Danbury, Conn. 06810
Release agents, degreasers, stripable protective film, et cetera.

Paper Corp. of America
630 Fifth Avenue
New York, N.Y. 10020
Part Wick, breathable, textured release paper.

Price Driscoll Corp.
75 Milbar Blvd.
Farmingdale, N.Y. 11735
Yellow Label mold release for general purpose release including wax. Epoxy ParFilm and Polyester ParFilm for those resin systems.

Rexco Chemical Co.
2956 Randolph Avenue
Costa Mesa, Calif.
Partall Paste #2.

Reynolds Metals Co.
6601 West Broad Street
Richmond, Va. 23230
Polyvinyl alcohol.

Silicone:

Ellen Products Co., Inc.
131 S. Liberty Drive
Stony Point, N.Y. 10980

General Electric Co.
Chemical Materials Dept.
1 Plastics Avenue
Pittsfield, Mass. 01201
Dry Film CS-87.

Specialty Products Corp.
15 Exchange Place
Jersey City, N.J. 17302

Urethane Release Agents:

Brulin & Co.
2920 Martindale Avenue
Indianapolis, Ind. 46205
Brulin's #731 for use before casting urethane foam. Also, 300M cleaner to clean mold.

Chem-Trend, Inc.
3205 E. Grand River
Howell, Mich. 48843
Liquid, sprayable release agent for urethane foam.

Dexter Corp.
Midland Division
Waukegan, Ill. 60085
or
Rock Hill, Conn. 06067
Barrier coat in a spray can for urethane casting.

Guardsman Chemical Coatings
1350 Steele, S.W.
Grand Rapids, Mich.
Guardsman's 72-365 clear basecoat primer for urethane masters and molds.

Miller-Stephenson Chemical Co., Inc.
1001 E. First St.
Los Angeles, Calif.
or
445 N. Lake Shore Drive
Chicago, Ill.
or
Route 7, Danbury, Conn.
Makes aerosol spray release agents for urethanes and epoxies.

Randolph Chemical
Carlstadt, N.J. 07072
Barrier coat in a spray can for urethane.

RESINS AND SHEETING

Acrylic:

American Acrylic Corp.
173 Marine Street
Farmingdale, L.I., N.Y. 11735
Decorative acrylic structural sheets in a variety of patterns; called LUMAsite; can be used as standard acrylic sheeting.

Cadillac Plastic
148 Parkway
Kalamazoo, Mich. 49006

J.B. Henriques, Inc.
420 Lexington Avenue
New York, N.Y. 10017

Imperial Chemicals Industries, Ltd.
Plastics Division
Welwyn Garden City
Herts, England

Industrial Plastics
324 Canal Street
New York, N.Y. 10013
Jobber of a large assortment of resins, sheetings, fillers, findings, et cetera.

Rohm & Haas Company
6th and Market Streets
Philadelphia, Pa. 19106
Plexiglas, MMA monomer, Rhoplex, AC-33, Acrysol GS, Tamol 731, Acryloid B7, B72.

Studio Plastique
W. H. Glover, Inc.
171 First Avenue
Atlantic Highlands, N.J.
Acrylic, polyester, epoxy, molds, and findings.

World of Plastics
1129 So. Elmora Avenue
Elizabeth, N.J. 07202
Sheeting and some findings. Supplies neon tubing and apparatus.

Epoxy:

Dow Chemical Co.
Midland, Mich. 48640
DER 332 and DOW.

Marblette Corp.
37–31 30th Street
Long Island City, N.Y. 11101
Marblette.

Reichhold Chemicals Inc.
RCI Building
White Plains, N.Y. 10602
Polylite.

Shell Chemical Co.
Industrial Chemicals Division
110 West 51st Street
New York, N.Y. 10020
Epon.

Thermoset Plastics
5101 E. 65th Street
Indianapolis, Ind. 46440

Union Carbide Corp.
Chemicals and Plastics Division
270 Park Avenue
New York, N.Y. 10010
ERL.

Flameproof Polyester Resin:

Hooker Chemical Corp.
Durez Plastics Division
North Tonawanda, N.Y. 14120
Hetron resins.

Packaged Casting Resin and Accessory Items:

California Titan Products
2501 S. Birch Street
Santa Ana, Calif. 92707
Polyester resin and contingent supplies sold in small quantities, et cetera.

Deep Flex Plastic Molds, Inc.
2740 Lipscomb Street
Fort Worth, Texas 76110

Industrial Plastics
324 Canal Street
New York, N.Y. 10013
Wide range of plastic materials, including resins, fillers, reinforcements, et cetera. All quantities available.

Sara Reid Designs
RR 1
Long Valley, N.J. 07853
MEK, acetate, polyethylene, RTV molds.

Seymour of Sycamore, Inc.
Sycamore, Ill. 60178
Spray Gel, unique spray can for a two-part system of polyester gel for fiberglass.

Taylor and Art Plastics
1710 E. 12th Street
Oakland, Calif. 94606

Valspar Corp.
200 Sayre Street
Rockford, Ill. 61101

Polyester-Epoxy Alloy:

FMC Corp.
633 Third Avenue
New York, N.Y. 10016
or
2121 Yates Avenue
Los Angeles, Calif. 90022

Polyester Resin:

American Cyanamid Co.
Plastics and Resins Division
Wallingford, Conn. 06492
Laminac.

Diamond Alkali Co. Eastern Representative:
Sara Reid Designs
RR 1
Long Valley, N.J. 07853
Diamond Alkali #6912.

Glidden Co.
Baltimore, Md. 21226
Glidpol.

Koppers Co., Inc.
Tar and Chemical Division
Koppers Building
Pittsburgh, Pa. 15219
Koppers.

PPG Industries
Coatings and Resins Division
One Gateway Center
Pittsburgh, Pa. 15222
Selectron.

Reichhold Chemicals Inc.
RCI Building
White Plains, N.Y. 10602
Polylite.

Studio Plastique
W.H. Glover, Inc.
171 First Avenue
Atlantic Highlands, N.J.
Acrylic, polyester, epoxy, molds, and findings.

Polyurethane (resins):

Flexible Products Co.
Marietta, Ga. 30060

Reichhold Chemicals Inc.
RCI Building
White Plains, N.Y. 10602
Polyurethane 92-342.

Union Carbide Corp.
Chemicals and Plastics Division
270 Park Avenue
New York, N.Y. 10010

Upjohn Co.
CPR Division
Torrance, Calif. 90307

Polyvinyl Acetate:

Union Carbide Co.
Chemicals and Plastics Division
270 Park Avenue
New York, N.Y. 10010

Polyvinyl Chloride:

Borden Chemical Co.
350 Madison Avenue
New York, N.Y. 10017

B.F. Goodrich
Chemical Division
1144 E. Market Street
Akron, Ohio 44305

Pyroxylin:

American Pyroxylin Corp.
72–86 Second Avenue
Kearny, N.J. 07032

Woodhill Chemical Corp.
18731 Cranwood Parkway
Cleveland, Ohio 44128
Celastic.

Silicone Casting Resins:

Dow Corning Corp.
Midland, Mich. 48640
Dow Corning 93-500 space grade encapsulant, room temperature curing.

Emerson & Cuming, Inc.
Canton, Mass.
Eccosil 2 CN, water white RTV silicone.

General Electric Co.
1 River Road
Schenectady, N.Y. 12306
RTV 602.

Sprayable Vinyl (cocoon):

Essex Chemical
Clifton, N.J.

R.M. Hollingshead Corp.
728 Cooper Street
Camden, N.J. 08108

Water Extended Polyester:

Ashland Chemical Co.
Columbus, Ohio 43212

Water Filled Polyester:

Reichhold Chemicals Inc.
RCI Building
White Plains, N.Y. 10602
Polylite 32-180.

SOLVENTS

National Solvent Corp.
3751 Jennings Road
Cleveland, Ohio 44109

Shell Chemical Co.
110 West 51st Street
New York, N.Y. 10020
Acetone.

Union Carbide Co.
Chemicals and Plastics Division
270 Park Avenue
New York, N.Y. 10017

appendix c

SHOPS THAT WILL WORK WITH SMALL USERS AND MAKE PROTOTYPES

Cassina Center 2, Via Manzoni 20036 Meda (Milano) Italy	Does experimental work—prototypes, molds, and models.
Consolidated Molded Products Corp. Division of Republic Corp. Warner Street and Greenwood Avenue Scranton, Pa. 18501	Large corporation equipped with all kinds of plastics processing machines. Makes molds and produces for companies. Does a good deal of work in furniture, from the mold to the finished piece.
Dura Plastics of New York, Inc. 49 Richmondville Avenue Westport, Conn. 06880	Blow molding, forming, cementing, et cetera. Produces on contract.
Fiber Tec Manufacturing Corp. 175 Murray Street Rochester, N.Y. 14606	Manufactures fiberglass parts. Does spraying with fiberglass on contract.
W.H. Glover, Inc. 165 First Avenue Atlantic Highlands, N.J.	Specializes in acrylic displays, furniture and silk-screened parts.
Gomma Gomma spa Via Confalonieri 4 20036 Meda, (Milano) Italy	Experts in urethane foaming for the furniture trade.
Just Plastics 250 Dyckman Street New York, N.Y. 10034	Mold making of epoxy, wood, Hydrostone jigs of wood and metal; various kinds of forming and finishing operations. Specialists in working with acrylics.
Oppenheimer Italiana Via Achille Grandi 20037 Paderno Dugnano, (Milano) Italy	Experts in urethane molding for the furniture industry.
R & M Plastics 6 Wells Avenue Yonkers, N.Y. 10701	Fabricates parts from acrylic.

Sara Reid Designs R.R. 1 Long Valley, N.J. 07853	Expertise in polyester resins. Some work with acrylics.
G.T. Schjedahl Co. Northfield, Minn. 55057	Manufacturers of foam portable buildings.
Sargent Industries 2045 Evans Avenue San Francisco, Calif. 94124	Manufacturers of inflatable structures.
Stan Cook Machine Shop 617 South Lawrence Street Burlington, N.J. 08016	Completely equipped metal shop and also specializes in fabricating parts from acrylic.

appendix d

TRADE ASSOCIATIONS AND TRADE PUBLICATIONS

Trade Associations:

The Society of the Plastics Industry, Inc., 250 Park Avenue, New York, N.Y. 10017

The Society of Plastics Engineers, 65 Prospect Street, Stamford, Conn. 06901

American Society for Testing Materials, 1916 Race Street, Philadelphia, Pa. 19103

Manufacturing Chemists Assoc., Inc., 1825 Connecticut Avenue, N.W., Washington, D. C. 20009

National Trade Publications:

Modern Plastics, Breskin Publications, Inc., 770 Lexington Avenue, New York, N.Y. 10021

Modern Plastics Encyclopedia, Breskin Publications, Inc., 770 Lexington Avenue, New York, N.Y. 10021

Plasticos Mundiales Y Empagne, Cleworth Publication Co., 1 River Road, Cos Cob, Conn. 06807

Plastics Design & Processing, Lake Publishing Corp., Box 270, 311 East Park Avenue, Libertyville, Ill. 60048

Plastics Technology, Sales Management Inc., 630 Third Avenue, New York, N.Y. 10017

Plastics World, Cleworth Publishing Co., 1 River Road, Cos Cob, Conn. 06807

Progressive Plastics, A MacLean-Hunter Publication, 481 University Avenue, Toronto 2, Ontario, Canada

SPE Journal, Society of Plastic Engineers, Inc., 65 Prospect Street, Stamford, Conn. 06901

Western Plastics, Western Business Publications, 274 Brannan Street, San Francisco, Calif. 94107

Western Plastics Directory, Western Business Publications, 274 Brannan Street, San Francisco, Calif. 94107

Western Plastics Directory, Western Business Publications, 274 Brannan Street, San Francisco, Calif. 94107

International Trade Publications:

Germany (West):

Kunststoff Berater, Umschau Verlag, Stuttgaerter Str. 20, Frankfurt am Main

Kunststoffe Carl Hanser Verlag, Kolberger Str. 22, Munchen 27

Die Kunststoff-Industrie und Ihre Helfer, Industrieschau Verlagsgesellschaft mbH., Berliner Allee 8, Darmstadt

Kunststoff-Rundschau, Verlag Brunke Garrels, Schloss Str. 2, Hamburg Wandsbek

Kunststoffwelt, A. M. Schoenleitner Verlag, Johann Sebastian Bach Str. 1, Munchen 19

Great Britain:

British Plastics, Lliffe Industrial Publications, Limited, Dorset House, Stamford Street, London, S.E.1.

Rubber and Plastics Age, Rubber and Technical Press Ltd., Gaywood House, Great Peter St., London S.W.1.

France:

Officiel des Matieres Plastiques, 12 rue de l'sly, Paris (8^{e})

Industrie des Plastiques Modernes, 40 rue du Colisee, Paris (8e)

Plastiques Informations, 94 rue Saint-Lazare, Paris (9e)

Plastiques Batiment, 94 rue Saint-Lazare, Paris (9e)

Pensez-Plastiques, 11 Boulevard des Batignolles Paris (8e)

Revue Generale des Matieres Plastiques, 4 rue Lamblardie, Paris (12e)

Connaissance des Plastiques, 112 Boulevard Voltaire, Paris (11e)

glossary

Accelerators: chemicals that speed up the curing of a polymer. Also known as *promoters.* They can never start reaction. See *Catalyst.*

Acrylic: a synthetic resin prepared from acrylic acid or from a derivative of acrylic acid. Common trade names are *Plexiglas* or *Lucite.*

Actinic Screen: materials which shut out the short-wave lengths of light, such as ultraviolet, and their chemicals reactions.

Addition Polymerization (*or chain*): a reaction that causes one molecule to hook onto another like molecule by addition to form a chain.

Adhesive: a bonding agent that holds surfaces together through one or more of the following methods: 1. A mechanical bond such as sealing wax, tar, various glues, paste, dissimilar to the material being joined. 2. Heat or room-temperature cured thermosetting resins to join similar or dissimilar materials. 3. Solvent-based cements, usually with resin filler, identical to the materials being joined.

Aging: the effect of time on plastics exposed indoors in ordinary conditions of temperature and relatively clean air. See *Weathering.*

Air Film: microscopic air space found between overlays of dry materials.

Alkyd: polyester resins made with some fatty acids as a modifier.

Allylies: a synthetic resin formed by the polymerization of chemical compounds containing the group $CH_2{=}CH{-}CH_2{-}$. The principal commercial allyl resin is a casting material—allyl carbonate polymer.

Amino Plastics: Melamine and Urea: a synthetic resin made from reaction of melamine or urea (as the case may be) with formaldehyde or its polymers.

Annealing: exposure at a temperature near, but below, deformation temperature to permit stress relaxation—followed by controlled cooling.

Antioxidant: substance that slows down or prevents oxidation of material exposed to air.

Back Draft: an undercut area in a mold that tends to inhibit removal of the molded piece.

Bag Molding: application of pressure during bonding or molding, in which a flexible cover exerts pressure on the material to be molded, through the introduction of air pressure or drawing off of air by vacuuming.

Bakelite: the name given to the phenol-formaldehyde plastics made originally by Dr. Leo Baekeland. A trade name.

Base Coat: normally associated with metallizing, first coat applied to part, either first or second surface, to form a base for subsequent incorporation of a metallic layer.

Binding Medium: a liquid which carries the pigment particles and holds them together.

Bleed: to give up color when in contact with water or a solvent. Sometimes called *migration.*

Blow Molding: economical, high-volume method for forming hollow plastic parts. In extrusion blow molding, the most common method, resin is fed through an extruder die and formed into tube or parison. Air is blown into the tube, making it conform to walls of the mold.

Blowing Agents: see *Foaming Agents.*

Blushing: when patches of whiteness appear in the cured resin. Unlinked portions in the molecular chain readily absorb moisture, which causes whiteness or blushing.

Bond: to attach materials together mechanically, chemically, or through the use of high-frequency welding, conventional, or suitable adhesives. See *Adhesive.*

Boss: a projection on a molded plastic piece to strengthen the piece, to facilitate its alignment, or to aid in its assembly.

Bubble: internal void or a trapped globule of air or other gas.

Calendering: method for forming thermoplastics into continuous sheet and film. Also used for coatings. Plastic compound passed between series of large, heated, revolving rollers which squeeze material into sheet or film.

Casting: to form an object by pouring a fluid resin into a mold.

Catalyst: chemicals often used to initiate polymerization. They are supposed to act by their presence and not to be affected by the chemical reaction which they induce. Acid and alkaline catalysts, for example, will influence the reaction between phenol and formaldehyde in the production of phenolic plastics.

Cavity: the depression in a mold which will form the outer surface of the molded piece.

Cellular Plastic: a plastic containing numerous voids throughout its mass created by Foaming Agents.

Cellulose: first modern plastic. A mixture of solid camphor and nitrocellulose under heat and pressure.

Cellulosic: a natural high-polymeric carbohydrate found in the fibrous matter of woody plants. Cotton fibers are one of the present forms of cellulose.

Cementing: joining of plastics by means of solvents or adhesives. See *Adhesive.*

Center Gated Mold: in injection molding, a mold in which each cavity is fed through an orifice at the center of the cavity.

Chemical Resistance: resistance of a plastic to chemical-acids and alkylines.

Clamping Pressure: pressure needed to keep a mold closed during the molding operation.

Coating: both thermoplastics and thermosets are widely used to coat many materials. Coating processes vary. Methods include: roller coating, which is similar to the calendering process; spread coating, which employs a blade in front of the roller to place resin on material; or brushing, spraying, dipping.

Cocoon Molding: a web of sprayable styrene is sprayed over a network to be followed by coatings of sprayed vinyl.

Coefficient of Linear Thermal Expansion: this term deals with the amount of growth which occurs in a material when it is heated and is normally expressed in terms of in./in./°F. Visualize a mercury thermometer where a few degrees temperature rise causes a very substantial growth in the column of mercury but has no visual effect on the glass container.

Cold Flow: see *Creep*

Cold Molded Plastics: plastics shaped by molding or stamping under high pressure at room temperature.

Collapse: the failure of voids in a cellular plastic resulting in a more dense material.

Colloidal Dispersion: a finely divided mixture that is not chemically combined.

Compression Molding: most common method for fabricating thermosetting resins. Fully automated machines have speeded up this process. Plastic, often compounded with fillers, is placed into the open cavity portion of a two-part mold. The plunger part of the mold closes and heat is applied.

Compressive Strength: the maximum load which a plastic can bear before its density begins to increase. Compressive strength is not always the best measure of a plastic's strength; tensile strength is often more useful.

Copolymerization: polymerizing one plastic with another at the same time to vary or add properties to the base plastic. A copolymer is polyester which is copolymerized with styrene or methyl methacrylate.

Core: the male element in a mold.

Crazing: fine cracks, which may extend, in a network, on or under the surface or through a layer of a plastic material, caused by internal and/or environmental stresses.

Cream Time: of polyurethane foam, the time interval from the start of mixing to the first sign of foaming action.

Creep: the dimensional change or distortion of plastic which takes place under continuous stress. Creep at room temperature is sometimes called *Cold Flow.*

Cross-Linking: setting up of chemical links between molecular chains. When extensive, as in thermosetting resins, cross-linking makes one infusible super-molecule of all chains.

Cure: change of physical properties by chemical reaction; usually accomplished by heat and/or catalysts, with or without pressure. Another term is *Set.*

Cut-Off: the line where the two halves of a mold come together.

Cycle: the complete molding operation, from the first step to the last.

Deflashing: one of a number of processes used to remove the flash on a molded piece left by spaces between the mold cavity edges.

Deformation Under Load: tells us what percentage of deformation will occur in a material under a given load in a given period of time. The time element is critical. While 1% deformation might be indicated under the standard 24- or 48-hour exposure, leaving the sample under the load for a 2-week period may show substantially higher deformation. The lower the deformation under load, the more likeihood there is that the value will not change with increased time.

Delamination: the separation of the layers in a laminate.

Density: weight per unit volume of a substance, expressed in grams per cubic centimeter, pounds per cubic foot, et cetera. A 4″ cube is normally specified, accurately cut, and its weight in pounds multiplied by 27 to give the density in lbs./ft.^{3}.

Design: is a visual language that combines human, physical, and functional properties with man's scientific, technologic, and aesthetic prowess into an expressive form.

Die: a metal block containing an orifice through which plastic is extruded to produce a desired profile.

Dielectric: any material that will resist the passage of an electric current, such as a good insulator. Most plastics are good insulators.

Dielectric Strength: differs from tensile or compressive strength in that the force is applied electrically rather than mechanically. This electrical force, rather than acting on the entire mass, acts upon portions of the molecules. Dielectric strength is expressed in volts per mil and represents the number of volts required to cause an electrical breakthrough of the sample. As the voltage increases, the molecule approaches a failure point; portions of the molecule fly off and carry a charge or a current.

Diffuser: any transmitting materials placed between the light source and the observer which tends to scatter the light rays passing through.

Dimensional Stability: ability of a plastic object to retain its precise shape when subject to environmental changes.

Dip Coating: applying a plastic coating by dipping an article into a tank of melted or liquid resin, and then chilling or rapidly curing the adhering plastic.

Dispersion: finely divided particles of a material in suspension in another substance.

Draft: the degree of taper of a side wall or the angle of clearance in a mold, to facilitate removal of parts from a mold.

Elasticity: the tendency of a substance to return to its original size or shape after having been stretched or compressed. Polymers such as vulcanized rubber display this characteristic to a remarkable degree, due to the folded shape of their long molecules. This type of elasticity is described as rubberlike elasticity.

Elastomer (*elastomeric*): an elastic, rubberlike substance, as natural or synthetic rubber.

Elongation: is the increase in original length at fracture, expressed as a percentage.

Embedment: enclosure of an object in resin for presentation and display. The resin is always colorless and transparent.

Emulsion: a suspension of fine droplets of one liquid in another.

Encapsulating: see *Potting.*

Entering Edge: the surface constituting the thickness of a part. In edge lighting, that surface nearest the light source and generally perpendicular to the light pipe.

Epoxy: a straight-chain thermosetting plastic.

Exotherm: either the temperature/time curve of a chemical reaction giving off heat, or the amount of heat given off in a reaction.

Extrusion: a widely used operation in which powdered or granular material is fed through a hopper, heated, worked through a revolving screw that forces the hot melt forward through a heating chamber and out through a shaping orifice, like the chopping of meat in a grinder. The extrusion is fed onto a conveyor for cooling.

Fatigue Endurance Limit: reversed stress, tension and compression; is the level where failure will not occur.

Fatigue Strength: the maximum cyclic stress a material can withstand for a given number of cycles before failure occurs.

Fiber Optics: generic term for light piping with extruded clear fibers, coated with a material of sufficiently dissimilar refractive index to maintain all light rays within the filament. Fibers may or may not be aligned and contained to be able to pipe an image half-tone in the form of dot segments. More normally used simply to distribute light from a central source.

Fiberglass: a material made of spun, woven, matted, or chopped fibers of glass made by spinning melted glass into filaments.

Fibonacci Series: each number is the sum of the two which precede it: 1, 1, 2, 3, 5, 8, 13, 21, 34, 55, 89, 144 . . . If you divide any number by the preceding one, the result approaches the Golden Section, ϕ.

Filament: a long, thin, threadlike material; often extruded plastic made into yarn such as nylon.

Filler: an inert substance added to a plastic to make it less costly, improve physical properties such as hardness, stiffness, and impact strength. The particles are usually smaller than reinforcements.

Film: an optional term for sheeting with nominal thickness not greater than .010″.

Fines: very small particles (under 200 mesh).

Finishing: the removal of defects or the development of desired surface characteristics on plastic products.

Flame Retardants: chemicals used to reduce or eliminate a plastic's tendency to burn.

Flame Spraying: the application of a coating of plastic such as polyethylene by spraying the plastic through a flame. The plastic melts as it passes through a flame, hardening into a coherent coating.

Flash: excess plastic attached to a molded piece along the parting line.

Flash Point: the lowest temperature at which a sufficient quantity of vapors are given off to form an ignitable mixture of vapor and air immediately above the surface of the liquid.

Flexible Molds: are molds made of rubber, elastomers, or flexible thermoplastics. Used for casting thermosetting plastics, especially when undercuts are present.

Flexible Strength: the capacity of a material to flex without permanent distortion or breaking.

Flexibilizer: sometimes a plasticizer, is an additive that makes a resin or rubber more flexible.

Flexural Modulus or Strength: is also expressed in pounds per square inch and is the same type of force applied in folding a sheet of paper. A sheet of paper slightly extended over the edge of a table assumes a slight downward curvature while a piece of cloth will hang almost perpendicular. This means the flexural strength of the paper is considerably higher since it resists bending under its own weight.

Flow: fluidity of plastic material, usually during the molding process.

Flow Lines: observable flow patterns indicating direction or path taken by material in injection molding. See *Weld Lines.*

Flow Marks: a wavy surface appearance in a molded part, caused by improper flow of the plastic into the mold.

Foaming Agents: chemicals added to plastics and rubbers that generate inert gases on heating, causing the resin to assume a cellular structure.

Foam-in-Place: use of a foaming machine at the work location.

Friability: generally denotes a foam which lacks toughness and is easily crumbled. In good quality foams, this is normally confined to outer surfaces and is evident only during a state of undercure.

Froth Foaming: the liquid components of the foam resin system are dispensed from the froth foam machine mixing head as a partially expanded, soft, and wet, froth, much like shaving cream from a pressure can.

FRP: fiberglass reinforced polyester.

Gate: in injection and transfer molding, the channel through which the molten plastic flows into the mold cavity.

Gel Coat: a thin, outer layer of resin, sometimes containing pigment, applied to reinforced plastic molding usually as a cosmetic.

Glazing: thin or transparent painting allowing dry underlayers to show through.

Golden Section (*Golden Mean, Golden Number, Golden Ratio Divine Proportion*) or ϕ (phi): if you start with a ϕ rectangle and attach a square on its longer side, the square, taken together with the original ϕ rectangle, will create a new larger ϕ rectangle. Side A is to side B as B is to the sum of A and B. The new ratio between the sides is about 1.618:1.

Gradient Filters: a material processed so as to have an infinitely variable light transmission factor throughout its entire area.

Granular Structure: nonuniform appearance of finished plastic usually due to incomplete fusion of particles within the mass.

Hardener: a substance which brings about the curing of a plastic. Formaldehyde is a hardener for casein. It links the casein molecules together to form a network structure.

Haze: the cloudy appearance of an otherwise transparent plastic.

Heat Sealing: a method of joining plastic films by simultaneous application of heat and pressure to areas in contact. Heat my be supplied conductively or it may be applied dielectrically.

Heat Softening Point (Heat Distortion Point): the temperature at which a standard test bar deflects 0.010″ under a stated load.

Heat Stability: the ability of a plastic to maintain its properties under increased temperatures.

Homo-Polymerization: the polymer is built by additive combination of the monomer. For example, styrene by itself is a monomer; when combined with itself it becomes a polymer—polystyrene; that is, homogeneous or a homopolymer.

Hot-Stamping: the transference of pigment by heat and pressure from a tape or film (acetate, cellophane, polyester) to a jigged plastic part. The tapes may be specific by color, material application, and applied surface, whether first or second. In addition to pigment, the tapes usually carry, on the same surface, top and bottom clear coats which hold pigment as a laminate and are transferred with the pigment during stamping to supply both protective and adhesive coats.

Hyperbolic Paraboloid (HP): a parabola is moved into the path of another parabola in such a manner that it remains upright throughout.

Ignitible Mixture: mixture within lower and upper limits of the flammable range that is capable of propagation of flame away from the source of ignition when lighted.

Impact Strength: ability of a material to withstand shock.

Index of Refraction: the ratio of light velocity in a vacuum to light velocity in the plastic, or the ratio of the sine of the angle of light refraction measured from a perpendicular to the plastic surface.

Infrared: radiant heat (in the sense used here).

Inhibitors: chemicals that retard a reaction.

Injection Molding: most widely used method for fabricating thermoplastics. Preferred, economical technique for making complex, solid plastic parts in medium and high-volume ranges. Many machines automated. Resin fed into hopper, then into heating chamber. Plunger or reciprocating screw pushes melted plastic into closed mold. After cooling, mold opens and finished piece is ejected.

Inorganic Pigments: natural or synthetic metallic oxides, sulfides, and other salts, calcined during processing at about 1,200° to 2,100°F.

Interface: the exposed surfaces forming the boundary of a molded part.

Irradiation: as applied to plastics, refers to bombardment with a variety of subatomic particles. Atomic irradiation has been used to initiate polymerization and copolymerization of plastics and in some cases to bring about changes in the physical properties of a plastic. For example, polyethylene can be cross-linked and rendered resistant to heat by irradiation with gamma rays.

Isocyanate Resins: based on combinations with polyols (such as polyesters, polyethers, et cetera). The reaction joins members through the formation of the urethane linkage.

Jig: tool or fixture for holding component parts to be assembled during the manufacturing process, or for holding other tools.

Laminated Plastic: material formed by impregnating paper or fabric with a resin, followed by the bonding together of many layers into a consolidated structure. Phenolic resins are commonly used for this purpose, the laminated material being heated under pressure until the resin has set. Low-pressure laminates are made by lamination with resins which set at low temperatures and without forming water or similar by-products (e.g., polyester resins cured with styrene).

Latices (singular, latex): a suspension in water of fine particles of rubber.

Layup: process of placing reinforcing material into position in a mold.

Light Box: usually a five-sided enclosure containing a light source whose energy is directed through a transparent or diffusing image panel that completes the enclosure.

Light Piping: ability of optical glass or polished acrylic to pass light from one end to another with little loss, even around bends.

Light Source: a source of energy; fluorescent, incandescent, luminescent, or some other.

Lubricant Bloom: an irregular, cloudy patch on a plastic surface.

Luminaire: pertaining to light; an artificial light source.

Mask: in edge lighting—the opaque element used to shield and thus eliminate stray light rays.

Masking: usually in the form of metal stencils for covering certain areas of the plastic part during spray painting, Four basic types include: *a.* Lip Mask—for spraying depressed designs, or letters. *b.* Cap Mask—for spraying background areas with unpainted raised designs. *c.* Plug Mask—sometimes formed from a low-melting-point alloy poured directly on the part to be decorated then trimmed for use. This mask is used to protect depressed designs, usually on the second surface. Plugs must be produced for each mold cavity. *d.* Block Mask—usually used to limit the area of spray in "spray-and-wipe" operations.

Mass Polymerization: polymerization which takes place in the absense of any solvent or other diluent. Polymethyl methacrylate sheets are made, for example, by the mass polymerization of methyl methacrylate.

Melamine: see *Amino Plastics.*

Melt Index: describes the flow behavior of a polymer at a specified temperature (374°F or 190°C). under specified pressure.

Melting Strength: strength of a plastic while in a molten state.

Metamerism: two colors which appear to be a visual match under one light source but a mismatch under another.

Methyl Methacrylate: see *Acrylic.*

Modulor: Le Corbusier's system of measurements and proportions based on the height of the average man and related to the Golden Section, ϕ.

Modulus: the term may be applied to either tensile, compressive, flexural, or torsional actions. It defines the number of pounds per square inch required to cause deformation, elongation, flexure, et cetera, in material. In other words, it represents stiffness. Imagine a rubber band and a piece of string 4″ long. Placing a 1 lb. weight on the rubber band will cause stretching or elongation, whereas, the same weight on the string would cause little or no visual elongation. Assume the rubber band stretched to double its original length. The relative modulus of this material is found by dividing the 1 lb. force by the elongation in terms of percentage. We therefore have 1 lb. over percentage in decimal form of 1/1 = 1. Assume the 4″ length of string has stretched .040″. This represents an elongation of 1%. Dividing this into the 1 lb. load we have 1/.01=100. The relative modulus of the string is 100 times higher than that of the rubber band. In actual practice, the modulus would be expressed in p.s.i.; consequently, the modulus for a material like string might be about 100,000 p.s.i. In determining the compressive or flexural modulus, the same types of units are involved except that we are dealing with compressive deformation and flexural displacement.

Modulus of Elasticity: ratio of stress to strain in a material that is elastically deformed.

Mold: a shaper, usually in the form of a hollow die, in which plastic is forced, for example, by a shaped plunger, to take up a desired form.

Mold Release: a material used to coat a mold cavity to prevent the molded part from sticking to it and to aid in the removal of the molded part.

Molding:
A. Blow Molding—the placing of a hollow body of material, referred to as a *parison* formed either by extrusion or injection molding, into a simple mold which generally pinches off the parison at one or both ends. Air is introduced at medium pressure to inflate the parison, forming the part as the material cools against the inside surface of the die cavity.
B. Compression Molding—a thermoset molding process in which, generally, preheated molding compound is placed in an open mold. Subsequent closing of the mold, with the application of heat and pressure, results in a curing of the material as a replica of the die cavity.
C. Extrusion—a thermoplastic process whereby pellets, granules, or powder are melted and conveyed by one or more screws in a heated barrel and forced through a die to form a particular shape.
D. Injection Molding—a molding procedure whereby a heat-softened plastic material is forced from a cylinder into a relatively cool cavity which gives the article the desired shape.

Molecular Weight: the relative average weight of a molecule of a substance, expressed by a number in a scale on which the weight of the oxygen atom was arbitrarily set at 16.

Monomer: relatively simple compound which can react to form a polymer. Ethylene, for example, is the monomer from which the polymer polyethylene is made.

Neon: a colorless, inert gaseous element occurring in air. Captured in an electric discharge tube, it is ignited by an electrical charge.

Notch Resistance: in general, the ability of a material to accept sharp contour changes without exhibiting undue stress concentration failures at these points. Antonym: *notch sensitivity*.

Nylon: generic term for all synthetic polyamides.

Optical Properties: amount of transmission of visible ultraviolet light, haze, angular deviation, minor defects in a transparent material.

Orange-Peel: unintentionally rough surface, usually in injection moldings.

Organic Pigments: are usually characterized by good brightness and brilliance although not so permanent as inorganic pigments. They are divided into two classes, toners, and lakes.

Organosol: a suspension of finely divided resin in a *volatile* organic liquid. The resin dissolves mainly at elevated temperatures. At that point the liquid evaporates and the residue becomes a homogeneous plastic mass when cool. Plasticizer can be dissolved in the volatile liquid.

Over-Spray: that unwanted portion of a coating requiring removal from a decorated part.

Packing: a condition resulting from the use of an excess of material over and above the amount required to fill a restrained mold or cavity.

Parting Agent: see *Mold Release.*

Parting Line: mark on a mold or cast where halves of mold meet in closing.

Pearlescent (*nacreous*): pigmentation of the shellfish variety used to create unusual visual effects; as Mother-of-Pearl, mostly synthetically produced now.

Phenolic: a thermosetting resin produced by condensation—a phenol-formaldehyde type.

Plastic: a substance which can be permanently formed or deformed under external stress or pressure, usually accelerated by the application of heat. The newly created form retains its shape by cooling, chemical action, or the removal of a solvent through evaporation.

Plastic Flow: deformation of a polymer which takes place under the effect of a sustained force. Plastic flow is accompanied by movement of the long molecules relative to one another; they take up new positions corresponding to a change in the shape of the mass of the polymer as a whole.

Plastic Memory: quality of some thermoplastics to return to their original form after reheating.

Plastic Welding: the joining of two or more pieces of plastic by fusion of adjacent parts, either with or without the addition of plastic from another source.

Plasticity: the property of being susceptible to deformation under stress, and to the retention of the deformation on removal of the stress; the quality of being able to be shaped by plastic flow.

Plasticizer: a plasticizer is a solvent that does not evaporate and makes the plastic material pliable, softer, easier to mold into shapes; any material capable of combining, but not chemically. Camphor is a plasticizer for nitrocellulose; the camphor molecules penetrate between the nitrocellulose molecules, enabling them to move more readily relative to one another.

Plastisol: a colloidal dispersion (a suspension of finely divided resin) of a synthetic resin in a plasticizer. The resin usually dissolves at elevated temperatures. A homogeneous plastic mass results when the mixture cools.

Platen: a steel plate used to transmit pressure and heat to a mold assembly in a press.

Polyamide: a thermoplastic material—often fiber forming.

Polyester: a thermosetting resin of a sirupy consistency.

Polyethylene: a thermoplastic material formed under elevated temperatures and high pressures.

Polymer: the combined or joined smaller molecules that have formed larger molecules.

Polymerization: the process whereby long molecules are formed as a result of the linking together of many small molecules of similar structure. (Chain or addition polymerization—see *Addition Polymerization.*) (Also condensation of two or more dissimilar molecules, e.g., urea and formaldehyde.)

Polystyrene: a water-white thermoplastic.

Pot-Life: the period during which a compound, after mixing with a catalyst, solvent, or other compounding material, remains suitable for its intended purpose.

Promoters: chemical, itself a feeble catalyst, that greatly increases the activity of a given catalyst.

Properties: physical qualities of plastics that enable differentiation of different plastics among themselves and other materials.

Prototype: a first or original model fabricated from a variety of materials for test or from a specified material by simulated production methods to determine validity of design.

Puddle: in decoration—overly heavy concentration of coating in a depression.

Purging: cleaning out one color or type of material before using a second color or material.

Pyrolysis: chemical decomposition by heat.

Quench: process of shock-cooling thermoplastic materials from the molten state.

Reflection: Diffuse—A beam of light is diffusely reflected from rough or matte surface which breaks up the beam and scatters it in all directions. Specular—A beam of light is specularly reflected from a smooth surface such as a mirror which does not change the shape of the beam. The angle of reflection of the beam to the mirror is equal to the angle of incidence.
Spread—A combination of diffuse and specular reflection.

Reflex: a contoured rear surface made up of corners of cubes capable of reflecting all light coming through the first surface, redirecting it back toward the light source.

Refraction: the change in direction of a ray of light in passing obliquely from one medium into another in which its speed is different.

Reinforcing: little or no pressure required. Both open and closed molds used; thermosets and thermoplastics employed. Fillers are glass fibers or other materials. Different techniques used. Generally plastic applied to filler inside mold. Chemical catalyst hardens material. Pressure sometimes used to improve texture, speed process.

Relief: the projection of an ornament or plane area from a principal surface.

Resiliency: ability of a plastic quickly to regain its original shape after having been strained or distorted.

Resin: a solid or semisolid complex, amorphous component or mixture of organic substances having no definite melting point and showing no tendency to crystallize.

Retarder: see *Inhibitor.*

Rise Time (*rate of rise*): the time interval from the beginning of foaming action to the first apparent end of foam growth. It is normally taken as the time when the rising foam crown in a control pour appears horizontally level with the open top of the cavity. (Rate of rise is usually reported in inches per minute.)

Rockwell Hardness: the surface hardness of a material determined by applying an ever-increasing load to a point on the surface.

Roll Coating: paint applied to flat or near flat surfaces as well as tops of raised lettering and markings with a paint covered roller.

Rotational Molding: a process generally used for forming large hollow articles. Plastic, in powder or liquid dispersion form, is placed in a mold which is sealed, heated, and rotated about two axes. The molten plastic clings to the walls of the mold and is distributed evenly.

Runner: in injection molding, the feed channel that connects the sprue with the cavity gate.

Sandwich Heating: a method of heating a thermoplastic sheet prior to forming which consists of heating both sides of the sheet simultaneously.

Screen Printing: porous stencil printing. Rubber blade or squeegee forces pigmented coating (ink or paint) through unblocked print areas of screen; nonprint areas are blocked or closed to coating passage. Although generally relegated to flat or single curved surfaces, may be tooled to print on complex surfaces.

Scumbling: opaque instead of transparent colors are painted thinly over a dry undercoat. Surplus color is wiped off, stippled, dabbled, or rubbed with brush, rag, or fingers.

Second Surface (*especially transparent*): the surface normally obscured to the observer or the surface farthest away from the observer; bottom or back surface.

Sensitizer: a chemical that after primary irritation causes sensitivity to the chemical from exposures which are so slight that they would have had little effect before sensitization. Its reaction is much like exposure of allergic people to poison ivy.

Separator: see *Mold Release.*

Set: converting liquid resin to a solid state by curing, by evaporation of a solvent, suspending medium, or gelling.

Setting Temperature: temperature needed to set a liquid resin to a solid state.

Shear Strength: the point at which a material fails due to shear stress.

Shelf-Life: the amount of time a material can be stored at room temperature before it polymerizes without a catalyst.

Shore Hardness: the hardness of a material determined by the size of an indentation made by an indenting tool under a fixed load, or the load needed to force the indentor to penetrate the material to a predetermined depth.

Shot: the yield from one complete molding cycle, including scrap.

Shot Capacity: the maximum amount of material that can be delivered to an injection mold by one stroke of the ram.

Shrinkage: thermal contraction, continuing polymerization or cure, relaxing of strains that reduce the volume size of plastic.

Silicone: highly heat-resistant resin derived from silica.

Silk Screen Printing: a printing process which uses a taut woven fabric as a stencil.

Sink Mark: a shallow depression on the surface of an injection molded piece. May be caused by local internal shrinkage.

Sintering: process of holding powders or pellets at just below the melting point for about half an hour. Particles are fused or sintered but not melted.

Slurry: mixture of a liquid and fine material.

Slush Molding: similar to rotational method. Used to form small, hollow plastic parts. Requires liquid dispersion. Dispersed resins poured into one- or two-piece mold, which is then vibrated or spun. Heating brings final fusion of resin. The excess resin is drained off; the mold is cooled and the form is stripped out.

Solution Polymerization: polymerization which takes place in a solvent.

Solvent: liquid in which a substance dissolves.

Specific Gravity: the specific gravity of water is 1. When a resin has a lower specific gravity than water, less than 1, it is lighter, will float, and has more volume for less weight; conversely, a resin with a higher specific gravity than water is heavier, will not float, and less volume for equal weight.

Specific Heat: the amount of heat required to raise a unit of mass one unit of temperature. In the C.G.S. system the units are in terms of calories per gram per degree centigrade; in English units, British thermal units per pound per degree fahrenheit. Specific heat will vary with temperature and with phase changes in the material.

Specular: a highly reflective, mirrorlike, high-polish surface.

Spray Coating: formation of a protective coating by spraying a solution of plastic onto a supporting structure, e.g., of webbing.

Spraylat: a trademark for an inert plasticized resin, readily removable, that can be applied as a mask or stencil to parts to be decorated by spray, roll painting, or vacuum metallizing.

Spray-Up: spray gun used as a processing tool to disperse or deposit plastic, color, or fibers.

Sprue: in injection or transfer molding, the main feed channel that connects the mold filling orifice with the runners leading to each cavity gate. May also refer to the piece of plastic formed in this channel. Also *sprue gate.*

Stabilizers: ingredients added to compounding resins in order to maintain the desired chemical and physical properties at, or close to, their original values.

Step-Wise Cures: one catalyst initiates the polymerization; another one takes over above the temperature range of the first catalyst. This eliminates stresses from one temperature to another.

Stereospecific Polymerization: selectively directs the formation of polymeric chains having a predetermined steric configuration (among the many stereoisomers theoretically possible in a given reaction). (*Steric*—relating to atoms in space: spatial.)

Storage Life: see *Shelf-Life.*

Stress: the load exceeds the strength of the material. Expressed as load in pounds per square inch.

Stretch Forming: a forming technique where heated thermoplastic sheet is stretched over a mold and subsequently cooled.

Sweating: exuding of droplets of liquid, usually plasticizer, on the surface of a plastic sheet.

Tack-Free Time: this is the time interval from the start of mixing until the foam crown becomes tack-free.

Tensile Strength: the measure of resistance against being pulled apart.

Thermal Conductivity: the rate at which heat is transferred through a material.

Thermal Degradation (*thermal shock*): a deleterious change in the chemical structure of the plastic due to excess heat.

Thermal Expansion: the fractional change in dimension of a material for a unit change in temperature.

Thermal Movement: the growth and shrinkage of thermoplastic parts due to changes in atmospheric conditions; usually in terms of thermal coefficient of expansion.

Thermal Stress Cracking: crazing and cracking of some thermoplastic resins which result from overexposure to elevated temperatures.

Thermoforming: a process used to form plastic sheets into three-dimensional shapes. The sheet is clamped to a frame, heated to make it soft, then pressure is applied (either mechanical or vacuum) to force the sheet to conform to a mold located beneath the sheet.

Thermoplastic: plastic that can be softened by heating and hardened by cooling, repeatedly.

Thermosetting Plastic: a substance that becomes a solid when heated; setting is brought about by the establishment of a molecular network structure; the molecules can no longer move relative to one another.

Thixotropic: the ability of a liquid to resist the pull of gravity. Materials that are gel-like at rest and fluid when agitated.

Top Coat: in relation to first-surface metallized parts; the clear or tinted transparent coating placed over the metallized film as a protective surface.

Transfer Molding: related version of compression molding. Automated system called injection molding of thermosetting plastics. Plastic is heated in cylinder or pot and then transferred to open cavity mold for compression.

Translucent: letting light pass through but diffusing it so that objects on the other side cannot be distinguished.

Transmission:

Diffuse—A beam of light is diffusely transmitted through a material when the shape of the beam is lost and the light scattered in all directions. The form of an object cannot be seen through a diffusing material.

Transparent—A beam of light is transmitted through a transparent material without being diffused or changed in shape. Objects can be seen clearly through transparent materials.

Ultraviolet: zone of invisible radiations beyond the violet end of the spectrum. Their short wave lengths have enough energy to initiate chemical reactions and degrade some plastics and colors.

Undercut: an indentation or protuberance in a mold that tends to impede removal of the finished piece from the mold.

Urea: see *Amino Plastics.*

Urethane: see *Isocyanate Resins.*

Urethane Foaming Systems: the polyisocyanate or isocyanate-containing prepolymer component of a formulated foam system is designated as Component "A" and the resin or polyol component of a formulated foam system is designated as Component "B".

U-V Stabilizer (*ultraviolet*): a chemical compound that when mixed with a thermoplastic selectively absorbs ultraviolet light rays.

Vacuum Forming: method of forming thermoplastic sheet material, e.g., acrylic, by using a vacuum to draw the heated plastic into a hollow mold.

Vacuum Metallization: the deposition of a minute aluminum coating on the surface of plastic parts to simulate a bright plated finish. This is accomplished by vaporizing pure aluminum within a vacuum chamber, which causes the vapor to radiate from the vaporizer source, usually a tungsten wire heater. Plastic parts held, masked, and sometimes revolved within this atmosphere receive line-of-sight coatings of aluminum particles. Normally, basic to this process is the application to the parts, prior to metallizing, of a base coat of lacquer or organic resin to provide an ultra-smooth surface for the metal film and to act as one layer in the final lamination of the aluminum. The back or top coat of lacquer is applied after metallization to provide the top laminate layer and to give overall protection to the finish. Either the base coat or top coat, depending on the application, may be color tinted to simulate golds, bronzes, or coppers.

Venting: refers to the incorporation of gas and air escape ports in the mold or cavity. Vents are placed in locations where required to prevent air traps and in areas where final mold filling occurs, e.g., corners, edges, et cetera. For foam molding the mold should be foam-tight but not air-tight. Vents can be used for completeness-of-fill checks, and ⅛" holes are sufficient for wire probe. Larger holes are not recommended because of poor foam characteristics around such vent holes.

Vinyl: a thermoplastic material formed through addition polymerization. Polyvinyl chloride, acetate, alcohol, butyral are members of this group.

Viscosity: the internal fluid resistance of a substance causing molecular attraction which makes it resist a tendency to flow. Low viscosity flows faster, like heavy cream. High viscosity flows more slowly, like molasses.

Viscosity Index: based upon empirical number such as Pennsylvania Oil as 100, or an asphaltic oil as 0. It indicates the effect of temperature change on the flow of oil.

Volume Resistivity: the volume resistivity of a material is its ability to impede the flow of electricity expressed in ohms per centimeter. This measurement is always made on a 1 cm. cube. Wire, for example, is a conductor having negligible volume resistivity.

Warpage: dimensional distortion in a plastic object.

Weathering: the exposure of plastics outdoors, and the effects of exposure.

Wedge Lighting: light entering the thickest edge of an acrylic wedge is reflected back and forth between the principal surfaces as in normal edge lighting. However, at each reflection the angle of incidence becomes progressively closer to normal until it is less than 42.2° and the light passes through the air interface to escape from the part at both faces of the wedge. The distance from the thick entering edge to where light begins to escape through the principal surfaces is a function of the wedge angle thickness of the part, and the extremes of angularity with which light enters. The convergence angle should be generally 2° to 6°.

Weld Line: a hairline mark indicating areas of cool material closure, e.g., material moving within a die cavity along two or more paths, cooling slightly as they converge by good molding practice.

Welding: joining of thermoplastic pieces by means of heat and/or pressure.

Wetting Agent: a chemical that cuts the surface film of a material and makes it possible for color to adhere more tenaciously.

Working Life: see *Pot Life.*

Working Stress: dividing the yield point by a number from 4 to 10 known as the safety factor.

Yield Point: there are various types of yield points—compressive, tensile, flexural, and torsional. The terms simply mean the point at which material under compression, tension, et cetera, will no longer return to its original dimensions after removal of stress. You can visualize yield point by taking a wooden matchstick and gently bending until a slight fracture occurs. Prior to this fracture, for all visual purposes, a matchstick will return to its original straightness. In actual practice, plastic materials under tension, compression, et cetera, show a small degree of fracture at the yield point. They consequently will not return to their original dimensions because the internal physical structure has now been slightly modified. Below the stress (yield point) the material is elastic; above it, it is viscous.

bibliography

A. BOOKS

Cook, J. Gordon. *Your Guide to Plastics.* Watford, Herts, England: Merrow Publishing Co., Ltd., 1968.

Dreyfus, Henry. *The Measure of Man.* New York: Whitney Publications, Inc.

Garrett, Lillian. *Visual Design: A Problem Solving Approach.* New York: Reinhold Publishing Co., 1967.

Gloag, John. *Plastics and Industrial Design.* London: George Allen and Unwin Ltd., 1945.

Horn, Milton B. *Acrylic Resins.* New York: Reinhold Publishing Co., 1960.

Kaufman, John E., ed. *IES Lighting Handbook.* Fourth edition, second printing. New York: Illuminating Engineering Society, 1968.

Kepes, Gyorgy, ed. *The Nature and Art of Motion.* New York: George Braziller, 1965.

Kinzey and Sharp. *Environmental Technologies in Architecture.* Englewood Cliffs, N.J.: Prentice-Hall, Inc., 1963.

Kobayashi, Akira. *Machining of Plastics.* New York: McGraw-Hill Book Co., 1967.

Kohler, Walter, and Wassili Luckhardt. *Lighting in Architecture.* New York: Reinhold Publishing Co., 1959.

Kovaly, Kenneth A. *Handbook of Plastic Furniture Manufacturing.* Stamford, Conn.: Technomic Publishing Co., Inc., 1970.

Lawrence, John R. *Polyester Resin.* New York: Reinhold Publishing Co., 1960.

Lubin, George, ed. *Handbook of Fiberglass and Advanced Plastics Composites.* New York: Van Nostrand Reinhold Company, 1969.

Meyer, Jerome S. *Prisms and Lenses.* New York: The World Publishing Co., 1959.

Moholy-Nagy, L. *Vision in Motion.* Chicago: Paul Theobald, 1947.

Nelson, George. *Problems of Design.* New York: Whitney Publications, Inc., 1957.

Newman, Thelma R. *Plastics as an Art Form.* Revised edition. Philadelphia: Chilton Book Company, 1969.

Oleesky, Samuel S., and J. Gilbert Mohr. *Handbook of Reinforced Plastics of the Society of the Plastics Industry, Inc.* New York: Reinhold Publishing Co., 1964.

Plastics Engineering Handbook of the Society of the Plastics Industry, Inc. New York: Reinhold Publishing Corp., 1960.

Rowland, Kurt. *The Shapes We Need.* London: Ginn and Company, Ltd., 1965.

B. PUBLICATIONS OF GOVERNMENT, LEARNED SOCIETIES AND OTHER ORGANIZATIONS

A Plastic Presence, sponsored by Philip Morris, Inc., Milwaukee Art Center, Milwaukee, Wisconsin, 1969.

Acrylite Fabrication Manual, Wakefield, Mass.: American Cyanamid Company, 1967.

American Chemical Society, *Papers Presented at Toronto Meeting,* Vol. 30, #1. Toronto: American Chemical Society, 1970.

Illuminating Engineering Society. *Design of Light Control.* New York, 1959.

Illuminating Engineering Society. *Laboratory Activities with Light.* New York, 1959.

Illuminating Engineering Society. *Recommended Levels of Illumination.* New York, 1966.

Kaminsky, S., and J. A. Williams. *Handbook for Welding and Fabricating Thermoplastic Materials.* Norwood, Mass.: Kamweld Products Co., 1964.

Kaufman, Edgar Jr. *What is Modern Design?* New York: The Museum of Modern Art, 1950.

LaPietra, Ugo. *Il Systema Disequilibrante.* Milan: Galleria Toselli, 1970.

Light Measurement and Control, 1965. General Electric Co., Large Lamp Department.

Lucite Design Handbook, E. I. du Pont de Nemours & Co. (Inc.), 1968.

"Mobay Design Report for Duromer and Duroflex." Mobay Chemical Company, Pittsburgh, Pa. 15205

Morris, G. E. "Condensation Plastics: Their Dermatological and Chemical Aspects." *AMA Archives of Industrial Health,* #5, 1952.

National Board of Fire Underwriters. *Fire and Explosion Hazards of Organic Peroxides.* New York, 1956.

Plastic as Plastic. New York: The Museum of Contemporary Crafts, 1969.

Plastics—General Methods of Testing, Nomenclature. Philadelphia: American Society for Testing and Materials, 1970.

Plastics Safety Handbook. New York: The Society of the Plastics Industry, Inc., 1959.

Plastics—Specifications: Methods of Testing Pipe, Film, Reinforced and Cellular Plastics. Philadelphia: American Society for Testing and Materials, 1970.

Product Environment, St. Louis City Art Museum, 1970.

Rohm & Haas Company. *Fabrication of Plexiglas.* Philadelphia: Rohm & Haas Company, 1970.

Rohm & Haas Company. *Plexiglas Design and Fabrication Data.* Philadelphia: Rohm & Haas Company, 1968.

Rugger, George R., and Joan B. Titus. *Weathering of Glass Reinforced Plastics.* Dover, New Jersey: Plastics Technical Evaluation Center, 1966.

Wood, Elizabeth A. *Crystals and Light.* Princeton, N.J.: D. Van Nostrand Co., Inc., 1964.

Wood, Elizabeth A. *Experiments with Crystals and Light.* New Jersey: Bell Telephone Laboratories, 1964.

C. PERIODICALS

Bennett, Richard A. "The Performance and Economy of Large Ground and Polished Acrylic Optical Components," *Optical Spectra,* Vol. 1, #3, Third quarter 1967.

"How To Work With Gel Coats," *Modern Plastics,* June 1962, pp. 120–121.

"Japanese Crystal Palace," *Architectural Forum,* Vol. 130, #4, May 1969.

Lynton, John. "Three Sculptors," *The New Statesman,* February 28, 1964.

McGarry, Frederick J. "Structural Considerations," *Progressive Architecture,* June 1960, pp. 169–171.

Mangiarotti, Angelo. "Razionalitá," *Casabella,* 325-1967.

Moholy-Nagy, L., "Painting With Light—A New Medium of Expression," *The Penrose Annual,* Vol. 41. London: Lund Humphries and Co., Ltd., 1939.

"New Art Forms in Plexiglas," *Rohm and Haas Reporter,* 24:2:4, March/April 1966.

Plastics Technology: Processing Handbook, Vol. 14, #11, October 1968.

"Polymer Chemistry: A Plastics Engineering Primer," *SPE Journal,* June 1970.

Popper, Frank. "Light and Movement," *Art and Artists,* Vol. 1, #9.

Progressive Architecture, October 1970.

Sawyer, E. B. "Art and Electricity," *Art,* Vol. 2, #28, December 1955.

Scheichenbauer, N. "Progettare con le materie plastiche," *Casabella,* 313-1967.

Schwartz, Louis. "Dermatitis from Synthetic Resins," *Journal of Investigative Dermatology,* August 1945, 6:4:247.

Scientific American, Vol. 219, #3, September 1968.

Snoke, J. Howard. "Materials Guide: Three Foams of Urethane," *Industrial Design,* Vol. 14, #7, September 1967.

Wilson, Rex H., and William E. McCormick. "Plastics," *AMA Archives of Industrial Health,* 21:56/536, June 1960.

D. ENCYCLOPEDIA ARTICLES

Encyclopedia of Polymer Science and Technology. "Fine Arts," Thelma R. Newman, 6:795. New York: John Wiley & Sons, Inc., 1967.

Modern Plastics Encyclopedia 1967, Vol. 44, #1A. New York: Breskin Publications, 1966.

Modern Plastics Encyclopedia 1968, Vol. 45, #1A. New York: Breskin Publications, 1967.

Modern Plastics Encyclopedia 1969, Vol. 46, #10A. New York: Breskin Publications, 1969.

Modern Plastics Encyclopedia 1970, Vol. 47, #10A. New York: Breskin Publications, 1970.

E. UNPUBLISHED MATERIALS

Walker, Ian M. "Functional Forms in Polyurethane Foam." Unpublished Master's thesis, Rochester Institute of Technology, 1968.

index

ABS, 5, 92, 66, 94, 95, 153, 167
Abrasion, resistance to, 238
Absorption aspects, 73
"A Case Study: The Plastic House," *see* Winfield, Armand
Acetal, 31, 58, 167
Acetate, 31
Acetone, 58, 162, 167, 272-273, 274
Acid treating, 58
Acrylic, 52, 63, 66, 73, 95, 125, 127, 128, 150, 154, 155, 158, 161, 162, 163, 167, 187, 190-2, 265, 274
  cast, 96
    fiber, heat-reacting qualities of, 13
    in light machines, 130, 145
    light-piping qualities of, 207
    in sculpture, 130
    resins, casting with, 304-307
    shapes, 51
    sheets, 96
    transparent, 5, 193
Acrylonitrile, 31
Acrylonitrile-butadiene-styrene, *see* ABS
Additive extraction, 58
Additives, 63, 218
Adhesion properties, 63
Adhesive quality, 274
Adhesives,
  anchor, 59
  Latex, 239
  mastic-type, 239
  resin, 239
  solvent-based, 58
  water-based, 238
Aesthetics, 5, 72-3
  functional, 68
  and light, 122
Alcohol, polyvinyl, 273
Aliphatic naphtha, 167
Allergies, 64
Alloys, structural, 92
Aluminum, 63, 83
  vacuum-coated, 264
American Celluloid and Chemical Corporation, 30
Amine products, 239
Anaerobics, 52
Annealing, 161, 187
Antioxidants, 63
Arch, hyperbolic vertical, 78
Architects, 70
Architect-designers, Italian, 2, 3, 5, 193
Architecture, Greek, 70
Arithmatic relationships, 69-70
Artist, designer as, 96
Art Journal, see Busch, Julia
Asbestos, 30
Atmosphere, and light, 122
Atomizing, 221
Atoms, 31
Attack, chemical, 73

Baekeland, Dr. Leo, 31
Bakelite, 31
Balloon lamps, 84
Band saw, 155
  vertical, 234
Bayer, Otto, 218
Beauty, 67
  and function, 89
Bismuth, 63
Bleeding, 145
Blood disorders, *see* Plastics, potentially hazardous
Blowing, 266
  agent, 218, 220, 239
Boeri, Cini, 95
Bonding, 52, 58, 158, 167, 228, 239
  and joining techniques, 52
  electromagnetic, 167
  thermal, 52, 167
Bosses, 74
Boundaries, hyperbolic, 78
Breakage, 89
  resistance, 153
Brookhaven National Laboratories, 311
Brushing, 63, 238
Buildings, 70, 83
Busch, Julia, quoted, 145
Butadiene, 31
Buttress, inclined, 83

Cadon, 265
Calendering, 41
Camphor, molten, 30
Cantilever, 83, 96
Capillary action, 166
Carbon dioxide, 64, 218
Carbon monoxide, 64
Car interiors, 63
Casein, 30
Castiglioni, Archille, 95
Casting, 43, 153, 269, 274, 275
  hand, cream time, 221
  silicone, crystal-clear, 284
  using syrup, 305
Castings, water-white, 274
Catalyst, 218, 269, 275, 280, *see also* Plastics, potentially hazardous
Catalyst-Resins, proportions of, 279-80
Caulking guns, 58
Caution,
  MEK peroxide, 280
  methyl Methacrylate, 307
Celanese Corporation, 30
Cell structure, 218, 220
Cellophane, 48, 63
Cellosolve acetate, 238
Celluloid, 30
Cellulose, 30
Cellulosics, 153
Cementing, 161-167, 265
Chairs, 70
  tube, 74
Chamber, cooling, 43
Champ-leve, 51
Cleaning, chemical, 58
Chemicals, toxic, 64
Chemists,
  German, *see* Spitteler, Adolf and Krische, W.
  organic, 31
"Chinese Puzzle," *see* Massin, Eugene
Chloride, 31, 221
Chroma, 63
Clarity, 67-68

Clay, 30, *see* Nature's materials
Cleaning,
abrasive, 58
and polishing, 74
Clothing, 63
Coat, barrier, 267, 273
Coating, 43, 63, 218
foam, 237
plastic, 51
release, 268
roller, 238
Cold flow, 89
Collography, *see* Graphics techniques
Colombo, Gianni, 96
Colombo, Joe C., 92, 96, quoted, 4
Color, 145, 153, 212
and light, 122
and transparency, 5
selection of, 89
variability of, 83
Colorants, 63
Coloring, 63
and decorating, 74
Combustion, 64
Compression, 90
structural, 83
Computers, 51
Concrete,
plastic irradiated, 311
reinforced, 83
Cone, 78
Construction, 83-84
Construction system,
cocoon, 84
Coolants, 156
Copolymers, 66
Copper, 63
Corona discharge, 58
Cost, 58
mold, 269
Cotton, 51, 59
Covering, 221
Cracking, 95
and crazing, 73
Crazing, 58, 95, 161, 187
Creativity, 70, 84, 214, 217
Creep, 89
material, 83
properties, consideration of, 89
Cross-linking agents, 239
Curing, 40, 43, 48, 84, 167, 274, 275
agents, 284
foam, 237
room-temperature, 266
time, 63, 269, 272, 273
Curves, 123, 125, 193
forms, 5
surfaces, 149
use of, 92
Curvilinear insulated spaces, 5
Cutting, 154
Cyanoacrylates, 167

Dacron, 51, 273
Decomposing, 66
Decoration, 74, 187
Deformation, 89, 273
properties, consideration of, 89
Defraction,
light, 122
Degradation, environmental, 63
Degrading, 66
Degreasing, 58
Demolding, 221
Density, 73, 221
Design, 67, 266, 274
aesthetic-functional, 68
and environment, 4
and imitation, 96
and technology, 5
applications, 218
approach, 84
avant-garde, 68
concepts, future, 309-311
considerations, 189-190
economic limitations in, 89
failure, 96
flexibility, 59, 153
general considerations, 70-83, 89
heritage, 68
human activity, relative to, 67
human properties in, 84
imaginative, 217
material limitations, 89
parameters, 70, 89-94, 96
performance, 73
practicability, 95
processes, 67, 70
rules for, 73
specifics, 189
specifications, 72
standards, 67
structural possibilities, 96
technical aspects of, 72
Designing with casting and laminating liquids, 266-307
Designer-artists, 130
Dielectrics, 52
Dies,
hot, 265
metal, 63
Dietz, Professor Albert G. H., 2
Dimensions, 72
and light, 122
Dimensional changes with time, 73
Dispensing, 221
Distortion, 154
Divine Proportion, the, 69
Doctor blades, *see* Filament winding
Dome, *see* Macro-modular units
Draft radii, 74
Drape molding, *see* Hand Layup
Drill,
variable-speed, 221
Drilling, 156, 159
Drury, Professor Ralph, 5
Dry colors, 63
Dry process, 63
Durethane 100, 63
Durethane 600, 63, 238
Dye, 63
alcohol-soluble, 238
Dyeing, 187, 189

Ecology,
and plastics, 308-309
Economics, 74, 89, 92, 153
and building, 83
limitations in design, 89, *see* Maintenance costs
Edge lighting, 96
Etched effect, 149
Elasticity,
material, 83
Elastic memory, 92
Elastomer,
cured, 273
polyurethane, 273
urethane, 221, 273
Electromagnetic bonding, 167
Elongation, 89, 238
and plastic memory, 92
Enamel,
elastomeric thermosetting, 63, 238
transparent, 51
Enameling, 51
Environment, 58, 66, 67, 70, 193
and mobility, 5
audio-visual, 128
biological, 65
degradation of polymers in, 65
effects of, 89
future, 217
Environmental effects, 89 and light, 123
Epoxy, 43, 51, 63, 66, 83-84, 167, 221, 239, 266, 268, 274
catalyzing, 64
fiberglass-reinforced, 266
liquid, 283-284
putty-type, working with, 283
reinforced, 59
resin, 43
Equipment,
dispensing, 220
hand-held, 265
metalworking, 153
Etching, *see* Graphics techniques
solution, 58
Ethylene dichloride, 162
Euclid,
formulas, 70
Evaporation rate, 238
Expansion, 220
Experimentation, 74
Exposure,
environmental, 73
Exotherm, 221
Extenders, 62
Extrusion, 40, 153, 218

Fabric, 63
liquid-saturated, 266
thermoformed, 265
vinyl-coated beta fiberglass, 264
Fabrication, 96, 153, 210, 265
Failure,
causes of, 72
material, 90
see Maintenance costs
Farbigem, 193
Fasteners, mechanical, 52
Fastening,
bonding and, 158
Fatigue, 73, 89
endurance limit, 90
Federal regulations, 63, 64

Fiber, 59, 95, 217
cotton-dacron, 264
modacrylic, 59
Monsanto, 265
straw, 59
synthetic, 13, 51, 256
see Nature's materials
Fiberglass, 43, 48, 217, 221, 266, 274
polyester, 268
vinyl-coated, 264
Fibonacci series, 70
Filament,
continuous glass, 43
Filler, 59, 96, 218, 274
Fillets,
use of, 92
Film, 52, 217, 264-265
plastic, 51
Finish, 238
Finishing, 74, 182
Fire hazard, 239
Flame,
retardants, 63
treating, 58
Flammability, 58, 63
Flammable Fabrics Act, 63
Flexibility, 89, 90, 217, 274
Flexural strength, 89, 94
Flow pattern, 74, 92
Fluorocarbons, 58, 66
fluorocarbon II, 239, 264
Foam,
density of, 234
flexible, 234
fluorocarbon-blown, 221
polystyrene, 238
polyurethane, 238
rigid urethane, 217
urethane, 5, 221, 238
urethane, flexible, 217
urethane, flexible expanded, 63
Foaming,
direct, 217
pour-in-place, 220
Form, 72, 123
and light, 122
blow-up, 5
curved, 5
structural, 78, 83
temporal, and light, 122
thin-walled, 122
three-dimensional, 122
transparent-translucent, 123
without materiality, 123
Formaldehyde, 30, 31
Formica, 5
Forming, 43, 170, 217
methods, 73-74
techniques, 154
"Four Yellow Balls," *see* Massin, Eugene
Fracturing, 89
Freon, 64, 122
Friction,
molecular, 265
Frosting, 189
Frothing, 220
Fumes,
volatile, 220
Furniture, 63
acrylic, 204
Furniture (continued)
blow-up, 5, 74
collapsible, 5
design of, 95
integral skin foam, 234
knock-down, 5
packable, 5
plastic, design in, 204
redisposable, 5
size, variations in, 70
stackable, 5
take-apart, 5

Ganging, 92
Gate,
location of, 74
see Molding process
Gel,
catalyzed polyester, 273
moisture-resistant, 274
properties of, 63
Gestalt, 70
Glass, 59
Farbigem, 193
laminated, 193
resin, 48
Golden Mean, the, 68
Golden Number, the, 69
Golden Ratio, 69
Golden Section, the, 69-70
Goodyear, Charles, 30
Graphics,
techniques, 51
Greeks, 69-70
Gun cotton, 30

Hallman, Ted, 265
Hand-crafting, 70, 214
Hand layup, 43
Handsaw, 156
Hangings, 51
Harmony, 67-68, 70
Hazard, health,
MEK peroxide, 280
polymer spraying, 64
HB units, 78
Healing, 63
Health hazard, 239, 280
methyl methacrylate, 307
Heat, 63
bath, 265
distortion, 73
frictional, 153, 167
sealing, 265
Heating, 153
chamber, 43
strip, 96
see Injection molding
Hopper, *see* Injection molding
Horizontals,
use of, 92
Hot air, 52
Hot melts, 52
Hot stamping, 63, 167, 189
HP,
segment, 79, 83
structures, 83
Hue, 63
Human element, 4, 68, 70
material relationships and contiguities, 5
Humidity, 58, 66, 220
Hyatt, John Wesley, 30
Hydrocarbons, 238
low molecular weight, 64
Hydrogen chloride, 64
Hydrogen cyanide, 64
Hyperbolic paraboloid (HP)
form, 83
segment, 78-79, 83
structure, 83
surface, 83

Illusions,
and light, 122
Image, 125, 127, 128
Imagination,
and design, 84
Imitation,
design and, 96
Impact
energy, 90
strength, selection of, 89
Index,
refractive, 274
Induction, 52
heating, 167
Industrial Revolution, 30
Infra-modular unit, 79
Injection
methods, 74
molding, 5, 59, 92, 94, 95, 153
Inorganic pigment, 63
Insideness, 123
Insulating, 221
Integral skin, 96
Integrity,
and beauty, 67
Internal-external relationships, 123
Intuition,
and design, 67
Inventor,
designer as, 96
Isocyanate, 221, 239
polymer, 218
Italy,
designers of, 193
design renaissance in, 2

Japan World Exposition, 264
Johnson, Jude, 149
Joining, 58, 161
Joints,
movable and immovable, 52

Kartell, 92, *see* Von Bohr, Olaf
Knock-down furniture, 5
Koch and Lowy, 96
Krische, W., 30

Lacquer, 30, 238
elastomeric, 228
elastomeric urethane, 63
Laminating,
liquids and castings, designing with, 266-307
Lamination, 274
Lamps, balloon, 84
LaPietra, Ugo, 13, 128, 193
Large molecules, 31, 39, 65
Latex, 30, 51, 221
Layering, 217, 266
Lead compound, 63
Le Corbusier,
and Euclid's formula, 70
Lenses, 125, 127

Leonardo da Vinci,
and Euclid's formula, 70
Light, 5, 122-150
control, 125
forms, and pattern, 13
liquid, 128
machines, acrylic, 130
moving, 122
pattern in, 207
rays, 127, 130
texture in, 207
transmission of, 122
waves, 128, 145
Lightness,
material, 83
Linear thermal expansion,
coefficient of, 94
Liquids,
casting and laminating, designing with, 266-308
Liquid-filled plastics, 123, 125
Load,
excessive, 73
factor, 89
indent, 221
varying, 90
Loading rate, 89
Lost wax process, 51
Low compression molding,
see Hand Layup
Lubrication,
surface, 63
Luminescence, 122

Macro-modular units, 79, 83
Magistreti (architect), 4
Magnesum, 63
Man-forms, 4
Manganese, 63
"Man in Purple, Green, and Red Progression,"
see Massin, Eugene
Manufacture, 70
Marble,
opacity of, 274
Massachusetts Institute of Technology, *see* Dietz, Professor Albert G. H.
Massin, Eugene, 145, 149
Master,
preparation of, 266
Mastics, 52, 238
Mat,
fiberglass, reinforcing, 59
Material, 70, 73, 74
cladding, 84
coating, 238
corrosion resistant, 83
decisions about, 74
demountable, 265
elastomeric, 266
expansion coefficient of, 83-84
fatigue in, 73
flexible, 92
heat-sealable, 264
life of, 73
limitations of, 5
maximum capabilities of, 123
nature's, 63
portable, 265
resistance to fire of, 83
stability of, 95
strength of, 73, 89-90

Material (continued)
structural, 83
surfacing, 234
traditional, treatment of, 193
water-based, 52
woven, 51
see also Design Standards
Mathematical models, 68, 70
Mating,
surface, 52
Matting,
fiberglass polyester, 274
Melamine, 48
Melt point, 63
Mercury, 94
Metal, 221, *see* Nature's materials
Metalizing,
vacuum, 187, 189
see also Coating
Metallic flakes, *see* Special effect pigments
Metering, 220-221
Methanol, 221
Methylene chloride solution, 273
Miller, Henry, 84
Mineral oil,
as a filler, 125
Mirrored surface, 145
Mirrors, 264
and light, 122, 125, 128
image, 145, 149
Mixing, 220-221
Mobay, 95
see Models, experimentation and development of
Mobility,
furniture for, 5
Models,
experimentation and development of, 16
Modern Plastics, see "What *is* Plastics' Image Anyway?"
Modulor, the, 70
Modulus of elasticity, 89
Moholy-Nagy, L., quoted, 130
Moiré effects, 128
Moisture,
effect of, 89
level of, 273
material reaction to, 73
Moldability, 31, 83, 153
Molding,
blow, 43
compression, 40, 51, 59, 218
injection, 5, 59, 92, 94, 95, 153, 214, 218
powders, micropulverized, 43
process of, 74
rotational, 43
solvent, 51
Molds, 43, 221, 266
definition of, 273
designing for, 92
duplication of, 269
forming from, 269
making, 266-274
open-ended, 43, 268
plaster, 51
release coating, 268-269
rigid, 266
transfer, 95
Molecules, large, 31, 39, 65
Monofilaments, 41

Monomers, 31, 65
Monsanto, *see* Models, experimentation and development of
Movement,
and light, 123
in design, 149
Mylar, 48, 145, 149, 150
light and, 149
see Polyester film

Neon,
and design with light, 128
Nerve disorders, *see* Plastics, potentially hazardous
Nitrocellulose, *see* Plastics, history of
Nonplanar areas, 5
Nylon, 51, 59, 95, 153, 167
cord, 96
second-generation, 265

Odor,
consideration of, 73, 89
Opacity,
pigment, 63
Opion Research Corp.,
survey, 3
Optical illusions and curved surfaces, 149
Organic pigment, 63
Organ malfunctioning, *see* Plastics, potentially hazardous
Outsideness, 123
Oxidation,
resin resistance to, 63

Packaging, 221
Paint,
acrylic-based, 264
Painting, 187
Paper,
breathable-type, 221
Paraboloid, 78
Paraffin wax–xylene, 273
Parkes, Alexander, 30
Parkesine, 30
Parting agent,
see Release agent
Pastes, 63
Patch,
body, 274
Pattern, 73, 123, 266
Pentagon, 69
Permanence,
color, 63
Phenol, 31
Phenolic plastic, *see* Bakelite
Phi, 69
ratios, 70
Photography,
and light, 122
Physical laws, 70
Physical properties, 84
Pigment, 63, 218
alcohol-soluble, 238
Plasticity, 31
Plastics,
and ecology, 308-309
as art and design form, 130
combinations of, 31, 64
families of, 31-33, 89
fiberglass-reinforced, 83-84
flame-retardant, 64

Plastics (continued)
history of, 30
in industry, 3
in structure-space sense, 3
liquid-filled, designing with, 266-308
man-made, 31, 51
metalized, 149
nature's, 30
physical properties of, 59
potentially hazardous, 64
synthetic, 31
technology of, 31
thermosetting, 59
transparent, light aspects of, 123
Plasticizers, *see* Plastics, potentially hazardous
Plastisols, 63
Platen, 265
Plexiglas, 96
Plexima, 96
Plique-à-jour, 51
Plunger mold, 95
Polishing,
cleaning and, 74
Polyacrylates, 167
Polycarbonates, 5, 153, 167
Polyester, 51, 59, 63, 64, 73, 83, 96, 145, 238, 265, 269, 274, 275
a focus on, 143
catalyzed thermosetting, 43
fiberglass-reinforced, 66, 266, 273
film, 63, 149
reinforced (RP), 284-304
resin, 63, 84, 128, 145
thermosetting, 43
transparent, 5
"Unfulfilled" working with, 274
water-extended (WEP), 284
Polyethylene, 43, 51, 58, 153, 167, 264, 266
sheet, vacuum-formed, 221
Polyhydroxyl, 218
Polyisocyanate, 218
Polymer chemistry, 31
Polymeric mixtures, 31
Polymerization, 31, 167, 306-307
Polymers, 31, 59, 62, 63, 64, 153
potentially toxic, 65
Polymethyl methacrylate, 153, 266
Polypropylene, 59, 153
Polyolefins, 58
Polystyrene, 51, 58, 66, 153
Polysulphones, 167
Polyurethane, 51, 153, 218-264
foam, high-density, 63
see also Isocyanate polymer
Polyvinyl acetate, 221
Polyvinyl chloride, 153, 266
Polyvinyl fluoride, 264
Polyvinyls, 31
Possibilities, design, potential range of, 153
Postcure, 273
Potassium dichromate, 58
Pouring, 266
Powder,
molding, 218
Prefabrication, 83
Preparation, surface, 237
Pressure, 40, 63, 264
in bonding, 58
Prisms, 125
Properties, 84
Process,
considerations, 73
decisions, 74
requirements, 74
Processes,
a primer of, 39
Processing,
conditions, and material property changes, 94
equipment, selection of, 89
techniques, 39
injection molding, 39
Product, 74
Production,
aspects, 72
techniques, 67
Progressive Architecture, see Winfield, Armand
Proportion, 70
Prototype,
evaluation of, 74
Pulp,
molding, 51
Pulp-resin, 51
PVC, 63, 167

Radiation,
ultraviolet, 63
Rayon, 59
Reactions,
chemical, 73
Reaction rate, 218
Reflections, 123, 125, 128
Reflectors, 130
Refraction,
angle of, 78
light, 122, 125
Refractive index, 145, 274
Reichold, *see* Models, experimentation and development of
Release agent,
mold, 273
silicone, 269
Reliability, 89
Remoldability, 153
Renaissance, 69
Repertoire,
processing, 210
Repetition,
and light, 122
Resiliency, 95, 154
Resin, 48, 62, 63, 73, 221, 273, 274, 275
Resistance,
chemical, 274
Retardant,
flame, 218
Rigidity, 274
"Rocking Men," *see* Massin, Eugene
Rohm and Haas, *see* Models, experimentation and development of
Rolling, 63
Roof,
construction of, 83
Routing, 156
RTC, 266
RTV, 266
catalyzed, 272
RTV, 266 (continued)
silicone, 269, 274
Rubber cement, 221
Rubber,
foam, 218
silicone, 221, 269, 273
Rugs,
synthetic yarns for, 265
Runners, 92
Rhythm,
and light, 122

Saddle-shape, 83
Safety factor, 89, 239
St. Thomas Aquinas, 67
Sandblasting,
see Cleaning, abrasive
Sanding, 266
Sandpaper, 266
see Cleaning, abrasive
Sansegundo, Carlos, 5
Saran, 59
Saw,
buck, 234
crosscut, 156
Danish, 224
jig, 156
sabre, 156
traveling, 155
Sawing,
circular, 154
Schonbein, Dr. Friedrich, 30
Science Museum (London), 30
Scriber, 156
Sculpture,
acrylic, 130
Sealing, 265
Seating, 70
Seggio, 92
Shadows,
and light, 122
Shape, 73, 90
post and lintel, 96
Shaping, 153, 156
Shear,
material, 83
Sheet,
pressure-sensitive, 52
vacuum-formed, 266
Shell,
common, 78
Shellac, 30
Shrinkage,
postmolding, 74
Silicone, 51, 63, 269
casting, crystal-clear, 284
Silk-screen, 51
Silk-screening, 187
Sink marks, 74
Skin disorders,
see Plastics, potentially hazardous
Slate powder, 30
Slip agents, 63
Small, Neal, 13
Softening, 153
Solvents, 52, 265
see Plastics, potentially hazardous
Spaces,
curvilear insulated, 5
Spanning, 83, 84
Spider, 43

Spitteler, Adolf, 30
Spray booth, 64
Spraying, 5, 63, 84, 221, 238, 239
Spring, 95
Sprue hole, 273
Stability,
  dimensional, 59
Stabilizers, 63
Stacking, 92
Steel, 92
Stiffness,
  degree of, 89
Straight edges, 132
Strength, 73
  stitch and tear, 59
  structural, 92
  tensile, 92
Stress, 58, 73
  properties, consideration of, 89
  reversed, 90
  tensile, 89
Stress-strain data, 89-90
Stripping, 272
Structure,
  air-supported, 264
  cantilever, 83
  frame-supported, 264
  saddle-shape, 83
Structural aspects, 72
Styrene, 31, 64
  webbing agent, 84
Substrate,
  copolymer, 264
  impregnated, 48
Sulfuric acid, 58
Sulphur, 30
Sunlight,
  effect of, 89
Supports, 92
Surfaces,
  curved, 140
  metalized, and light, 122
  reflective, and light, 122
Surfactants, 218
Systems,
  double layer, 83
  processable, 273
  single layer, 83
  supporting, 83
  water-blown, 221

Taste, consideration of, 89
Techniques, fabrication,
  bonding and joining, 52
  manual, 58
Technology, 52, 63, 83, 96, 274
  and potential, 311-314
  new, 265
Temperature, 40, 58, 89, 154, 161, 220, 221, 234, 269, 272, 273
  changes in, 66
  pyrolizing, 64
  range, useful, selection of, 89
  room, 266
Tension, 90
  and motions, 73
  structural, 83
Terpolymers, 66
Testing, 74
Tetrafluoroethylene (TFE), 153, 167
Texture, 73, 123, 153, 212
  and transparency, 5
Texture (continued)
  variability of, 83
Textured surfaces,
  and light, 122
TFE, 153, 167
"The Plastic House," *see* Winfield, Armand
Thermoforming, 96
  see Forming
Thermoplastics, 31, 43, 51, 52, 59, 92, 128, 153, 167, 218
  acrylic, 43
  cellulose nitrate, 40, 41,
Thermosets, 31, 40, 48, 94
Thermosetting, 218
Thickness, 123
  wall, 74
Thinness, 123
"Three Brown and Clear Blocks," *see* Massin, Eugene
"Three Rivers" project, *see* Drury, Professor Ralph
Time, 40
  and light, 122
  processing, 221
Tolerance, 92, 95
  and plastic memory, 92
  dimensional, 74
Tools,
  hand-powered, 154
  reaction to, 73
Toxic effects,
  precautions against, 65-66
Toxicity, 73
Toys, 3
Transparency, 5, 63, 122-150, 153
Troweling, 238
Tuluol, 238

Ultrasonics, 52
Umbrella, *see* Macro-modular unit
Undercutting, 74, 92, 266, 272, 273, 274
Unit,
  HB, 78
  infra-modular, 79
United States pavilion, 264
Upholstery, 59
Urethane, 51, 63, 66, 95, 96, 217, 220, 221, 238, 239, 264, 268

Value,
  colorant, 63
Vapor honing, *see* Cleaning, abrasive
Vaseline, 273
Vault, 78
  see Macro-modular unit
Ventillation, 64, 163, 188, 220, 239
Versatility, 217
Verticals, 92, 204
Vibrations,
  sound, high-frequency, 52
Vinyl, 43, 51, 58, 59, 63, 237
  cocoon system, 84
Vision in Motion, *see* Moholy-Nagy, L.
Visual echoes, 122, 123
Visual image, 123
Volume, 123
Von Bohr, Olaf, 5
Vulcanization, 30
Vulcanizing,
Vulcanizing (continued)
  room-temperature, 51, 266
Walker, Ian, 234
Wall hangings, synthetic
  yarns for, 265
Warpage, 92, 221, 274
Water, 66, 89
Wax, 51, 266, 268, 269, 273
Wear, 73
Weather, 265
  effects of, 89
Weatherability, 66, 238
Webbing agent, 84
Weight, 153
Weld,
  high-frequency, 264
Welding
  rod, 52
  ultrasonic, 167
Winding,
  filament, 43
Windows,
  epoxy, 193
  polyester-fiberglass, 193
  stained glass, 51
Window-lighting effects, 51
Winfield, Armand, "A Case Study: The Plastic House", *Progressive Architecture*, 2
Wood,
  laminated forms, 5
  opacity of, 274
Working stress, 89-90
Wrapper, 96

Yarn,
  acrylic, 265
  synthetic, 13
Yield point, 89

Zakas, Spiro, 13
Zeidman, Robert, quoted, 4
Zerning, John, 84
Zinc, 63
Znamierowski, Nell, 13, 265

thelma r. newman

Thelma R. Newman has become an internationally recognized authority on plastics since the publication of her definitive and monumental book, *Plastics As An Art Form.*

She received her bachelor's degree from the College of the City of New York, her Master of Arts degree from New York University, her doctorate from Columbia University, Teacher's College, and studied design engineering for plastics at Newark (N.J.) College of Engineering. She has also studied sculpture under Seymour Lipton at the New School for Social Research,. photography with Carlotta Corpron, Aaron Siskind and Josef Breitenbach, and stained glass window making as an apprentice to J. Gordon Guthrie at Durham and Sons.

Widely known as a lecturer and a recognized teacher and authority on plastic sculpture and other art forms, she has exhibited her work expressing light and transparency in cities throughout the country. She has her own studio in which, between teaching, consulting and lecturing, she constantly seeks new worlds to conquer in the realm of plastics.

She has taught at New Jersey State College, North Texas State College, Newark State College and for four years was director of art for the Union Township schools, Union, New Jersey.

In addition to *Plastics As An Art Form,* Thelma Newman is the author of *Wax As Art Form* and *Creative Candlemaking,* in addition to numerous articles in technical, art and learned journals. She is a senior member of the Society of Plastics Engineers and is president of and a plastics designer for Poly-Dec. Company, Inc.

A

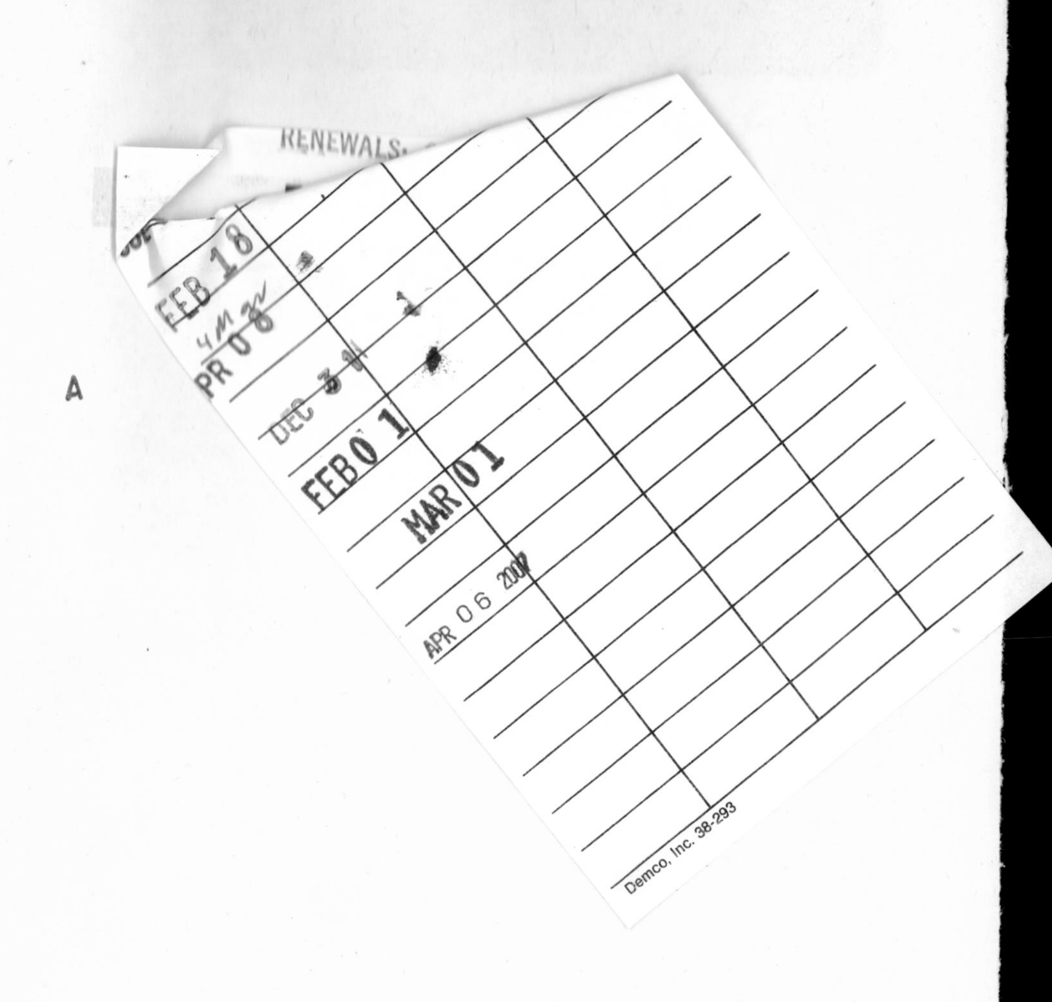